TABU SEARCH

TABU SEARCH

Fred Glover
Manuel Laguna

University of Colorado at Boulder

Kluwer Academic Publishers

Boston/Dordrecht/London

Distributors for North, Central and South America:
Kluwer Academic Publishers
101 Philip Drive
Assinippi Park
Norwell, Massachusetts 02061 USA
Telephone (781) 871-6600
Fax (781) 871-6528
E-Mail <kluwer@wkap.com>

Distributors for all other countries:
Kluwer Academic Publishers Group
Distribution Centre
Post Office Box 322
3300 AH Dordrecht, THE NETHERLANDS
Telephone 31 78 6392 392
Fax 31 78 6546 474
E-Mail services@wkap.nl>

Electronic Services <http://www.wkap.nl>

Library of Congress Cataloging-in-Publication

Glover, Fred.
Tabu search / Fred Glover, Manuel Laguna.
p. cm.
Includes bibliographical references and index.
ISBN 0-7923-9965-X Hardcover
ISBN 0-7923-8187-4 Paperback
1. Operations research. 2. Mathematical optimization.
3. Artificial intelligence. I. Laguna, Manuel. II. Title.
T57.6.G568 1997
0003'.3--dc21 97-24688
CIP

This printing is a digital duplication of the original edition.

Printed on acid-free paper.

Printed in Great Britain by IBT Global, London.

To Diane and Zuza,
who have brought us to appreciate
the tabus worth transcending.

CONTENTS

PREFACE

Faced with the challenge of solving hard optimization problems that abound in the real world, classical methods often encounter great difficulty — even when equipped with a theoretical guarantee of finding an optimal solution. Vitally important applications in business, engineering, economics and science cannot be tackled with any reasonable hope of success, within practical time horizons, by solution methods that have been the predominant focus of academic research throughout the past three decades (and which are still the focus of many textbooks).

The impact of technology and the advent of the computer age have presented us with the need (and opportunity) to solve a range of problems that could scarcely have been envisioned in the past. We are confronted with applications that span the realms of resource planning, telecommunications, VLSI design, financial analysis, scheduling, space planning, energy distribution, molecular engineering, logistics, pattern classification, flexible manufacturing, waste management, mineral exploration, biomedical analysis, environmental conservation and scores of others.

This book explores the meta-heuristic approach called tabu search, which is dramatically changing our ability to solve problems of practical significance. In recent years, journals in a wide variety of fields have published tutorial articles, computational studies and applications documenting successes by tabu search in extending the frontier of problems that can be handled effectively — yielding solutions whose quality often significantly surpasses that obtained by methods previously applied.

A distinguishing feature of tabu search, represented by its exploitation of adaptive forms of memory, equips it to penetrate complexities that often confound alternative approaches. Yet we are only beginning to tap the potential of adaptive memory strategies, and the discoveries that lie ahead promise to be as exciting as those made to date. The knowledge and principles that have currently evolved give a foundation to create practical systems whose capabilities markedly exceed those available earlier, and at the same time invite us to explore still untried variations that may lead to further advances.

We present the major ideas of tabu search with examples that show their relevance to multiple applications. Numerous illustrations and diagrams are used to elucidate principles that deserve emphasis, and that have not always been well understood or applied. Our goal is to provide "hands-on" knowledge and insight alike, rather than to focus exclusively either on computational recipes or on abstract themes. This book is designed to be useful and accessible to researchers and practitioners in management science, industrial engineering, economics, and computer science. It can appropriately be used as a textbook in a masters course or in a doctoral seminar. Because of its emphasis on presenting ideas through illustrations and diagrams, and on identifying associated practical applications, it can also be used as a supplementary text in upper division undergraduate courses.

The development of this book is largely self-contained, and (with the exception of a chapter on applying tabu search to integer programming) does not require prior knowledge of special areas of operations research, artificial intelligence or optimization. Consequently, students with diverse backgrounds can readily grasp the basic principles and see how they are used to solve important problems. In addition, those who have reasonable programming skills can quickly gain a working knowledge of key ideas that will enable them to implement their own tabu search procedures.

Discussion questions and exercises are liberally incorporated in selected chapters to facilitate the use of this book in the classroom. The exercises are designed to be "tutorial," to lead the reader to uncover additional insights about the nature of effective strategies for solving a wide range of problems that arise in practical settings.

A conspicuous feature of tabu search is that it is dynamically growing and evolving, drawing on important contributions by many researchers. Evidence of these contributions will be seen throughout this book. In addition, a chapter is included that underscores the impact of these contributions through a collection of real world applications, together with associated studies of prominent "classical problems."

There are many more applications of tabu search than can possibly be covered in a single book, and new ones are emerging every day. Our goal has been to provide a grounding in the essential ideas of tabu search that will allow readers to create successful applications of their own. We have also sought to provide an understanding of advanced issues that will enable researchers to go beyond today's developments and create the methods of tomorrow.

FRED GLOVER
MANUEL LAGUNA

1 TABU SEARCH BACKGROUND

The abundance of difficult optimization problems encountered in practical settings such as telecommunications, logistics, financial planning, transportation, and production has motivated the development of powerful optimization techniques. These techniques are usually the result of adapting ideas from a variety of research areas. The hope is to develop procedures that are efficient and are able to handle the complexity of today's optimization problems. The ideas that motivate the particular structure of each methodology sometimes come from unexpected connections.

Network flow programming, for example, shares a heritage with models exploiting ideas from electricity and hydraulics. Simulated annealing (SA) is based on a physical process in metallurgy. Genetic algorithms (GAs) seek to imitate the biological phenomenon of evolutionary reproduction, while ant systems simulate a colony of ants that cooperate in a common problem solving activity. The philosophy of tabu search (TS) is to derive and exploit a collection of principles of intelligent problem solving. In this sense, it can be said that tabu search is based on selected concepts that unite the fields of artificial intelligence and optimization.

The basic form of TS is founded on ideas proposed by Fred Glover. (See Section 1.8 for a brief sketch of historical origins.) The method is based on procedures designed to cross boundaries of feasibility or local optimality, which were usually treated as barriers. Early examples of these procedures (derived from surrogate constraint methods and cutting plane approaches) systematically imposed and released constraints to permit exploration of otherwise forbidden regions. Seminal related ideas were also developed by Pierre Hansen in what he labeled a *steepest ascent / mildest descent* method (see Figure 1.1). In addition, as will become clear in this book, many researchers have made important contributions to TS which are shaping the evolution of the method and are responsible for its growing body of successful applications.

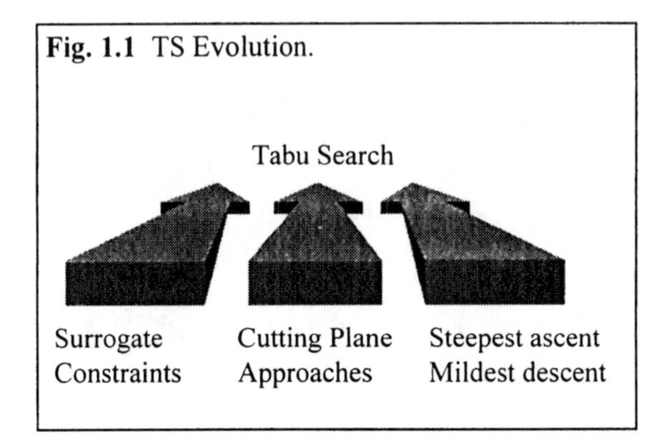

Fig. 1.1 TS Evolution.

Tabu search is a meta-heuristic that guides a local heuristic search procedure to explore the solution space beyond local optimality. (Section 1.9 provides a background on meta-heuristics in general.) The local procedure is a search that uses an operation called *move* to define the neighborhood of any given solution. One of the main components of TS is its use of adaptive memory, which creates a more flexible search behavior. Memory-based strategies are therefore the hallmark of tabu search approaches.

Together with simulated annealing and genetic algorithms, tabu search was evaluated in the widely referenced report by the Committee on the Next Decade of Operations Research (CONDOR 1988) to be "extremely promising" for the future treatment of practical applications. The subsequent rapid and sustained growth of tabu search applications has amply confirmed this evaluation. Pure and hybrid approaches have set new records in finding better solutions to problems in production planning and scheduling, resource allocation, network design and routing in telecommunications among many other areas. Chapter 8 describes numerous specific examples of such applications.

1.1 General Tenets

Webster's dictionary defines *tabu* or *taboo* as "set apart as charged with dangerous supernatural power and forbidden to profane use or contact ..." or "banned on grounds of morality or taste or as constituting a risk ..." Tabu search scarcely involves reference to supernatural or moral considerations, but instead is concerned with imposing restrictions to guide a search process to negotiate otherwise difficult regions. These restrictions operate in several forms, both by direct exclusion of search alternatives classed as "forbidden," and also by translation into modified evaluations and probability of selection. Restrictions are imposed or created by making reference to memory structures that are designed for this specific purpose.

Tabu search is based on the premise that problem solving, in order to qualify as intelligent, must incorporate *adaptive memory* and *responsive exploration.* A good analogy is mountain climbing, where the climber must selectively remember key elements of the path traveled (using adaptive memory) and must be able to make strategic choices along the way (using responsive exploration). (See Figure 1.2.) The adaptive memory feature of TS (whose importance is suggested by the analogy of the mountain climber who must analyze current alternatives in relation to previous ascents of similar terrain) allows the implementation of procedures that are capable of searching the solution space economically and effectively. Since local choices are guided by information collected during the search, TS contrasts with memoryless designs that heavily rely on semirandom processes that implement a form of sampling. Examples of memoryless methods include semi-greedy heuristics and traditional annealing and evolutionary approaches. (Non-traditional memory-based evolutionary approaches incorporating TS ideas are discussed in Chapter 9.) Adaptive memory also contrasts with rigid memory designs typical of branch and bound strategies.

Fig. 1.2 Adaptive memory and responsive exploration.

The emphasis on responsive exploration (and hence purpose) in tabu search, whether in a deterministic or probabilistic implementation, derives from the supposition that a bad strategic choice can yield more information than a good random choice. (In a system that uses memory, a bad choice based on strategy can provide useful clues about how the strategy may profitably be changed. Even in a space with significant randomness — which fortunately is not pervasive enough to extinguish all remnants of order in most real world problems — a purposeful design can be more adept at uncovering the imprint of structure, and thereby at affording a chance to exploit the conditions where randomness is not all-encompassing.)

Responsive exploration integrates the basic principles of intelligent search (i.e., exploiting good solution features while exploring new promising regions). Tabu search is concerned with finding new and more effective ways of taking advantage of the mechanisms associated with both adaptive memory and responsive exploration. The development of new designs and strategic mixes makes TS a fertile area for research and empirical study.

1.2 Use of Memory

The memory structures in tabu search operate by reference to four principal dimensions, consisting of recency, frequency, quality, and influence (Figure 1.3). *Recency-based* and *frequency-based* based memory complement each other, and are addressed in detail in Chapters 2 to 4. The *quality* dimension refers to the ability to differentiate the merit of solutions visited during the search. In this context, memory can be used to identify elements that are common to good solutions or to paths that lead to such solutions. Operationally, quality becomes a foundation for incentive-based learning, where inducements are provided to reinforce actions that lead to good solutions and penalties are provided to discourage actions that lead to poor solutions. The flexibility of these memory structures allows the search to be guided in a multi-objective environment, where the goodness of a particular search direction may be determined by more than one function. The tabu search concept of quality is broader than the one implicitly used by standard optimization methods. This characteristic is elaborated further in Chapter 2.

Fig. 1.3 Four TS dimensions.

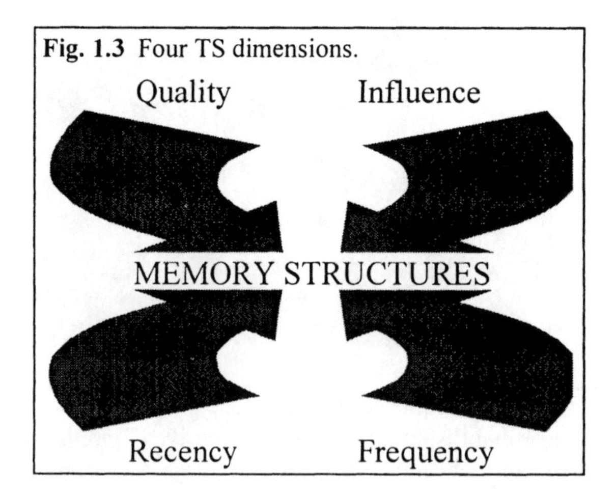

The fourth dimension, *influence*, considers the impact of the choices made during the search, not only on quality but also on structure. (In a sense, quality may be regarded as a special form of influence.) Recording information about the influence of choices on particular solution elements incorporates an additional level of learning. By contrast, in branch and bound, for example, the separation rules are prespecified and the branching directions remain fixed, once selected, at a given node of a decision tree. It is clear however that certain decisions have more influence than others as a function of the neighborhood of moves employed and the way that this neighborhood is negotiated (e.g., choices near the root of a branch and bound tree are quite influential when using a depth-first strategy). The assessment and exploitation of influence by a more flexible memory than embodied in tree searches is an important feature of the TS framework.

Some descriptions of tabu search emphasize the connection that views quality as one of the components of influence. Within this type of classification, *logic* is specified as the fourth dimension of the approach. Such a classification has the merit of underscoring that influence relies on a balance of multiple factors, and highlighting the fact that logic has a useful (often critical) role in intelligent processes. However, quality is such a special form of influence that we prefer to list it separately, and we may consider logic to be one of the elements of strategy generally, by which we organize and exploit the other dimensions indicated.

The memory used in tabu search is both *explicit* and *attributive.* Explicit memory records complete solutions, typically consisting of elite solutions visited during the search. An extension of this memory records highly attractive but unexplored neighbors of elite solutions. The memorized elite solutions (or their attractive neighbors) are used to expand the local search, as indicated in Chapter 3. In some cases explicit memory has been used to guide the search and avoid visiting solutions more than once. This application is limited, because clever data structures must be designed to avoid excessive memory requirements.

Alternatively, TS uses attributive memory for guiding purposes. This type of memory records information about solution attributes that change in moving from one solution to another. For example, in a graph or network setting, attributes can consist of nodes or arcs that are added, dropped or repositioned by the moving mechanism. In production scheduling, the index of jobs may be used as attributes to inhibit or encourage the method to follow certain search directions.

1.2.1 An Illustrative Preview

As a preview of how attributes can be used within a recency-based memory structure, consider the problem of finding an optimal tree (a subgraph without cycles) on the graph with nodes numbered 1 to 7, as shown in Figure 1.4. For the case where the objective function is linear the problem is very easy, and so we may suppose the problem has a more complex nonlinear objective. Such objectives are common in electrical power distribution and telecommunication network design problems, for example. To interpret Figure 1.4, assume that all possible edges that join pairs of nodes are available for composing the tree, and the three subgraphs illustrated for iterations *k*, *k*+1, and *k*+2 (where *k* is arbitrary) identify particular trees generated at different stages of solving the problem.

Fig. 1.4 Optimal tree problem with nonlinear objective.

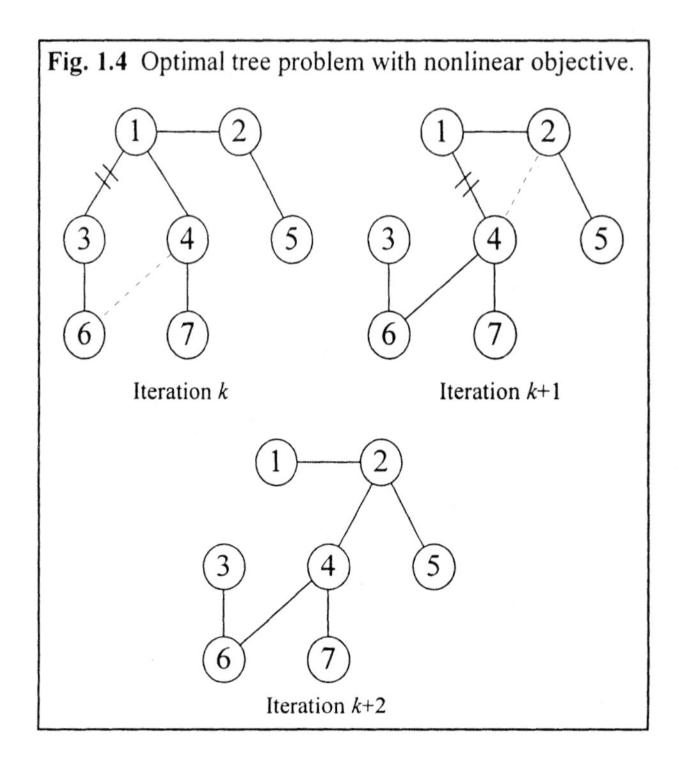

We stipulate for illustration purposes that the moves used for changing one tree into another consist of selecting an edge to be dropped and another to be added, so that the result remains a tree. Note that this is a design issue, given that the search does not have to be limited to the feasible region (i.e., the set of all possible trees). The notion of crossing the feasibility boundary from inside and outside forms the basis for a special kind of tabu search strategy that will be examined in detail later. For convenience in our present example, however, the edge dropped must lie on the unique cycle produced by introducing the edge added, or equivalently, the edge added must join the two separate subtrees created by removing the edge dropped. (This is illustrated by the trees shown for iterations *k* and *k*+1.)

The move applied at iteration *k* to produce the tree of iteration *k*+1 consists of dropping the edge (1,3), and adding the edge (4,6), as shown respectively by the edge marked with two crossed lines and the edge that is dotted in the upper left portion of Figure 1.4. The presence of the edge (1,3) and the absence of the edge (4,6) in the tree of the iteration *k* may be considered as two different solution attributes, which we denote by (1,3)*in* and (4,6)*out*. Since these are attributes that change as a result of the move, they qualify to be designated tabu-active, and to be used to define the tabu status of moves at future iterations. Assume for the moment we will classify a move to be tabu if any of its attributes is tabu-active. In a recency-based memory, for example, we can specify that (1,3)*in* should be tabu-active for 3 iterations, seeking to prevent edge (1,3) from being added back to the current tree for this duration. Therefore, the earliest that edge (1,3) can belong to the current tree will be iteration k+4. In a similar way, (4,6)*out* can be declared tabu-active for 1 iteration, seeking to prevent edge (4,6) from being removed from the current tree for this duration. These conditions effectively seek to avoid "reversing" particular changes created by the move at iteration k.

The indicated *tabu tenures* of 3 and 1 of course are very small, and we discuss how such tabu tenures may be chosen appropriately in subsequent chapters. However, it is useful in this preview to briefly mention the rationale for giving a larger tenure to (1,3)*in* than to (4,6)*out*. Specifically, in our illustration, 15 edges (all those not shown, which are not part of the current tree) can be added to the tree as part of a move to create a new tree, but only 6 edges can be dropped as part of such a move. Thus making (1,3)*in* tabu-active, which prevents edge (1,3) from being added, is much less restrictive than making (4,6)*out* tabu-active, which prevents edge (4,6) from being dropped. Stated differently, preventing an edge from being added excludes a smaller number of moves than preventing an edge from being dropped. In general, the tabu tenure of an attribute depends on the restrictiveness of the associated tabu condition.

In the context of our example, explicit memory can be used to guide the search from the current solution to an elite solution, forcing the search to visit solutions that are difficult to reach when guided solely by changes in the objective function value. Consider, for example, the solution at iteration k+2 depicted in Figure 1.4 and assume that a previously identified elite tree with edges (1,2), (2,5), (3,4), (3,6), (4,5), and (6,7) has been stored in explicit memory. The primary objective of the search can now be made to find a path between the current solution and the stored elite solution. This exploration allows the search to perform moves that may be considered highly unattractive by the objective function evaluation but which at the same time may be *essential* to reach better regions in the solution space. In this particular example, one possibility is to forbid moves that drop edges that are part of both the current tree and the target elite tree. The first iteration of the search stage will then forbid edges (1,2) and (2,5) from being removed. This use of explicit memory is a simple illustration of the path relinking process detailed in Chapter 4.

As illustrated above, explicit and attributive memory are complementary. While explicit memory expands the neighborhood during local search (by including elite solutions), attributive memory typically reduces it (by selectively screening or forbidding certain moves). A special type of memory is created by employing hash functions. These representations may be viewed as a semi-explicit memory that can be used as an alternative to attributive memory. More details about hashing are given in Chapter 7.

1.3 Intensification and Diversification

Two highly important components of tabu search are intensification and diversification strategies. Intensification strategies are based on modifying choice rules to encourage move combinations and solution features historically found good. They may also initiate a return to attractive regions to search them more thoroughly. Since elite solutions must be recorded in order to examine their immediate neighborhoods, explicit memory is closely related to the implementation of intensification strategies. As Figure 1.5 illustrates, the main difference between intensification and diversification is that during an intensification stage the search focuses on examining neighbors of elite solutions.

Fig. 1.5 Intensification and diversification.

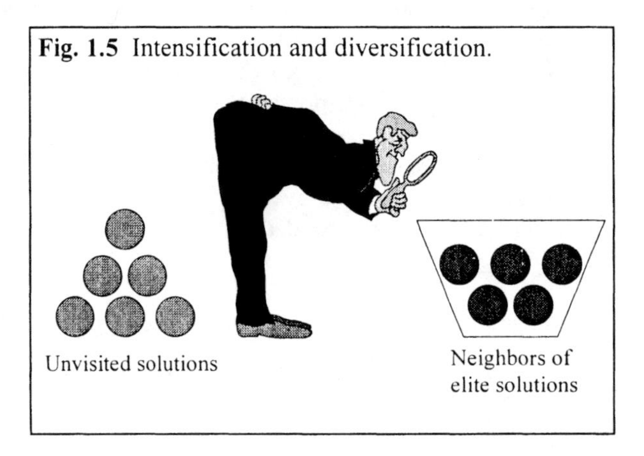

Here the term "neighbors" has a broader meaning than in the usual context of "neighborhood search." That is, in addition to considering solutions that are adjacent or close to elite solutions by means of standard move mechanisms, intensification strategies generate "neighbors" by either grafting together components of good solutions or by using modified evaluations that favor the introduction of such components into a current (evolving) solution. The diversification stage on the other hand encourages the search process to examine unvisited regions and to generate solutions that differ in various significant ways from those seen before. Again, such an approach can be based on generating subassemblies of solution components that are then "fleshed out" to produce full solutions, or can rely on modified evaluations as embodied, for example, in the use of penalty/incentive functions.

Intensification strategies require a means for identifying a set of elite solutions as basis for incorporating good attributes into newly created solutions. Membership in the elite set is often determined by setting a threshold that is connected to the objective function value of the best solution found during the search. However, considerations of clustering and "anti-clustering" are also important in generating such a set, and more particularly for generating subsets of elite solutions that may be used for a specific phase of intensification. In Chapter 4, we will examine forms of intensification and diversification approaches that make use of memory structures (e.g., frequency-based memory) or address the goal of intensifying and diversifying in a more direct way.

As a basis for applying the ideas discussed in this book, we would like to emphasize the value of gaining a working knowledge of the different aspects and strategies contained in the TS framework before tackling difficult problems. Incorporating only a couple of TS-related concepts into a search procedure may result in an inferior method and a frustrating experience. Many of the TS implementations in the literature do not incorporate several of the advanced features described here, due to the good (or bad) fortune of achieving immediate success with simple (and sometimes naive) designs. Our goal, however, is to present most of the strategic issues as early as possible in order to encourage future TS researchers and practitioners to incorporate these concepts into their applications.

1.4 Adaptive Memory Programming

Tabu search has become the focus of numerous comparative studies and practical applications in recent years, and fruitful discoveries about preferred strategies for solving difficult optimization problems have surfaced as a result.

However, sometimes the nature and implications of these discoveries have not been made entirely clear. The reason for this ambiguity is that tabu search has been presented with two faces in the literature, causing it to be viewed as two different methods — one simpler and one more advanced. The simpler method incorporates a restricted portion of the TS design, and is sometimes used in preliminary analyses to test the performance of a limited subset of its components — usually involving only short term memory. The more advanced method embodies a broader framework that includes longer term memory, with associated intensification and diversification strategies. This second approach, due to its focus on exploiting a collection of strategic memory components, is sometimes referred to as *Adaptive Memory Programming* (AMP).

More encompassing forms of tabu search that qualify for the AMP label often prove considerably more effective than the primitive forms. In many situations, attempts to rely predominantly on restricted short term memory is an evident handicap — one that invites comparison to attempts to solve a problem by disengaging (or removing) a part of the brain. Nevertheless, simplified TS approaches are sometimes surprisingly successful. Since they are also frequently quite easy to implement, these approaches will undoubtedly continue to appear in the literature. However, it is important to be aware that they can be strongly dominated by more complete TS methods. Relevant considerations underlying these differences are sketched in Chapters 5 and 7.

1.5 Is Memory Really a Good Idea?

Memory would seem unquestionably to be an integral component of any search that deserves to be called "intelligent." Yet, surprisingly, some of the methods widely hailed as innovations in artificial intelligence — as applied to optimization — are largely devoid of memory. The avoidance of memory is not as unreasonable as might be imagined. Memory, particularly in its adaptive forms, introduces too many degrees of freedom to be treated conveniently in "theorem and proof" developments. Researchers who prefer to restrict consideration to processes (and behavior) that can be characterized by rigorous proofs must focus their efforts in other directions.

Yet there are more subtle and valid reasons to be wary about the use of memory. Malleable forms of memory entail certain dangers — potential pitfalls that go hand in hand with the ability to provide valuable strategic opportunities. These dangers are the price to be paid for the evolution of "intelligent" mechanisms, including biological mechanisms embodied in a brain.

From an evolutionary standpoint, the emergence of memory may be viewed as posing a challenge comparable to the emergence of oxygen, whose corrosive properties (as evolutionary biologists are fond of telling us) caused considerable destruction until organisms adapted to take advantage of them. Analogous perils may well have been created by the emergence of memory, though today we only see the outcomes that survived and flourished. Characteristically, the blind alleys of poorly designed physical structure are conspicuously imprinted on the fossil record. But the blind alleys of poorly regulated mental adjustment — which may have affected survival in subtler ways — remain invisible to us.

It is noteworthy, however, that we are memory users whose evolutionary line has survived. Since we tend to endow our problem solving schemes with features that reflect our own disposition, such schemes tend to be protected (at least to a degree) from dangers otherwise presented by adaptive memory. Even so, hastily contrived uses of memory can lead to conspicuously undesirable outcomes.

Accordingly, in order to solve complex problems more effectively, TS approaches seek to uncover the potential gains of adaptive memory without being caught in the traps of ill-considered memory designs. This leads to a quest for "integrating principles," by which alternative forms of memory are appropriately combined with effective strategies for exploiting them. A novel finding is that such principles are sometimes sufficiently potent to yield effective problem solving behavior with negligible reliance on memory. Over a wide range of problem settings, however, strategic use of memory can make dramatic differences in the ability to solve problems.

1.6 Points of Departure

A starting premise of tabu search is that intelligent inhibition plays a critical role in making effective use of memory. This may be conceived as a reflection of an analogous supposition that appropriate forms of inhibition and restraint correspondingly are essential to survival, although we may not always think of such elements as survival tools. (The usual tenet of our culture is that inhibition represents something that must be overcome, rather than something that can provide important advantages when properly utilized.)

The connotation of the "tabu" term in tabu search carries an implication, as it does in other domains, of rules that are contextual and subject to change. This type of variability can range from narrowly confined interaction to highly complex coordination. The potential intricacy of managing such variation understandably may pose an obstacle to rules that are too rigidly constrained by the quest for mathematical precision, but there is a reverse danger of seeking to handle complexity by the expedient of simplistic rules, particularly those that rely heavily on randomization as a substitute for identifying strategic relationships.

Currently it is fashionable to base the design of search mechanisms on a level of organization represented by primitive organisms. But we may legitimately wonder whether intelligent behavior can be adequately encompassed within physical or biological processes that are distant precursors of the forms of organization embodied in our own brains. If there is value in having the capabilities we call human, then it seems questionable to aspire to mimic something less. (This issue is considered further in Section 1.9.)

There is of course no reason to limit consideration to forms of intelligence that match our own. Evolution presumably may have honed our skills to handle problems that have typically presented themselves in our surroundings. Our prowess may be less impressive for problems confronted in other settings, including problems created as a result of our own technology. A leading goal of TS research is to identify memory and strategy combinations that have merit in a wide range of contexts, not restricted to those we have commonly encountered by the accident of history. If this pursuit may yield insights into different types of intelligence, that would be a welcome bonus. Conceivably, by this perspective, the field of memory-based search methods may have something useful to contribute to the field of cognitive behavior. Up to now, the complementary realms of *search* and *psychology* have been largely isolated from each other. This situation may change, however, as findings about the connections between adaptive memory processes and improved problem solving become systematized.

1.7 Elements of Adaptive Memory

Adaptive memory involves an attribute-based focus, and as already emphasized, depends intimately on the four elements of recency, frequency, quality, and influence. This simple catalog disguises a surprising range of alternatives — as becomes apparent when the four basic elements are considered in combination, and differentiated for different attribute classes over varying regions and spans of time. The notion of influence, for example, characteristically refers to changes in structure, feasibility and regionality. In addition, the logical constructions used to interrelate these elements are not limited to a single dimension, but invite distinctions between "sequential logic" and "event driven logic," whose alternative forms appropriately give rise to different kinds of memory structures.

A number of key questions arise about the nature of these elements, which have important implications for designing search methods. A brief listing of some of these questions follows.

1) Which types of solution attributes can be most effectively exploited by adaptive memory? (What is the impact of different memory-exploiting strategies on selecting neighborhoods for conducting the search?)

2) What types of functions are useful for generating new attributes as combinations of others? (What implications do such functions have for *vocabulary building* methods in tabu search? (See Chapter 7.))

3) What are relevant measures of influence, as reflected in attribute changes caused by moving from one solution to another? (How can these measures assist in isolating characteristics of past trajectories that are relevant for designing current ones?)

4) What thresholds should be established to identify levels of recency, frequency, quality and influence? (What role should these levels play in determining tabu restrictions and aspiration criteria? How can thresholds be used to provide penalties and inducements for selecting particular moves, and for changing the phase of search?)

5) How should randomization be designed to take advantage of information provided by evaluative measures and thresholds? (Which search functions and domains should be governed by probabilistic variation and which should preferably be treated deterministically?)

6) How may memory be applied most effectively to coordinate the use of compound neighborhoods? (What forms of memory are most useful for ejection chain strategies, as a basis for concatenating component moves into more elaborate alternatives? (See Chapter 5.))

7) What clustering and pattern classification approaches are best suited to take advantage of the search history? (How can these approaches be

coordinated to improve intensification and diversification strategies in tabu search?)

8) Which special adaptations of memory and learning give best results for highly context-specific problems? (Conversely, which "generic" forms are most effective over wide ranges of problems whose structure is not predicated in advance?)

Basic considerations and research directions associated with these questions are the focus of the remainder of this book.

1.8 Historical Note on Tabu Search

The foundations of tabu search reflect the theme that good heuristics are motivated by many of the same concerns as good algorithms (exact methods). Moreover, heuristics and algorithms alike can derive benefit from principles distilled from the domains of artificial intelligence (AI) and operations research (OR). Reciprocally, the effective design of such methods can contribute to developing new and sharper versions of principles in areas of AI and OR that involve a quest for improved problem solving techniques.

Some of the popular treatments of tabu search have lost sight of the broader perspectives on which it is based, and it is useful to trace some of the historical background behind the development of the approach to understand its character and the influences that have shaped it. Today's researchers in artificial intelligence and operations research tend to forget that these two fields grew up together, and originally shared many common notions. Both started out with the goal of developing methods to solve challenging problems — problems which often came from common or closely related origins. Some of the early papers on heuristic decision making consciously adopted approaches that bridged AI and OR perspectives (e.g., Simon and Newell, 1958; Fisher and Thompson, 1963; Crowston, et al. 1963). Soon, however, the fields began to diverge, with OR (and optimization) focusing more strongly on mathematical results (especially related to convergence) and AI giving more attention to symbolic and qualitative analyses.

While this division was still in its infant stages, there were a few efforts to introduce non-traditional approaches into the optimization field, which broke away from the customary mathematical emphasis on monotonic convergence ideas, and from the associated somewhat narrowly conceived uses of memory. At the same time, efforts were made to introduce probabilistic and integrated design concepts into heuristic procedures. These developments — which fell somewhere between the dichotomized points of view emerging in the AI and OR communities, and offered the chance for bridges between them — were unfortunately submerged for some years. They provide the foundations for the ideas that resurfaced in the mid to late 1980s and became the source of strategies now at the heart of tabu search.

Four main early developments set the stage for tabu search: (1) strategies that combine decision rules based on logical restructuring and non-monotonic (variable depth) search, (2) systematic violation and restoration of feasibility, (3) flexible memory based on recency and frequency, and (4) selective processes for combining solutions, applied to a systematically maintained population.

The first of these developments comes from a study of decision rules for job shop scheduling problems. Traditional approaches antedating the 1960s had evolved a variety of decision criteria for generating schedules for job shop problems, and embedded these criteria in local decision rules that were applied in isolation from each other. Fisher and Thompson (1963) introduced an innovation of alternating between multiple rules at each decision node by a probabilistic strategy, which used reinforcement learning to amend the probabilities of choosing the rules according to the quality of schedules produced over multiple solution runs. This approach of seeking to benefit from multiple rules by intelligently alternating among them motivated consideration of a contrasting strategy (Glover, 1963), which sought to exploit a collection of decision rules by establishing a way to combine them to create new rules. This approach, whose ideas are woven into various parts of the fabric of tabu search, began by amending the structure of component rules to make them susceptible to being parameterized, according to a common metric. (The result constitutes one of the two main types of logical restructuring used in tabu search, as discussed in Chapter 3.) The method then generated a series of trial solutions by varying the parameters determining the integrated decisions. Instead of stopping at a local optimum of the series, the process generated a non-monotonic objective function surface by systematically continuing to vary the underlying parameters to produce additional trial solutions. The two key ideas — restructuring multiple decision rules for the purpose of combining them to yield new decisions not provided by any of the rules in isolation, and carrying search paths beyond simple local optimality — enabled the method to obtain better solutions than previous approaches, including the strategy of alternating between decision rules probabilistically. Additional implications of this early study are noted in Chapter 9, where consequences of these notions for the use of "intelligent agents" in AI are also examined.

The second development whose ideas have become imprinted on associated strategies of tabu search, was embodied in a method for solving integer programming problems by reference to corner polyhedral relaxations (Glover, 1966, 1969). The method endowed each variable with its own memory as a basis for creating restrictions to avoid the generation of "unproductive" solutions. These restrictions were keyed to recording limited attributes of the solutions, rather than recording full solutions explicitly. The approach also introduced a strategy of combining pairs of solutions, from a select subset of those generated, to provide new solutions as candidates for optimality. The mode of joining solutions used progressive simple summations, constituting a precursor of the scatter search approach (see Chapter 9). Again, the method did not stop at a local optimum relative to this sequence of trial solutions, but instead terminated the progression by selecting the best local optimum when the memory and associated bounding rules disallowed any variable from being used to extend the sequence.

Following terminology popularized by Papadimitriou and Steiglitz (1982), the approach constitutes an instance of what more recently has come to be called a "variable depth" method. Such methods are often emphasized as incorporating a design for going beyond local optimality, although this characterization is not accurate in all cases. As described by Papadimitriou and Steiglitz, drawing on ideas of Kernighan and Lin (1970), such methods terminate with a local optimum defined relative to successive variable depth passes, stopping when the solution obtained on a given variable depth pass is no better than the solution obtained on the immediate preceding pass. By contrast, in the form of the approach for general integer programming problems (Glover, 1966), successive variable depth passes generate cutting planes as well as trial solutions, yielding a coordinated process that ultimately assures a global optimum will be obtained. If a given pass is carried to completion rather than terminated heuristically, it also yields a global optimum for an underlying group theoretic integer program. (As shown by Denardo and Fox (1979), the method gives the best known computational bounds for this group theoretic problem.)

From the tabu search perspective, a key added contribution of this approach was the use of a flexible individualized memory, and associated conditional restrictions for controlling the solutions that can be admissibly generated. The fact that this method is not a heuristic but a convergent algorithm reinforces the premise that heuristic and algorithmic procedures share significant concerns in common.

The third development likewise involves an exact approach for integer programming problems. In this case the underlying procedure was founded on an augmentation of the simplex method for linear programming (LP), using cutting plane processes to introduce new constraints implied by the integer requirement for the variables. The typical LP design underlying cutting plane approaches of this type was to begin from a feasible starting point (in either a "primal" or "dual" sense), and then to maintain this feasibility condition while progressing monotonically to an optimal solution. However, combinatorial optimization poses different topological challenges than linear optimization, giving rise to primal and dual regions that are nonconvex and even disconnected. This fact motivated the creation of a *pseudo primal-dual* method (Glover, 1968) that purposely violated and then restored feasibility conditions, making it possible to "cut across" various regions of the solution space. Finite convergence to optimality was assured by rules that allowed feasibility always to be recovered with a net advance. The strategy of purposely visiting both infeasible and feasible regions in successive waves likewise has become a feature of one of the principal component strategies of tabu search.[1] This approach, called strategic oscillation, is treated at length in Chapter 4.

---

[1] As pointed out by Ralph Gomory, in an interview commemorating his invention of the first finitely convergent cutting plane method for integer programming, (Gomory, 1996), early

The last of the four developments involves the introduction of surrogate constraint methods for integer programming (Glover 1965). Such methods were based on the strategy of combining constraints to produce new ones, with the goal of yielding information not contained separately in the parent constraints. The information capturing function soon led to uses of recency and frequency memory both as a means to extract such information and as a way to exploit it within heuristic and algorithmic processes. These strategies are a focus of developments covered throughout this book, with specific consideration of surrogate constraint concepts in Chapter 4, and we will not elaborate further on them here. However, it is interesting to note an observation from a paper that sought to combine surrogate constraint procedures with heuristic methods for integer programming (Glover, 1977). The observation addresses the attitude toward heuristics that prevailed in optimization circles during the 1970s, by which time the schism between AI and OR had become quite strongly defined. The climate of the times (and a challenge to it) was expressed as follows.

> Algorithms have long constituted the more respectable side of the family [of optimization methods], assuring an optimal solution in a finite number of steps. Methods that merely claim to be clever, and do not boast an entourage of supporting theorems and proofs, are accorded a lower status. Algorithms are conceived in analytic purity in the high citadels of academic research, heuristics are midwifed by expediency in the dark corners of the practitioner's lair. [Yet we are coming to recognize that] algorithms are not always successful, and their heuristic cousins deserve a chance to prove their mettle.... Algorithms, after all, are merely fastidious heuristics in which epsilons and deltas abide by the dictates of mathematical etiquette. It may even be said that algorithms exhibit a somewhat compulsive aspect, being denied the freedom that would allow an occasional inconstancy or an exception to ultimate convergence. (Unfortunately, ultimate convergence sometimes acquires a religious significance: it seems not to happen in this world.) The heuristic approach, robust and boisterous, may have special advantages in terrain too rugged or varied for algorithms. In fact, those who are fond of blurring distinctions suggest that an algorithm worth its salt is one with "heuristic power."

Ironically, in spite of the suggestion made in this quote that a reconciliation of viewpoints was imminent, widespread recognition of the relevance of heuristics within optimization was not to occur in the mainstream of the OR field for nearly a decade more. The irony is heightened by the fact that the paper containing the preceding quote introduced several connections that have become part of tabu search, and which likewise were not to be pursued by researchers until the mid 1980s.

---

cutting plane methods and associated polyhedral strategies were abruptly abandoned by researchers in the early 1970's, without the benefit of appreciable testing, and may well warrant reconsideration.

Regardless of the fashions that may capture the attention of researchers in OR or AI, then or now, the theme that (good) algorithms and heuristics are intimately related continues to be a basic part of the perspective that underlies tabu search. Manifestations of this perspective occur throughout this book, and particularly in the application of tabu search to the general area of integer programming, as discussed in Chapter 6.

1.9 Historical Note on Meta-Heuristics

The term *meta-heuristic* (also written *metaheuristic*) was coined in the same paper that introduced the term *tabu search* (Glover, 1986), and has come to be widely applied in the literature, both in the titles of comparative studies and in the titles of volumes of collected research papers (see, e.g., Laporte and Osman, 1995; Gendreau, Laporte and Potvin, 1995; Al-Mahmeed, 1996; Charon and Hudry, 1996; Osman and Kelly, 1996).

A meta-heuristic refers to a master strategy that guides and modifies other heuristics to produce solutions beyond those that are normally generated in a quest for local optimality. The heuristics guided by such a meta-strategy may be high level procedures or may embody nothing more than a description of available moves for transforming one solution into another, together with an associated evaluation rule.

The contrast between the meta-heuristic orientation and the "local optimality" orientation is significant. For many years, the primary conception of a heuristic procedure (a conception still prevalent today) was to envision either a clever rule of thumb or an iterative rule that terminates as soon as no solutions immediately accessible could improve the last one found. Such iterative heuristics are often referred to as descent methods, ascent methods, or local search methods. (A sign of the times is that "local search" now sometimes refers to search that is not limited to being local in character.) Consequently, the emergence of methods that departed from this classical design — and that did so by means of an organized master design — constituted an important advance. Widespread awareness of this advance only began to dawn during the last decade, though its seeds go back much farther.

The evolution of meta-heuristics during the past half dozen years has taken an explosive upturn. Meta-heuristics in their modern forms are based on a variety of interpretations of what constitutes "intelligent" search. These interpretations lead to design choices that in turn can be used for classification purposes. However, a rigorous classification of different meta-heuristics is a difficult and risky enterprise, because the leading advocates of alternative methods often differ among themselves about the essential nature of the methods they espouse. This may be illustrated by considering the classification of meta-heuristics in terms of their features with respect to three basic design choices: (1) the use of adaptive memory, (2) the kind of neighborhood exploration used, and (3) the number of current solutions carried from one iteration to the next. We embed these options in a classification scheme of the form $x/y/z$, where the choices for x are A (if the meta-heuristic employs adaptive memory) and M (if the method is "memoryless"). The choices for y are N (for a method that employs some systematic neighborhood search either to select the next move or to improve a given solution) and S (for those methods relying on random sampling). Finally, z may be 1 (if the method moves from one current solution to the next after every iteration) or P (for a population-based approach with a population of size P). This simple 3-dimensional scheme gives us a preliminary basis of classification, which discloses that agreement on the proper way to label various meta-heuristics is far from uniform. We show this by providing classifications for a few well-known meta-heuristics in Table 1.1.

Table 1.1 Meta-heuristic classification.

Meta-heuristic	*Classification* 1	*Classification* 2
Genetic algorithms	M/S/P	M/N/P
Scatter search	M/N/P	A/N/P
Simulated annealing	M/S/1	M/N/1
Tabu search	A/N/1	A/N/P

Two different ways are given for classifying each of these procedures. The first classification most closely matches the "popular conception" and the second is favored by a significant (if minority) group of researchers. The differences in these classifications occur for different reasons, depending on the method. Some differences have been present from the time the methods were first proposed, while others represent recent changes that are being introduced by a subgroup of ardent proponents. For example, the original form of simulated annealing (as reported in Kirkpatrick, et. al, 1983) has come to be modified by a group that believes stronger elements of neighborhood search should be incorporated. A similar change came about in genetic algorithms, a few years before it was introduced in simulated annealing, in the mid 1980s. Still, it should be pointed out that not all the advocates of simulated annealing and genetic algorithms view these changes as appropriate.

On the other hand, among those examples where different classifications were present from the start, the foundation papers for tabu search included population-based elements in the form of strategies for exploiting collections of elite solutions saved during the search. Yet a notable part of the literature has not embraced such population-based features of tabu search until recently. Similarly, scatter search was accompanied by adaptive memory elements as a result of being associated with early tabu search ideas, but this connection is likewise only beginning to be pursued.

A few proponents of simulated annealing and genetic algorithms have recently gone farther in modifying the original conceptions than indicated in Table 1.1, to propose the inclusion of elements of adaptive memory as embodied in tabu search. Such proposals are often described by their originators as *hybrid* methods, due to their marriage of aspects from different frameworks. We will comment on these developments later.

1.9.1 Additional Meta-Heuristic Features

In addition to the three basic design elements used in our classification, meta-heuristics incorporate other strategies with the goal of guiding the search. A meta-heuristic may strategically modify the evaluation provided by a component heuristic (which normally consists of identifying the change in an objective function value produced by a move). For example, simulated annealing relies on a problem objective function to provide each evaluation, but then amends this evaluation based on the current solution. In the amended form, all improving moves are considered equally attractive, and any such move encountered is accepted. Disimproving moves are accepted or rejected by a probabilistic criterion that initially assigns a high probability (when the temperature is high) to accepting any move generated, regardless of its quality. However, a bias is incorporated that favors smaller disimprovements over larger ones, and over time this bias is increased, ultimately reducing the probability of accepting a nonimproving move to zero. The set of available moves can be taken from another heuristic, but classical SA pre-empts all other move generation processes to generate moves randomly from the proposed domain.

A meta-heuristic may also modify the neighborhood of moves considered to be available, by excluding some members and introducing others. This amended neighborhood definition may itself necessitate a change in the nature of evaluation. The strategic oscillation approach of tabu search (Glover, 1977) illustrates this intimate relationship between changes in neighborhood and changes in evaluation. A standard neighborhood that allows moves only among feasible solutions is enlarged by this approach to encompass infeasible solutions. The search is then strategically driven to cross the feasibility boundary to proceed into the infeasible region. After a selected depth is reached, the search changes direction to drive back toward feasibility, and upon crossing the feasibility boundary similarly continues in the direction of increased feasibility. (One-sided oscillations are employed in some variants to remain predominantly on a particular side of the boundary.) To guide these trajectories, the approach modifies customary evaluations to take account of the induced direction of movement and the region in which the movement occurs. The result generates a controlled behavior that exploits the theme of non-monotonic exploration.

The emphasis on guidance differentiates a meta-heuristic from a simple random restart procedure or a random perturbation procedure. However, sometimes these naive restarting and perturbation procedures are also classed as low-level meta-heuristics, since they allow an opportunity to find solutions that are better than a first local optimum encountered. "Noising" procedures, which introduce controlled randomized changes in parameters such as cost or resource availability coefficients, provide one of the popular mechanisms for implementing such approaches. Another popular mechanism is simply to randomly modify evaluations, or to choose randomly from evaluations that fall within a chosen window. Such randomized processes are also applied to selecting different types of moves (neighborhood definitions) at different junctures.

In contrast to an orientation that still often appears in the literature, the original conception of a meta-heuristic does not exclude consideration of constructive moves for generating initial solutions, but likewise allows these moves to be subjected to meta-heuristic guidance. (A popular orientation in the literature is to suppose that meta-heuristics are only used in connection with "transition" moves, which operate on fully constructed solutions.) From a broader perspective, a partial solution created by a constructive process is simply viewed as a solution of a particular type, and procedures for generating such solutions are natural candidates to be submitted to higher level guidance. This view has significant consequences for the range of strategies available to a meta-heuristic approach.

Strategic oscillation again provides an illustration. By the logical restructuring theme of tabu search, constructive moves are complemented by creating associated destructive moves, allowing the oscillation to proceed constructively to (and beyond) a stipulated boundary, and then to reverse direction to proceed destructively to various depths, in alternating waves. Transition moves permit refinements at varying levels of construction and destruction.

The perspective that restricts attention only to transition moves is gradually eroding, as researchers are coming to recognize that such a restriction can inhibit the development of effective methods. However, there remain pockets where this recognition is slow to dawn. (For example, methods that alternate between construction and transition moves — affording a simple subset of options provided by strategic oscillation — have recently been characterized in a segment of the literature as a "new development.")

1.9.2 Common Distinctions

Our earlier meta-heuristic classification, which differentiates between population-based strategies and adaptive memory strategies, is often taken to be a fundamental distinction in the literature. Population-based strategies manipulate a collection of solutions rather than a single solution at each stage. Such procedures are now often referred to as composing the class of *evolutionary methods.* A prominent subclass of these methods is based on strategies for "combining" solutions, as illustrated by genetic algorithms, scatter search and path relinking methods (see Chapter 4). Another prominent subclass consists of methods that are primarily driven by utilizing multiple heuristics to generate new population members. This incorporation of multiple heuristics for generating trial solutions, as opposed to relying on a single rule or decision criterion, is a very old strategy whose origins are probably not traceable. Some of the recent evolutionary literature occasionally cites work of the mid 1960s as embodiments of such ideas, but such work was clearly preceded by earlier developments, as noted in the observations of Simon and Newell (1958). The key to differentiating the contributions of such methods obviously rests on the novelty of the component heuristics and the ingenuity of the strategies for coordinating them. Such concerns are more generally the focus of parallel processing solution methods, and many "evolutionary" contributions turn out chiefly to be a subset of the strategies that are being developed to a higher level of sophistication under the parallel processing rubric.

The adaptive memory classification provides a more precise means of differentiation, although it is not without pitfalls. From a naive standpoint, virtually all heuristics other than complete randomization induce a pattern whose present state depends on the sequence of past states, and therefore incorporate an implicit form of "memory." Given that the present is inherited from the past, the accumulation of previous choices is in a loose sense "remembered" by current choices. This sense is slightly more pronounced in the case of solution combination methods such as genetic algorithms and scatter search, where the mode of combination more clearly lends itself to transmitting features of selected past solutions to current solutions. Such an implicit memory, however, does not take a form normally viewed to be a hallmark of an intelligent memory construction. In particular, it uses no conscious design for recording the past and no purposeful manner of comparing previous states or transactions to those currently contemplated. By contrast, at an opposite end of the spectrum, procedures such as branch and bound and A* search use highly (and rigidly) structured forms of memory — forms that are organized to generate all nondominated solution alternatives with little or no duplication.

Adaptive memory procedures, properly conceived, embody a use of memory that falls between these extremes, based on the goal of combining flexibility and ingenuity. Such methods typically seek to exploit history in a manner inspired by (but not limited to) human problem solving approaches. They are primarily represented by tabu search and its variations that sometimes receive the "adaptive memory programming" label. In recent years, as previously intimated, other approaches have undertaken to incorporate various aspects of such memory structures and strategies, typically in rudimentary form. Developments that produce hybrids of tabu search with other approaches at a more advanced level have become an important avenue for injecting adaptive memory into other methods, and constitute an active area of research.

Another distinction based on memory is introduced by neural network (NN) approaches. Such methods emphasize an associative form of memory, which has its primary application in prediction and pattern matching problems. Neural network procedures also implicitly involve a form of optimization, and in recent years such approaches have been adapted to several optimization settings. Performance is somewhat mixed, but researchers in optimization often regard neural networks as appropriate to be included within the meta-heuristic classification. Such an inclusion is reinforced by the fact that NN-based optimization approaches sometimes draw on standard heuristics, and produce solutions by transformations that are not limited to ordinary notions of local optimality. A number of initiatives have successfully combined neural networks with simulated annealing, genetic algorithms and, most recently, tabu search (see Chapters 8 and 9).

Meta-heuristics are often viewed as composed of processes that are intelligent, but in some instances the intelligence belongs more to the underlying design than to the particular character (or behavior) of the method itself. The distinction between intelligent design and intelligent behavior can be illustrated by considering present day interior point methods of linear programming. Interior point methods (and more general barrier function methods) exploit a number of ingenious insights, and are often remarkably effective for achieving the purposes for which they were devised. Yet it seems doubtful whether such methods should be labeled intelligent, in the sense of being highly responsive to varying conditions, or of changing the basis for their decisions over time as a function of multiple considerations. Similar distinctions arise in many other settings. It must be conceded that the line that demarks intelligent methods from other methods is not entirely precise. For this reason it is not necessary for a master procedure to qualify as intelligent in a highly rigorous sense in order to be granted membership in the category of meta-heuristics.

1.9.3 Metaphors of Nature

A popular thrust of many research initiatives, and especially of publications designed to catch the public eye, is to associate various methods with processes found in nature. This trend embodies a wave of "New Romanticism," reminiscent of the Romanticism of the 18th and 19th centuries (distinguished by their preoccupation with Nature with a capital "N"). The current fascination with natural phenomena as a foundation for problem-solving methods undoubtedly is fueled by our sense of mystery concerning the ability of such phenomena to generate outcomes that are still far beyond our comprehension. However, the New Romanticism goes farther, to suggest that by mimicking the rules we imagine to operate in nature (especially "rudimentary" processes of nature) we will similarly be able to produce remarkable outcomes.

Models of nature that are relied upon for such inspiration are ubiquitous, and it is easy to conjure up examples whose metaphorical possibilities have not yet been tapped. To take an excursion in the lighter side of such possibilities (though not too far from the lanes currently traveled), we may observe that a beehive offers a notable example of a system that possesses problem solving abilities. Bees produce hives of exceptional quality and complexity, coordinate diverse tasks among different types of individuals, perform spatial navigation, and communicate via multiple media. (It is perhaps surprising in retrospect that the behavior of bees has not been selected as a basis for one of the "new" problem solving methods.)

Nor is it necessary to look simply to sentient creatures for analogies that inspire templates for effective problem solving. The root system of a tree, for example, provides an intriguing model for parallel computation. In order to find moisture and nutrients (analogous to a quest for "solutions"), roots distribute themselves across different regions, sending out probes that multiply or atrophy according to the efficacy of their progress. The paths of such a system may cross, as different channels prove promising by virtue of the regions in which they lie and also according to the directions in which they are explored. Obstacles are effectively skirted, or over time are surmounted by longer range strategies — as by extending finer probes, which ultimately expand until the medium is broached. There exist some root systems, as in groves of aspen, where roots of one entity can merge with those of another, thus enlarging the potential sources of communication and contact available to each. (Such an interlinked community gives rise to the largest known organisms on the planet.)

These analogies to systems in nature invite us to ponder a key question. If we were allowed to place our bets on the probable success of a hive of bees or a grove of aspen, as opposed to that of a group of humans, when confronted with a challenging task that requires intelligence and the ability to learn from the past, how would we wager? Undoubtedly we would be drawn to reflect that our goals and problem structures may often be different than those to which "natural processes" apply. In addition, we ourselves — as products of a rather special and extended chain of natural developments — may incorporate capabilities not present in the processes that produced us.

Metaphors of nature have a place. They appear chiefly to be useful for spurring ideas to launch the first phases of an investigation. As long as care is taken to prevent such metaphors from cutting off lines of inquiry beyond their scope, they provide a means for "dressing up" the descriptions of various meta-heuristics in a way that appeals to our instinct to draw parallels between simple phenomena and abstract designs.

Invoking such parallels may sometimes appear to embody a primitive mysticism, akin to chanting about campfires in the night, but it gives us a foundation for connecting the new to the old, and for injecting passion into our quests. It is up to prudence to determine when the symbolism of the New Romanticism obscures rather than illuminates the pathway to improved understanding. Within the realm of meta-heuristic design, there is a great deal we have yet to learn. The issue of whether the analogies that underlie some of our models may limit or enhance our access to further discovery deserves careful reflection.

1.10 Discussion Questions and Exercises

1. Consider a tabu search for a sequencing problem that uses insert moves to transform one sequence into another (see Figure 1.6). What is the most likely form of an explicit memory structure? Describe at least two attributive memory structures that could be used in this method.

Fig. 1.6 Insert move.

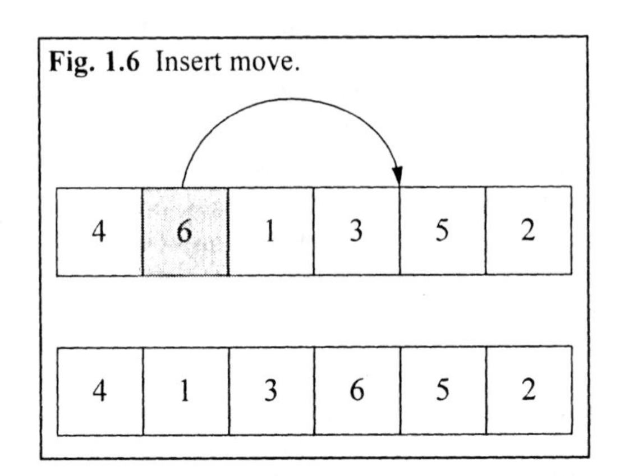

2. For the sequencing problem and move mechanism in question 1, discuss the interpretation of the four memory dimensions: recency, frequency, quality, and influence.

3. Elite solutions are used during an intensification stage to construct solutions that incorporate "good" attributes from these solutions. Discuss a form of intensification based on elite solutions that can be used for solving sequencing problems.

4. A variety of search methods rely on randomization as a means of achieving diversification. Discuss the advantages and disadvantages of this approach compared to a systematic form of diversification based on the use of memory. Can these two approaches be more nearly harmonized by a "strongly biased" randomization that uses probabilities influenced by memory?

2 TABU SEARCH FOUNDATIONS: SHORT TERM MEMORY

Tabu search can be applied directly to verbal or symbolic statements of many kinds of decision problems, without the need to transform them into mathematical formulations. Nevertheless, it is useful to introduce mathematical notation to express a broad class of these problems, as a basis for describing certain features of tabu search. We characterize this class of problems as that of optimizing (minimizing or maximizing) a function $f(x)$ subject to $x \in \mathbf{X}$, where $f(x)$ may be linear or nonlinear, and the set $\mathbf{X}$ summarizes constraints on the vector of decision variables x. The constraints may include linear or nonlinear inequalities, and may compel all or some components of x to receive discrete values. While this representation is useful for discussing a number of problem solving considerations, we emphasize again that in many applications of combinatorial optimization, the problem of interest may not be easily formulated as an objective function subject to a set of constraints. The requirement $x \in \mathbf{X}$, for example, may specify logical conditions or interconnections that would be cumbersome to formulate mathematically, but may be better be left as verbal stipulations that can be then coded as rules.

Tabu search begins in the same way as ordinary local or neighborhood search, proceeding iteratively from one point (solution) to another until a chosen termination criterion is satisfied. Each $x \in \mathbf{X}$ has an associated neighborhood $\mathbf{N}(x) \subset \mathbf{X}$, and each solution $x' \in \mathbf{N}(x)$ is reached from x by an operation called a *move*.

As an initial point of departure, we may contrast TS with a simple descent method where the goal is to minimize $f(x)$ (or a corresponding ascent method where the goal is to maximize $f(x)$). Such a method only permits moves to neighbor solutions that improve the current objective function value and ends when no improving solutions can be found. A pseudo-code of a generic descent method is presented in Figure 2.1. The final x obtained by a descent method is called a local optimum, since it is at least as good or better than all solutions in its neighborhood. The evident shortcoming of a descent method is that such a local optimum in most cases will not be a global optimum, i.e., it usually will not minimize $f(x)$ over all $x \in \mathbf{X}$.[1]

Fig. 2.1 Descent method.

1) Choose $x \in \mathbf{X}$ to start the process.
2) Find $x' \in \mathbf{N}(x)$ such that $f(x') < f(x)$.
3) If no such x' can be found, x is the local optimum and the method stops.
4) Otherwise, designate x' to be the new x and go to 2).

The version of a descent method called *steepest descent* scans the entire neighborhood of x in search of a neighbor solution x' that gives a smallest $f(x')$ value over $x' \in \mathbf{N}(x)$. Steepest descent implementations of some types of solution approaches (such as certain path augmentation algorithms in networks and matroids) are guaranteed to yield globally optimal solutions for the problems they are designed to handle, while other forms of descent may terminate with local optima that are not global optima. In spite of this attractive feature, in certain settings steepest descent is sometimes impractical because it is computationally too expensive, as where $\mathbf{N}(x)$ contains many elements or each element is costly to retrieve or evaluate. Still, it is often valuable to choose an x' at each iteration that yields a "good" if not smallest $f(x')$ value.

The relevance of choosing good solutions from current neighborhoods is magnified when the guidance mechanisms of tabu search are introduced to go beyond the locally optimal termination point of a descent method. Thus, an important first level consideration for tabu search is to determine an appropriate *candidate list strategy* for narrowing the examination of elements of $\mathbf{N}(x)$, in order to achieve an effective

[1] Descent methods are often viewed as applicable only to transition neighborhoods, as opposed to constructive neighborhoods. However, most constructive methods can be conceived to be descent methods, where the objective penalizes infeasibility, and each constructive step toward a more complete (and more nearly feasible) solution therefore is an improving step in the neighborhood under consideration. The final constructive step qualifies as a local optimum since going beyond a complete construction (as by adding additional edges to a tour) is either once again infeasible or conspicuously undesirable.

tradeoff between the quality of x' and the effort expended to find it. (Here quality may involve considerations beyond those narrowly reflected by the value of $f(x')$.)[2] To give a foundation for understanding the basic issues involved, we turn our attention to the following illustrative example, which will also be used as a basis for illustrating various aspects of tabu search in later sections.

Machine Scheduling / Job Sequencing Example

Suppose that a descent method is used to find a local optimum for the single machine scheduling problem that seeks an order for sequencing jobs on the machine to minimize total tardiness. In this problem, each job j ($j = 1, \ldots, n$) has a processing time p_j, a due date d_j, and a tardiness penalty w_j, and the objective is to minimize

$$T = \sum_{j=1}^{n} w_j \left[C_j - d_j\right]^+ ,$$

where C_j is the completion time of job j and for any value v, $[v]^+ = \mathrm{Max}(0, v)$. The completion time C_j equals the processing time p_j for job j, plus the sum of the processing times of all jobs that are sequenced before job j. Thus, our goal is to find a sequence of jobs that will minimize total weighted tardiness as expressed by T. The objective function is a "regular measure of performance" (in standard scheduling terminology), and it has been shown that there exists an optimal schedule in which there is no idle time between any two jobs scheduled. Therefore a sequence of jobs, which constitutes a permutation, completely defines a solution.

Pairwise exchanges (or *swaps*) are frequently used as one of the ways to define neighborhoods in permutation problems, identifying moves that lead from one sequence to the next. In this case, a swap identifies two particular jobs and places each in the sequence location previously occupied by the other. The complete "swap neighborhood" of a given sequence contains all combinations of n jobs taken two at a time. Consequently, there are $\frac{n(n-1)}{2}$ swaps in the neighborhood $\mathbf{N}(x)$, where x is a solution that corresponds to a current permutation . This means that even if each swap can be evaluated in one operation, finding the best swap at each step of the descent method is a process that can require considerable effort ($O(n^2)$, in computational complexity notation).

---

[2] If a neighborhood space is totally random, then of course nothing will work better than a totally random choice. (In such a case there is no merit in trying to devise an effective solution procedure.) Assuming that neighborhoods can be identified that are reasonably meaningful for a given class of problems, the challenge is to define solution quality appropriately so that evaluations likewise will have meaning. By the TS orientation, the ability to use history in creating such evaluations then becomes important for devising effective methods.

In reality it is not necessary to apply this "brute-force" process to find the next best move to make. By using context information, we can limit the local search to those moves that are more likely to improve the current solution. To illustrate this, suppose that a 6-job problem has processing times given by (6, 4, 8, 2, 10, 3), due dates specified by (9, 12, 15, 8, 20, 22), and tardiness penalties $w_j = 1$ for $j = 1, ..., 6$. Also suppose that the sequence (1, 2, 3, 4, 5, 6) with objective function value of 36 is used to start a descent method that employs swaps as the move mechanism. We define the *move value* of a move from a current solution x to a new solution x' as the net improvement (or disimprovement) in the objective function, i.e., as the value $f(x') - f(x)$.

Table 2.1 Swap neighborhood.

Jobs			*move*	*abs*
i	*j*	*T*	*value*	$(d_i - d_j)$
1	2	37	1	3
1	3	42	6	6
1	4	32	-4	1
1	5	57	21	11
1	6	40	4	13
2	3	39	3	3
2	4	30	-6	4
2	5	56	20	8
2	6	43	7	10
3	4	30	-6	7
3	5	40	4	5
3	6	30	-6	7
4	5	44	8	12
4	6	39	3	14
5	6	29	-7	2

Table 2.1 shows the evaluation of the neighborhood $\mathbf{N}(x_0)$ of the initial solution x_0 (15 swaps). It is clear from this table that there is a large variation in the quality of each swap in the neighborhood (measuring quality here by the move value). It also becomes apparent that some low quality swaps may be eliminated heuristically before calculating their move value, by a "first level filter" that considers the difference between the due dates of the exchanging jobs. In fact by implementing a rule which specifies that only swaps of jobs with a maximum absolute due date difference of 3 are considered, only 4 out of 15 swaps need to be evaluated. The reduced neighborhood is approximately 27% of the original one, and it contains the best swap (which consists of exchanging jobs 5 and 6 with a move value of -7). More sophisticated strategies for neighborhood reduction are possible, and they

become more attractive as the evaluation of the entire neighborhood becomes more time-consuming. Again, this consideration becomes magnified when we concern ourselves with aspects of tabu search that go beyond the realm of descent methods. We examine candidate list strategies to address such issues in Chapter 3.

Regardless of the approach used to examine the neighborhood **N**(x), descent methods by themselves have very limited success at solving hard combinatorial optimization problems. This is why several procedures have emerged that attempt to find improved solutions by allowing the search to generate alternatives other than those produced by applying the strategy of descent. Some of these procedures rely on randomization to sample the search space by various mechanisms until a termination criterion is met. Among these procedures are genetic algorithms, simulated annealing, and multistart methods such as GRASP and other forms of iterated descent. We discuss some connections and possibilities for creating hybrid methods in Chapter 9. Tabu search, in contrast to these methods, employs a somewhat different philosophy for going beyond the criterion of terminating at a local optimum. Randomization is de-emphasized, and generally is employed to facilitate operations that are otherwise cumbersome to implement or whose strategic implications are unclear. (Randomization in TS is strongly biased to reflect strategic evaluations, by introducing probabilistic variation to counter "noise" in these evaluations.) The main assumption is that intelligent search should be based on systematic forms of guidance, like those we are about to discuss.

2.1 Memory and Tabu Classifications

An important distinction in TS arises by differentiating between short term memory and longer term memory. Each type of memory is accompanied by its own special strategies. However, the effect of both types of memory may be viewed as modifying the neighborhood **N**(x) of the current solution x. The modified neighborhood, which we denote by $\mathbf{N}^*(x)$, is the result of maintaining a selective history of the states encountered during the search.

In the TS strategies based on short term considerations, $\mathbf{N}^*(x)$ characteristically is a subset of **N**(x), and the tabu classification serves to identify elements of **N**(x) excluded from $\mathbf{N}^*(x)$. In TS strategies that include longer term considerations, $\mathbf{N}^*(x)$ may also be expanded to include solutions not ordinarily found in **N**(x). Characterized in this way, TS may be viewed as a dynamic neighborhood method. This means that the neighborhood of x is not a static set, but rather a set that can change according to the history of the search. This feature of a dynamically changing neighborhood also applies to the consideration of selecting different component neighborhoods from a collection that encompasses multiple types or levels of moves, and provides an important basis for parallel processing. Characteristically, a TS process based strictly on short term strategies may allow a solution x to be visited more than once, but it is likely that the corresponding reduced neighborhood $\mathbf{N}^*(x)$ will be different each time. With the inclusion of longer term considerations, the likelihood of duplicating a previous neighborhood upon revisiting a solution, and more generally of making choices that repeatedly visit only a limited subset of **X**, is all but nonexistent. From a practical standpoint, the method will characteristically identify an optimal or near optimal solution long before a substantial portion of **X** is examined.

A crucial aspect of TS involves the choice of an appropriate definition of $\mathbf{N}^*(x)$. Due to the exploitation of memory, $\mathbf{N}^*(x)$ depends upon the trajectory followed in moving from one solution to the next (or upon a collection of such trajectories in a parallel processing environment). As a starting point for examining short term memory concerns, consider a form of memory embodied in a tabu list **T** that explicitly contains t different solutions, that is, $\mathbf{T} = \{x_1, x_2, \ldots, x_t\}$, where $\mathbf{N}^*(x) = \mathbf{N}(x) \setminus \mathbf{T}$, i.e., $\mathbf{N}^*(x)$ consists of the solutions of $\mathbf{N}(x)$ not included on the tabu list.

Suppose we apply this structure to a TS procedure that uses swaps to search the solution space of a 4-job sequencing problem as illustrated by the Machine Scheduling / Job Sequencing Example of the preceding section. If the current solution is given by $x = (1,\ 2,\ 3,\ 4)$ and the tabu list consists of

$$\mathbf{T} = \{(1,\ 3,\ 2,\ 4), (3,\ 1,\ 2,\ 4), (3,\ 2,\ 1,\ 4)\},$$

Fig. 2.2 Reduced neighborhood.

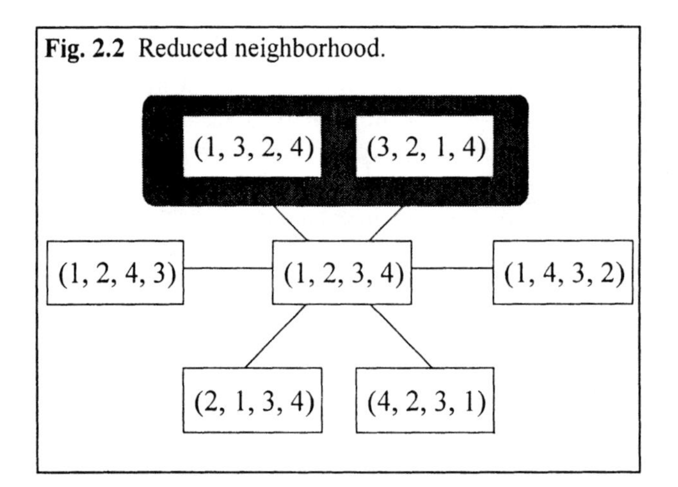

then the modified neighborhood $\mathbf{N}^*(x)$ contains only 4 out of the 6 solutions in $\mathbf{N}(x)$, as depicted in Figure 2.2. This figure shows that only two out of the three solutions in **T** belong to the neighborhood of x. If the solutions in **T** are a subset of those visited in the past, this memory approach can be used to prevent cycles of length less than or equal to $|\mathbf{T}|$ from occurring in the search trajectory. As a first approximation, this is the basis for a TS memory approach discussed in the following subsection.

The approach of storing complete solutions (*explicit* memory) generally consumes an enormous amount of space and time when applied to each solution generated. A scheme that emulates this approach with limited memory requirements is given by the use of hash functions, as discussed in Chapter 7. (Also, as will be seen, explicit memory has a valuable role when selectively applied in strategies that record and analyze certain "special" solutions.) Regardless of the implementation details, short term memory functions provide one of the important cornerstones of the TS methodology. These functions give the search the opportunity to continue beyond local optima, by allowing the execution of nonimproving moves coupled with the modification of the neighborhood structure of subsequent solutions. However, instead of recording full solutions, these memory structures are generally based on recording attributes (*attributive* memory), as illustrated in the "preview example" of Chapter 1. In addition, short term memory is often based on the most recent history of the search trajectory.

2.2 Recency-Based Memory

The most commonly used short term memory keeps track of solutions attributes that have changed during the recent past, and is called *recency-based* memory. This is the kind of memory that is included in most short descriptions of tabu search in the literature (although a number of its aspects are often left out by popular summaries).

To exploit this memory, selected attributes that occur in solutions recently visited are labeled *tabu-active*, and solutions that contain tabu-active elements, or particular combinations of these attributes, are those that become tabu. This prevents certain solutions from the recent past from belonging to $\mathbf{N}^*(x)$ and hence from being revisited. Other solutions that share such tabu-active attributes are also similarly prevented from being visited. Note that while the tabu classification strictly refers to solutions that are forbidden to be visited, by virtue of containing tabu-active attributes (or more generally by violating certain restriction based on these attributes), we also often refer to moves that lead to such solutions as being tabu. We illustrate these points with the following example.

Minimum k-Tree Problem Example

The *Minimum k-Tree* problem seeks a tree consisting of k edges in a graph so that the sum of the weights of these edges is minimum (Lokketangen, et al. 1994). An instance of this problem is given in Figure 2.3, where nodes are shown as numbered circles, and edges are shown as lines that join pairs of nodes (the two "endpoint" nodes that determine the edge). Edge weights are shown as the numbers attached to these lines. As noted in the preliminary TS illustration of Chapter 1, a tree is a set of edges that contains no cycles, i.e., that contains no paths that start and end at the same node (without retracing any edges).

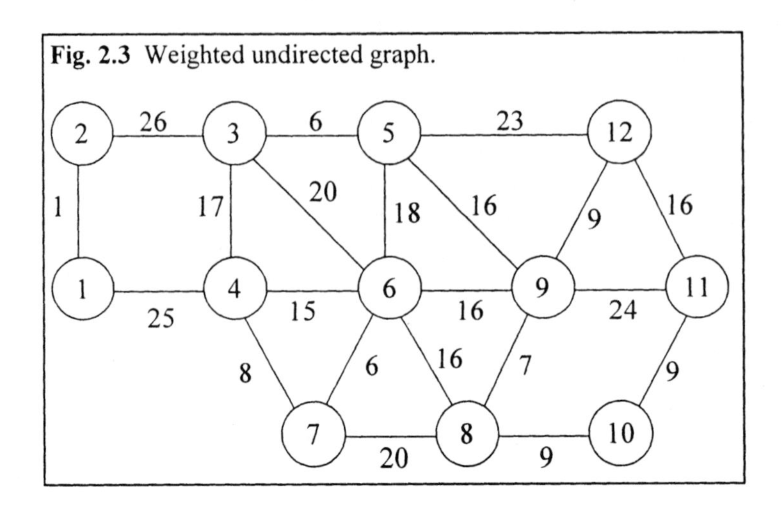

Fig. 2.3 Weighted undirected graph.

Assume that the move mechanism is defined by edge-swapping, as subsequently described, and that a greedy procedure is used to find an initial solution. The greedy construction starts by choosing the edge (i, j) with the smallest weight in the graph, where i and j are the indexes of the nodes that are the endpoints of the edge. The remaining k-1 edges are chosen successively to minimize the increase in total weight at each step, where the edges considered meet exactly one node from those that are endpoints of edges previously chosen. For $k = 4$, the greedy construction performs the steps in Table 2.2.

Table 2.2 Greedy construction.

Step	Candidates	Selection	Total Weight
1	(1,2)	(1,2)	1
2	(1,4), (2,3)	(1,4)	26
3	(2,3), (3,4), (4,6), (4,7)	(4,7)	34
4	(2,3), (3,4), (4,6), (6,7), (7,8)	(6,7)	40

The construction starts by choosing edge (1,2) with a weight of 1 (the smallest weight of any edge in the graph). After this selection, the candidate edges are those that connect the nodes in the current partial tree with those nodes not in the tree (i.e., edges (1,4) and (2,3)). Since edge (1,4) minimizes the weight increase, it is chosen to be part of the partial solution. The rest of the selections follow the same logic, and the construction ends when the tree consists of 4 edges (i.e., the value of k). The initial solution in this particular case has a total weight of 40.

The swap move mechanism, which is used from this point onward, replaces a selected edge in the tree by another selected edge outside the tree, subject to requiring that the resulting subgraph is also a tree. There are actually two types of such edge swaps, one that maintains the current nodes of the tree unchanged (static) and one that results in replacing a node of the tree by a new node (dynamic). Figure 2.4 illustrates the best swap of each type that can be made starting from the greedy solution. The added edge in each case is shown by a heavy line and the dropped edge is shown by a dotted line.

The best move of both types is the static swap of Figure 2.4, where for our present illustration we are defining *best* solely in terms of the change on the objective function value. Since this best move results in an increase of the total weight of the current solution, the execution of such move abandons the rules of a descent approach and sets the stage for a tabu search process. (The feasibility restriction that requires a tree to be produced at each step is particular to this illustration, since in general the TS methodology may include search trajectories that violate various types of feasibility conditions.)

Fig. 2.4 Swap move types.

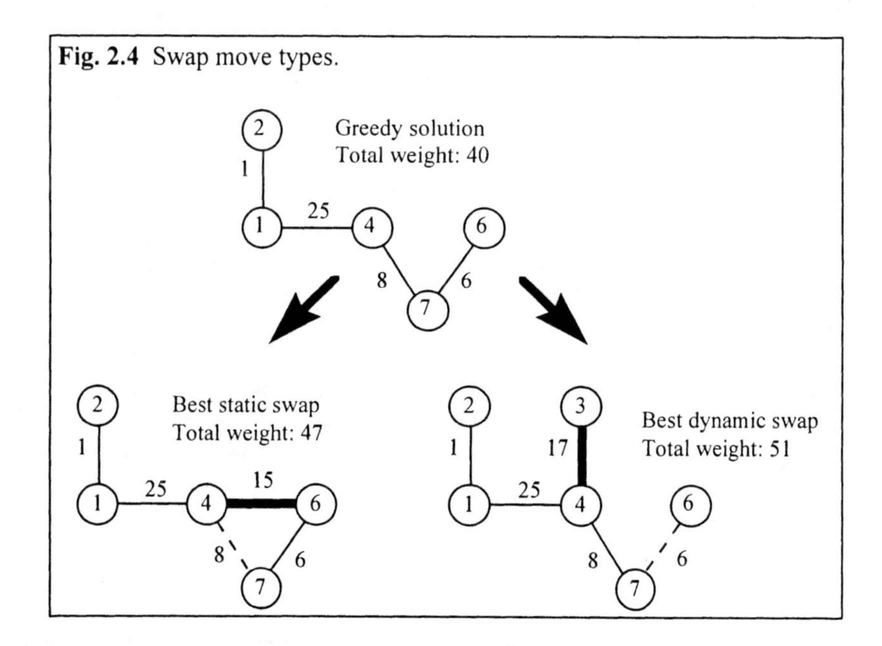

Given a move mechanism, such as the swap mechanism we have selected for our example, the next step is to choose the key attributes that will be used for the tabu classification. Tabu search is very flexible at this stage of the design. Problem-specific knowledge can be used as guidance to settle on a particular design. In problems where the moves are defined by adding and deleting elements, the labels of these elements can be used as the attributes for enforcing tabu status. This type of labeling was used in the example of Chapter 1. The terminology used there can be relaxed to simply refer to the edges as attributes of the move, since the condition of being *in* or *out* is always automatically known from the current solution.

Choosing Tabu Classifications

Tabu classifications do not have to be symmetric, that is, the tabu structure can be designed to treat added and dropped elements differently (as we also did in Chapter 1). Suppose for example that after choosing the static swap of Figure 2.4, which adds edge (4,6) and drops edge (4,7), a tabu status is assigned to both of these edges. Then one possibility is to classify both of these edges tabu-active for the same number of iterations. The tabu-active status has different meanings depending on whether the edge is added or dropped. For an added edge, tabu-active means that this edge is not allowed to be dropped from the current tree for the number of iterations that defines its tabu tenure. For a dropped edge, on the other hand, tabu-active means the edge is not allowed to be included in the current solution during its tabu tenure. Since there are many more edges outside the tree than in the tree, it seems reasonable to implement a tabu structure that keeps a recently dropped edge tabu-active for a longer period of time than a recently added edge. Notice also that for this problem the tabu-active period for added edges is bounded by k, since if no added edge is allowed to be dropped for k iterations, then within k steps all available moves will be classified tabu.

The concept of creating asymmetric tabu classifications can be readily applied to settings where add/drop moves are not used. For example, in the sequencing problem introduced at the beginning of this chapter, the jobs participating in a swap may receive different tabu-active tenures based on their contribution to the objective function value. In the neighborhood of the initial solution (1, 2, 3, 4, 5, 6), the best move is to exchange the positions of jobs 5 and 6. Their current contributions to the objective function value are respectively 10 and 11. After the swap is performed, the contribution of job 5 increases to 13, while the contribution of job 6 decreases to 1 (recall that the move value of this swap is -7). Since 6 is clearly in a better position, one strategy may be to forbid 6 from moving to a later position in the sequence for a longer time than 5 is being prevented from moving to an earlier position in the sequence.

Illustrative Tabu Classifications for the Min k-Tree Problem

We return to the Min k-Tree example to demonstrate additional features of tabu search. As previously remarked, the tabu-active classification may in fact prevent the search from visiting solutions that have not been examined yet. We illustrate this phenomenon as follows. Suppose that in the Min k-Tree problem instance of Figure 2.3, dropped edges are kept tabu-active for 2 iterations, while added edges are kept tabu-active for only one iteration. (The number of iterations an edge is kept tabu-active is called the *tabu tenure* of the edge.) Also assume that we define a swap move to be tabu if either its added or dropped edge is tabu-active. If we examine the full neighborhood of available edge swaps at each iteration, and always choose the best that is not tabu, then the first three moves are as shown in Table 2.3 below (starting from the initial solution found by the greedy construction heuristic). The move of iteration 1 is the static swap move previously identified in Figure 2.4. Diagrams showing the successive trees generated by these moves, starting with the initial greedy solution, are given in Figure 2.5.

Table 2.3 TS iterations.

Iteration	Tabu-active net tenure		Add	Drop	Weight
	1	2			
1			(4,6)	(4,7)	47
2	(4,6)	(4,7)	(6,8)	(6,7)	57
3	(6,8), (4,7)	(6,7)	(8,9)	(1,2)	63

The net tenure values of 1 and 2 in Table 2.3 for the currently tabu-active edges indicate the number of iterations that these edges will remain tabu-active (including the current iteration).

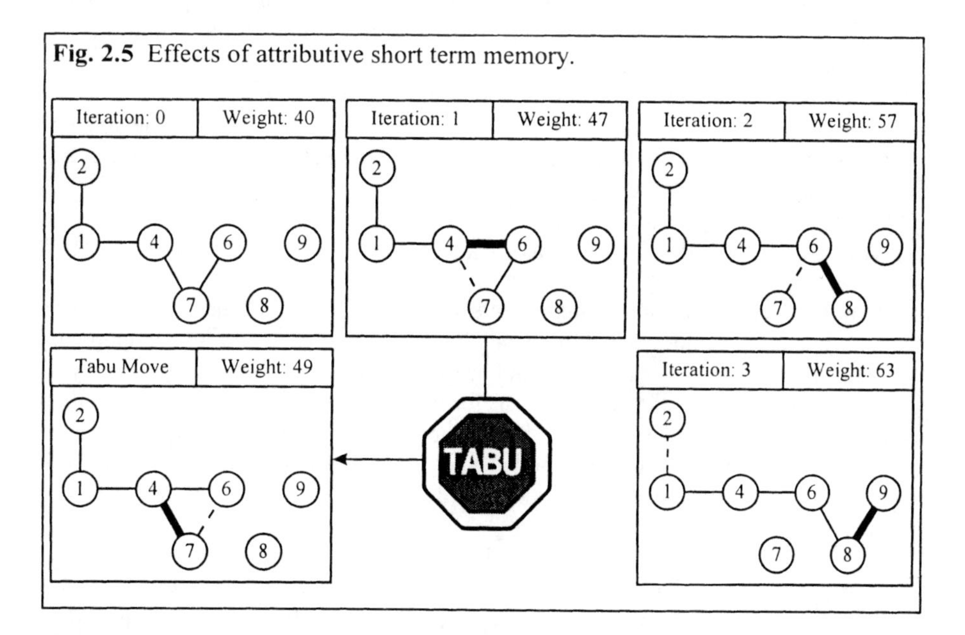

Fig. 2.5 Effects of attributive short term memory.

At iteration 2, the reversal of the move of iteration 1 (that is, the move that now adds (4,7) and drops (4,6)) is clearly tabu, since both of its edges are tabu-active at iteration 2. In addition, the move that adds (4,7) and drops (6,7) is also classified tabu, because it contains the tabu-active edge (4,7) (with a net tenure of 2). This move leads to a solution with a total weight of 49, a solution that clearly has not been visited before (see Figure 2.5). The tabu-active classification of (4,7) has modified the original neighborhood of the solution at iteration 2, and has forced the search to choose a move with an inferior objective function value (i.e., the one with a total weight of 57). In this case, excluding the solution with a total weight of 49 has little effect on the quality of the best solution found (since we have already obtained one with a weight of 40).

In other situations, however, additional precautions must be taken to avoid missing good solutions. These strategies are known as aspiration criteria and are the subject of Section 2.7. For the moment we observe simply that if the tabu solution encountered at the current step instead had a weight of 39, which is better than the best weight of 40 so far seen, then we would allow the tabu classification of this solution to be overridden and consider the solution admissible to be visited. The aspiration criterion that applies in this case is called the *improved-best* aspiration criterion. (It is important to keep in mind that aspiration criteria do not compel particular moves to be selected, but simply make them available, or alternately rescind evaluation penalties attached to certain tabu classifications.)

One other comment about tabu classification deserves to be made at this point. In our preceding discussion of the Min *k*-Tree problem we consider a swap move tabu if either its added edge or its dropped edge is tabu-active. However, we could instead stipulate that a swap move is tabu only if both its added and dropped edges are tabu-active. In general, the tabu status of a move is a function of the tabu-active attributes of the move (i.e., of the new solution produced by the move). The relevance of various alternative ways of generating restrictions is examined in Exercises 1 and 2 at the end of the chapter.

2.3 A First Level Tabu Search Approach

We now have on hand enough ingredients for a first level tabu search procedure. Such a procedure is sometimes implemented in an initial phase of a TS development to obtain a preliminary idea of performance and calibration features, or simply to provide a convenient staged approach for the purpose of debugging solution software. While this naive form of a TS method omits a number of important short term memory considerations, and does not yet incorporate longer term concerns, it nevertheless gives a useful starting point for demonstrating several basic aspects of tabu search.

We start from the solution with a weight of 63 as shown previously in Figure 2.5 which was obtained at iteration 3. At each step we select the least weight non-tabu move from those available, and use the improved-best aspiration criterion to allow a move to be considered admissible in spite of leading to a tabu solution. The reader may verify that the outcome leads to the series of solutions shown in Table 2.4, which continues from iteration 3, just executed. For simplicity, we select an arbitrary stopping rule that ends the search at iteration 10.

Table 2.4 Iterations of a first level TS procedure.

Iteration	Tabu-active net tenure		Add	Drop	Move Value	Weight
	1	2				
3	(6,8), (4,7)	(6,7)	(8,9)	(1,2)	6	63
4	(6,7), (8,9)	(1,2)	(4,7)	(1,4)	-17	46
5	(1,2), (4,7)	(1,4)	(6,7)	(4,6)	-9	37*
6	(1,4), (6,7)	(4,6)	(6,9)	(6,8)	0	37
7	(4,6), (6,9)	(6,8)	(8,10)	(4,7)	1	38
8	(6,8), (8,10)	(4,7)	(9,12)	(6,7)	3	41
9	(4,7), (9,12)	(6,7)	(10,11)	(6,9)	-7	34*
10	(6,7), (10,11)	(6,9)	(5,9)	(9,12)	7	41

The successive solutions identified in Table 2.4 are shown graphically in Figure 2.6 below. In addition to identifying the dropped edge at each step as a dotted line, we also identify the dropped edge from the immediately preceding step as a dotted line which is labeled 2*, to indicate its current net tabu tenure of 2. Similarly, we identify the dropped edge from one further step back by a dotted line which is labeled 1*, to indicate its current net tabu tenure of 1. Finally, the edge that was added on the immediately preceding step is also labeled 1* to indicate that it likewise has a current net tabu tenure of 1. Thus the edges that are labeled with tabu tenures are those which are currently tabu-active, and which are excluded from being chosen by a move of the current iteration (unless permitted to be chosen by the aspiration criterion).

As illustrated in Table 2.4 and Figure 2.6 the method continues to generate different solutions, and over time the best known solution (denoted by an asterisk) progressively improves. In fact, it can be verified for this simple example that the solution obtained at iteration 9 is optimal. (In general, of course, there is no known way to verify optimality in polynomial time for difficult discrete optimization problems, i.e., those that fall in the class called NP-hard. The Min *k*-Tree problem is one of these.)

Fig. 2.6 Graphical representation of TS iterations.

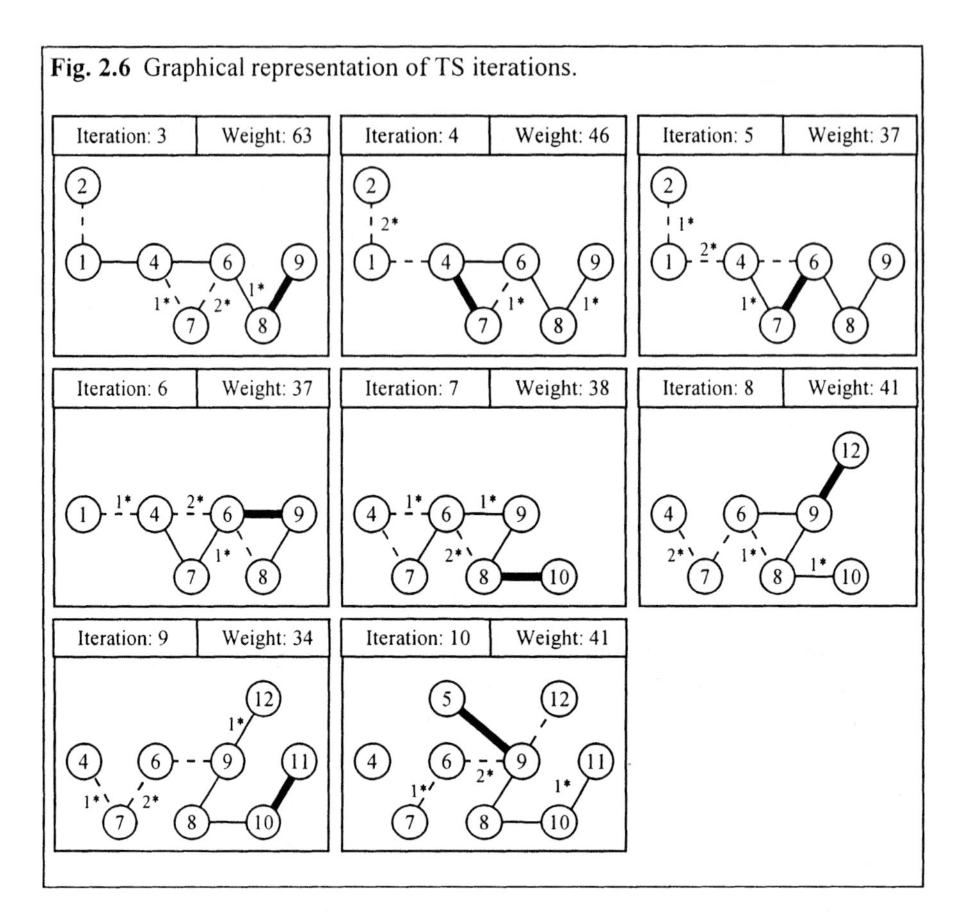

It may be noted that at iteration 6 the method selected a move with a move value of zero. Nevertheless, the configuration of the current solution changes after the execution of this move, as illustrated in Figure 2.6.

The selection of moves with certain move values, such as zero move values, may be strategically controlled, to limit their selection as added insurance against cycling in special settings. We will soon see how considerations beyond this first level implementation can lead to an improved search trajectory, but the non-monotonic, gradually improving, behavior is characteristic of TS in general. Figure 2.7 provides a graphic illustration of this behavior for the current example.

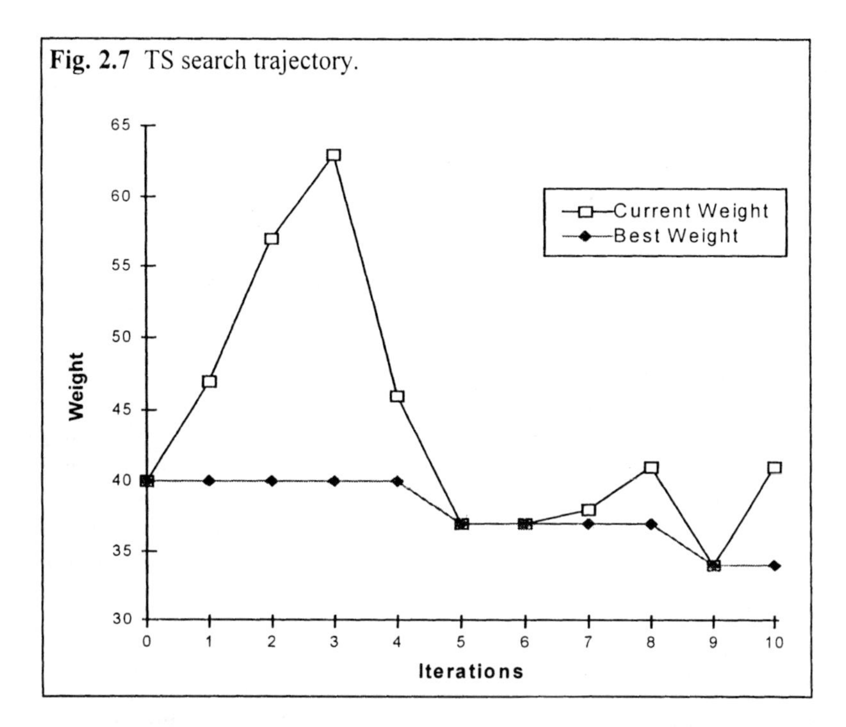

Fig. 2.7 TS search trajectory.

We have purposely chosen the stopping iteration to be small to illustrate an additional relevant feature, and to give a foundation for considering certain types of longer term considerations. One natural way to apply TS is to periodically discontinue its progress, particularly if its rate of finding new best solutions falls below a preferred level, and to restart the method by a process designated to generate a new sequence of solutions.

Classical restarting procedures based on randomization evidently can be used for this purpose, but TS often derives an advantage by employing more strategic forms of restarting. We illustrate a simple instance of such a restarting procedure, which also serves to introduce a useful memory concept.

2.3.1 Critical Event Memory

Critical Event memory in tabu search, as its name implies, monitors the occurrence of certain *critical events* during the search, and establishes a memory that constitutes an aggregate summary of these events. For our current example, where we seek to generate a new starting solution, a critical event that is clearly relevant is the generation of the previous starting solution. Correspondingly, if we apply a restarting procedure multiple times, the steps of generating all preceding starting solutions naturally qualify as critical events. That is, we would prefer to depart from these solutions in some significant manner as we generate other starting solutions.

Different degrees of departure, representing different levels of *diversification*, can be achieved by defining solutions that correspond to critical events in different ways (and by activating critical event memory by different rules). In the present setting we consider it important that new starting solutions not only differ from preceding starting solutions, but that they also differ from other solutions generated during previous passes. One possibility is to use a blanket approach that considers each complete solution previously generated to represent a critical event. The aggregation of such events by means of critical event memory makes this entirely practicable, but often it is quite sufficient (and, sometimes preferable) to isolate a smaller set of solutions.

For the current example, therefore, we will specify that the critical events of interest consist of generating not only the starting solution of the previous pass(es), but also each subsequent solution that represents a "local TS optimum," i.e. whose objective function value is better (or no worse) than that of the solution immediately before and after it. Using this simple definition we see that four solutions qualify as critical (i.e., are generated by the indicated critical events) in the first solution pass of our example: the initial solution and the solutions found at iterations 5, 6 and 9 (with weights of 40, 37, 37 and 34, respectively).

Since the solution at iteration 9 happens to be optimal, we are interested in the effect of restarting before this solution is found. Assume we had chosen to restart after iteration 7, without yet reaching an optimal solution. Then the solutions that correspond to critical events are the initial solution and the solutions of iterations 5 and 6. We treat these three solutions in aggregate by combining their edges, to create a subgraph that consists of the edges (1,2), (1,4), (4,7), (6,7), (6,8), (8,9) and (6,9). (Frequency-based memory, as discussed in Chapter 4, refines this representation by accounting for the number of times each edge appears in the critical solutions, and allows the inclusion of additional weighting factors.)

To execute a restarting procedure, we penalize the inclusion of the edges of this subgraph at various steps of constructing the new solution. It is usually preferable to apply this penalty process at early steps, implicitly allowing the penalty function to decay rapidly as the number of steps increases. It is also sometimes useful to allow one or more intervening steps after applying such penalties before applying them again.

For our illustration, we will use the memory embodied in the subgraph of penalized edges by introducing a large penalty that effectively excludes all these edges from consideration on the first two steps of constructing the new solution. Then, because the construction involves four steps in total, we will not activate the critical event memory on subsequent construction steps, but will allow the method to proceed in its initial form.

Applying this approach, we restart the method by first choosing edge (3,5), which is the minimum weight edge not in the penalized subgraph. This choice and the remaining choices that generate the new starting solution are shown in Table 2.5.

Table 2.5 Restarting procedure.

Step	Candidates	Selection	Total Weight
1	(3,5)	(3, 5)	6
2	(2,3), (3,4), (3,6), (5,6), (5,9), (5,12)	(5, 9)	22
3	(2,3), (3,4), (3,6), (5,6), (5,12), (6,9), (8,9), (9,12)	(8, 9)	29
4	(2,3), (3,4), (3,6), (5,6), (5,12), (6,8), (6,9), (7,8), (8,10), (9,12)	(8, 10)	38

Beginning from the solution constructed in Table 2.5, and applying the first level TS procedure exactly as it was applied on the first pass, generates the sequence of solutions shown in Table 2.6 and depicted in Figure 2.8. (Again, we have arbitrarily limited the total number of iterations, in this case to 5.)

Table 2.6 TS iterations following restarting.

Iteration	Tabu-active net tenure		Add	Drop	Move Value	Weight
	1	2				
1			(9,12)	(3,5)	3	41
2	(9,12)	(3,5)	(10,11)	(5,9)	-7	34*
3	(3,5), (10,11)	(5,9)	(6,8)	(9,12)	7	41
4	(5,9), (6,8)	(9,12)	(6,7)	(10,11)	-3	38
5	(9,12), (6,7)	(10,11)	(4,7)	(8,10)	-1	37

It is interesting to note that the restarting procedure generates a better solution (with a total weight of 38) than the initial solution generated during the first construction (with a total weight of 40). Also, the restarting solution contains 2 "optimal edges" (i.e., edges that appear in the optimal tree). This starting solution allows the search trajectory to find the optimal solution in only two iterations, illustrating the benefits of applying an critical event memory within a restarting strategy. As will be seen in Chapter 4, related memory structures can also be valuable for strategies that drive the search into new regions by "partial restarting" or by directly continuing a current trajectory (with modified decision rules).

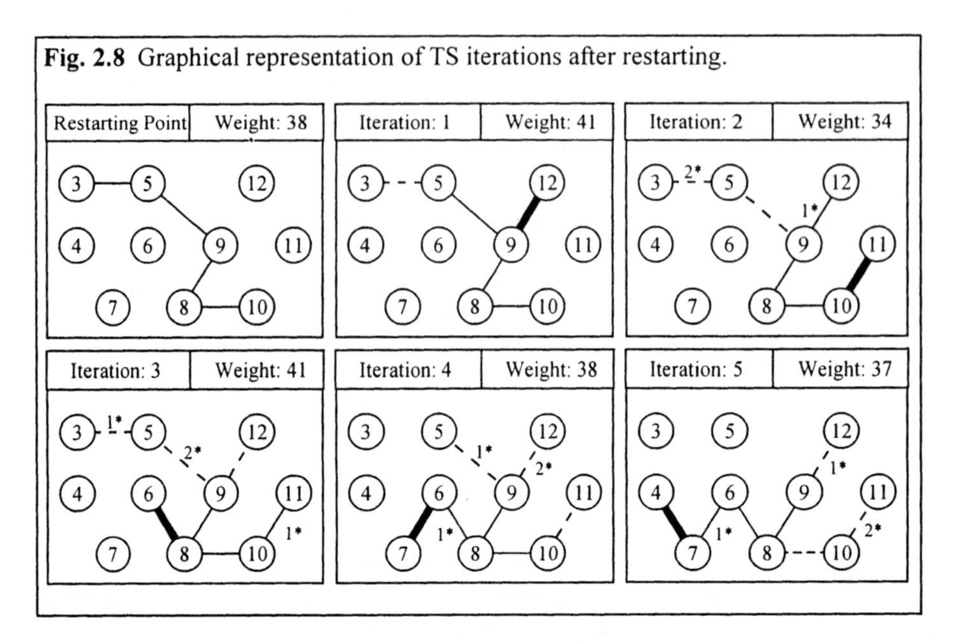

Fig. 2.8 Graphical representation of TS iterations after restarting.

Now we return from our example to examine elements of TS that take us beyond these first level concerns, and open up possibilities for creating more powerful solution approaches. We continue to focus primarily on short term aspects, and begin by discussing how to generalize the use of recency-based memory when neighborhood exploration is based on add/drop moves. From these foundations we then discuss issues of logical restructuring, tabu activation rules and ways of determining tabu tenure. We then examine the important area of aspiration criteria, together with the role of influence.

2.4 Recency-Based Memory for Add / Drop Moves

To understand procedurally how various forms of recency-based memory work, and to see their interconnections, it is useful to examine a convenient design for implementing the ideas illustrated so far. Such a design for the Min k-Tree problem creates a natural basis for handling a variety of other problems for which add/drop moves are relevant. In addition, the ideas can be adapted to settings that are quite different from those where add/drop moves are used.

As a step toward fuller generality, we will refer to items added and dropped as *elements*, though we will continue to make explicit reference to edges (as particular types of elements) within the context of the Min k-Tree problem example. (Elements are related to, but not quite the same as, solution attributes. The difference will be made apparent shortly.) There are many settings where operations of adding and dropping paired elements are the cornerstone of useful neighborhood definitions. For example, many types of exchange or swap moves can be characterized by such operations. Add/drop moves also apply to the omnipresent class of *multiple choice* problems, which require that exactly one element must be chosen from each member set from a specified disjoint collection. Add/drop moves are quite natural in this setting, since whenever a new element is chosen from a given set (and hence is "added" to the current solution), the element previously chosen from that set must be replaced (and hence "dropped"). Such problems are represented by discrete *generalized upper bound* (GUB) formulations in mathematical optimization, where various disjoint sets of 0-1 variables must sum to 1 (hence exactly one variable from each set must equal 1, and the others must equal 0). An add/drop move in this formulation consists of choosing a new variable to equal 1 (the "add move") and setting the associated (previously selected) variable equal to 0 (the "drop move").

Add/drop moves further apply to many types of problems that are not strictly discrete, that is, which contain variables whose values can varying continuously across specified ranges. Such applications arise by taking advantage of *basis exchange* (pivoting) procedures, such as the simplex method of linear programming. In this case, an add/drop move consists of selecting a new variable to enter (add to) the basis, and identifying an associated variable to leave (drop from) the basis. A variety of procedures for nonlinear and mixed integer optimization rely on such moves, and have provided a useful foundation for a number of tabu search applications. Additional related examples will be encountered throughout the course of this book.

2.4.1. Some Useful Notation

The approach used in the Min *k*-Tree problem can be conveniently described by means of the following notation. For a pair of elements that is selected to perform an add/drop move, let *Added* denote the element that is added, and *Dropped* the element that is dropped. Also denote the current iteration at which this pair is selected by *Iter*. We maintain a record of *Iter* to identify when *Added* and *Dropped* start to be tabu-active. Specifically, at this step we set:

$$TabuDropStart(Added) = Iter$$
$$TabuAddStart(Dropped) = Iter.$$

Thus, *TabuDropStart* records the iteration where *Added* becomes tabu-active (to prevent this element from later being dropped), and *TabuAddStart* records the iteration where *Dropped* becomes tabu-active (to prevent this element from later being added).

For example, in the Min *k*-Tree problem illustration of Table 2.3, where the edge (4,6) was added and the edge (4,7) was dropped on the first iteration, we would establish the record (for *Iter* = 1)

$$TabuDropStart(4,6) = 1$$
$$TabuAddStart(4,7) = 1$$

To identify whether or not an element is currently tabu-active, let *TabuDropTenure* denote the tabu tenure (number of iterations) to forbid an element to be dropped (once added), and let *TabuAddTenure* denote the tabu tenure to forbid an element from being added (once dropped). (In our Min *k*-Tree problem example of Section 2.2, we selected *TabuAddTenure* = 2 and *TabuDropTenure* = 1.)

As a point of clarification, when we speak of an element as being tabu-active, our terminology implicitly treats elements and attributes as if they are the same. However, to be precise, each element is associated with two different attributes, one where the element belongs to the current solution and one where the element does not. Elements may be viewed as corresponding to variables and attributes as corresponding to specific value assignments for such variables. There is no danger of confusion in the add/drop setting, because we always know when an element belongs or does not belong to the current solution, and hence we know which of the two associated attributes is currently being considered.

We can now identify precisely the set of iterations during which an element (i.e., its associated attribute) will be tabu-active. Let *TestAdd* and *TestDrop* denote a candidate pair of elements, whose members are respectively under consideration to be added and dropped from the current solution. If *TestAdd* previously corresponded to an element *Dropped* that was dropped from the solution and *TestDrop* previously corresponded to an element *Added* that was added to the solution (not necessarily on the same step), then it is possible that one or both may be tabu-active and we can check their status as follows. By means of the records established on earlier iterations, where *TestAdd* began to be tabu-active at iteration *TabuAddStart*(*TestAdd*) and *TestDrop* began to be tabu-active at iteration *TabuDropStart*(*TestDrop*), we conclude that as *Iter* grows the status of these elements will be given by:

TestAdd is tabu-active when:

$$Iter \leq TabuAddStart(TestAdd) + TabuAddTenure$$

TestDrop is tabu-active when:

$$Iter \leq TabuDropStart(TestDrop) + TabuDropTenure$$

Consider again the Min *k*-Tree problem illustration of Table 2.3. As previously noted, the move of Iteration 1 that added edge (4.6) and dropped edge (4,7) was accompanied by setting the *TabuDropStart*(4,6) = 1 and *TabuAddStart*(4,7) = 1, to record the iteration where these two edges start to be tabu-active (to prevent (4,6) from being dropped and (4,7) from being added). The edge (4,6) will then remain tabu-active on subsequent iterations, in the role of *TestDrop* (as a candidate to be dropped), as long as

$$Iter \leq TabuDropStart(4,6) + TabuDropTenure.$$

Hence, since we selected *TabuDropTenure* = 1 (to prevent an added edge from being dropped for 1 iteration), it follows that (4,6) remains tabu-active as long as

$$Iter \leq 2.$$

Similarly, having selected *TabuAddTenure* = 2, we see that the edge (4,7) remains tabu-active, to forbid it from being added back, as long as

$$Iter \leq 3.$$

An initialization step is needed to be sure that elements that have never been previously added or dropped from the solutions successively generated will not be considered tabu-active. This can be done by initially setting *TabuAddStart* and *TabuDropStart* equal to a large negative number for all elements. Then, as *Iter* begins at 1 and successively increases, the inequalities that determine the tabu-active status will not be satisfied, and hence will correctly disclose that an element is not tabu-active, until it becomes one of the elements *Added* or *Dropped.* (Alternately, *TabuAddStart* and *TabuDropStart* can be initialized at 0, and the test of whether an element is tabu-active can be skipped when it has a 0 value in the associated array.)

2.4.2 Streamlining

The preceding ideas can be streamlined to allow a more convenient implementation. First, we observe that the two arrays, *TabuAddStart* and *TabuDropStart*, which we have maintained separately from each other in to emphasize their different functions, can be combined into a single array *TabuStart*. The reason is simply that we can interpret *TabuStart*(*E*) to be the same as *TabuDropStart*(*E*) when the element *E* is in the current solution, and to be the same as *TabuAddStart*(*E*) when *E* is not in the current solution. (There is no possible overlap between these two states of *E*, and hence no danger of using the *TabuStart* array incorrectly.) Consequently, from now on, we will let the single array *TabuStart* take the role of both *TabuAddStart* and *TabuDropStart*. For example, when the move is executed that (respectively) adds and drops the elements *Added* and *Dropped*, the appropriate record consists of setting:

$$TabuStart(Added) = Iter$$
$$TabuStart(Dropped) = Iter.$$

The *TabuStart* array has an additional function beyond that of monitoring the status of tabu-active elements. (As shown in Chapter 4, this array is also useful for determining a type of frequency measure called a *residence frequency*.) However, sometimes it is convenient to use a different array, *TabuEnd*, to keep track of tabu-active status for recency-based memory, as we are treating here. Instead of recording when the tabu-active status starts, *TabuEnd* records when it ends. Thus, in place of the two assignments to *TabuStart* shown above, the record would consist of setting:

TabuEnd(Added) = *Iter* + *TabuDropTenure*
TabuEnd(Dropped) = *Iter* + *TabuAddTenure*.

(The element *Added* is now available to be dropped, and the element *Dropped* is now available to be added.) In conjunction with this, the step that checks for whether a candidate pair of elements *TestAdd* and *TestDrop* are currently tabu-active becomes:

TestAdd is tabu-active when:
Iter ≤ *TabuEnd(TestAdd)*
TestDrop is tabu-active when:
Iter ≤ *TabuEnd(TestDrop)*.

This is a simpler representation than the one using *TabuStart*, and so it is appealing when *TabuStart* is not also used for additional purposes. (Also, *TabuEnd* can simply be initialized at 0 rather than at a large negative number.)

As will be discussed more fully in the next section, the values of *TabuAddTenure* and *TabuDropTenure* (which are explicitly referenced in testing tabu-active status with *TabuStart*, and implicitly referenced in testing this status with *TabuEnd*), are often preferably made variable rather than fixed. The fact that we use different tenures for added and dropped elements discloses that it can be useful to differentiate the tenures applied to elements of different classes. This type of differentiation can also be based on historical performance, as tracked by frequency-based measures. Consequently, tenures may be individually adjusted for different elements (as well as modified over time). Such adjustment can be quite effective in some settings (e.g., see Laguna, et al. 1995). We show how these basic considerations can be refined to create effective implementations and also can be extended to handle additional move structures in Chapter 3.

2.5 Tabu Tenure

In general, recency-based memory is managed by creating one or several tabu lists, which record the tabu-active attributes and implicitly or explicitly identify their current status. Tabu tenure can vary for different types or combinations of attributes, and can also vary over different intervals of time or stages of the search. This varying tenure makes it possible to create different kinds of tradeoffs between short term and longer term strategies. It also provides a dynamic and robust form of search.

The choice of appropriate types of tabu lists depends on the context. Although no single type of list is uniformly best for all applications, some guidelines can be formulated. If memory space is sufficient (as it often is) to store one piece of information (e.g., a single integer) for each solution attribute used to define the tabu activation rule, it is usually advantageous to record the iteration number that identifies when the tabu-active status of an attribute starts or ends as illustrated by the add/drop data structure described in Sections 2.3 and 2.4. This typically makes it possible to test the tabu status of a move in constant time. The necessary memory space depends on the attributes and neighborhood size, but it does not depend on the tabu tenure.

Depending on the size of the problem, it may not be feasible to implement the preceding memory structure in combination with certain types of attributes. In general, storing one piece of information for each attribute becomes unattractive when the problem size increases or attribute definition is complex. Sequential and circular tabu lists are used in this case, which store the identities of each tabu-active attribute, and explicitly (or implicitly, by list position) record associated tabu tenures.

Effective tabu tenures have been empirically shown to depend on the size of the problem instance. However, no single rule has been designed to yield an effective tenure for all classes of problems. This is partly because an appropriate tabu tenure depends on the strength of the tabu activation rule employed (where more restrictive rules are generally coupled with shorter tenures). Effective tabu tenures and tabu activation rules can usually be determined quite easily for a given class of problems by a little experimentation. Tabu tenures that are too small can be recognized by periodically repeated objective function values or other function indicators, including those generated by hashing, that suggest the occurrence of cycling. Tenures that are too large can be recognized by a resulting deterioration in the quality of the solutions found (within reasonable time periods). Somewhere in between typically exists a robust range of tenures that provide good performance.

Once a good range of tenure values is located, first level improvements generally result by selecting different values from this range on different iterations. (A smaller subrange, or even more than one subrange, may be chosen for this purpose.) Problem structures are sometimes encountered where performance for some individual fixed tenure values within a range can be unpredictably worse than for other values in the range, and the identity of the isolated poorer values can change from problem to problem. However, if the range is selected to be good overall then a strategy that selects different tenure values from the range on different iterations typically performs at a level comparable to selecting one of the best values in the range, regardless of the problem instance.

Short term memory refinements subsequently discussed, and longer term considerations introduced in later chapters transform the method based on these constructions into one with considerable power. Still, it occasionally happens that even the initial short term approach by itself leads to exceptionally high quality solutions. Consequently, some of the TS literature has restricted itself only to this initial part of the method.

In general, short tabu tenures allow the exploration of solutions "close" to a local optimum, while long tenures can help to break free from the vicinity of a local optimum. These functions illustrate a special instance of the notions of *intensification* and *diversification* that will be explored in more detail later. Varying the tabu tenure during the search provides one way to induce a balance between closely examining one region and moving to different parts of the solution space.

In situations where a neighborhood may (periodically) become fairly small, or where a tabu tenure is chosen to be fairly large, it is entirely possible that iterations can occur when all available moves are classified tabu. In this case an *aspiration-by-default* is used to allow a move with a "least tabu" status to be considered admissible. Such situations rarely occur for most problems, and even random selection is often an acceptable form of aspiration-by-default. When tabu status is translated into a modified evaluation criterion, by penalties and inducements, then of course aspiration-by-default is handled automatically, with no need for to monitor the possibility that all moves are tabu.

There are several ways in which a *dynamic* tabu tenure can be implemented. These implementations may be classified into *random* and *systematic* dynamic tabu tenures.

2.5.1 Random Dynamic Tenure

Random dynamic tabu tenures are often given one of two forms. Both of these forms use a tenure range defined by parameters t_{min} and t_{max}. The tabu tenure t is randomly selected within this range, usually following a uniform distribution. In the first case, the chosen tenure is maintained constant for $\alpha\ t_{max}$ iterations, and then a new tenure is selected by the same process. The second form draws a new t for every attribute that becomes tabu at a given iteration. The first form requires more bookkeeping than the second one, because one must remember the last time that the tabu tenure was modified.

Either of the two arrays *TabuStart* or *TabuEnd* discussed in Section 2.4 can be used to implement these forms of dynamic tabu tenure. For example, a 2-dimensional array *TabuEnd* can be created to control a dynamic recency-based memory for the sequencing problem introduced at the beginning of this chapter. As in the case of the Min k-Tree problem, such an array can be used to record the time (iteration number) at which a particular attribute will be released from its tabu status. Suppose, for example, that $t_{min} = 5$ and $t_{max} = 10$ and that swaps of jobs are used to move from one solution to another in the sequencing problem. Also, assume that *TabuEnd*(j,p) refers to the iteration that job j will be released from a tabu restriction that prevents it from being assigned to position p. Then, if at iteration 30, job 8 in position 2 is swapped with job 12 in position 25, we will want to make the attribute (8,2) and (12,25) tabu-active for some number of iterations to prevent a move that will return one or both of jobs 8 and 12 from re-occupying their preceding positions. If t is assigned a value of 7 from the range $t_{min} = 5$ and $t_{max} = 10$, then upon making the swap at iteration 30 we may set *TabuEnd*(8,2) = 37 and *TabuEnd*(12,25) = 37.

This is not the only kind of *TabuEnd* array that can be used for the sequencing problem, and we examine other alternatives and their implications in Chapter 3. Nevertheless, we warn against a potential danger. An array *TabuEnd*(i,j) that seeks to prevent jobs i and j from exchanging positions, without specifying what these positions are, does not truly refer to attributes of a sequencing solution, and hence entails a risk if used to determine tabu status. (The pair (i,j) here constitutes an attribute of a move, in a lose sense, but does not serve to distinguish one solution from another.) Thus, if at iteration 30 we were to set *TabuEnd*(8,12) = 37, in order to prevent jobs 8 and 12 from exchanging positions until after iteration 37, this still might not prevent job 8 from returning to position 2 and job 12 from returning to position 25. In fact, a sequence of swaps could be executed that could return to precisely the same solution visited before swapping jobs 8 and 12.

Evidently, the *TabuEnd* array can be used by selecting a different t from the interval (t_{min}, t_{max}) at every iteration. As remarked in the case of the Min k-Tree problem, it is also possible to select t differently for different solution attributes.

2.5.2 Systematic Dynamic Tenure

Dynamic tabu tenures based on a random scheme are attractive for their ease of implementation. However, relying on randomization may not be the best strategy when specific information about the context is available. In addition, certain diversity-inducing patterns can be achieved more effectively by not restricting consideration to random designs. A simple form of *systematic* dynamic tabu tenure consists of creating a sequence of tabu search tenure values in the range defined by t_{min} and t_{max}. This sequence is then used, instead of the uniform distribution, to assign the current tabu tenure value. Suppose it is desired to vary t so that its value alternately increases and decreases. (Such a pattern induces a form of diversity that will rarely be achieved randomly.) Then the following sequence can be used for the range defined above:

$$\{ 5, 8, 6, 9, 7, 10 \}.$$

The sequence may be repeated as many times as necessary until the end of the search, where additional variation is introduced by progressively shifting and/or reversing the sequence before repeating it. (In a combined random/systematic approach, the decision of the shift value and the forward or backward direction can itself be made random.) Another variation is to retain a selected tenure value from the sequence for a variable number of iterations before selecting the next value. Different sequences can be created and identified as effective for particular classes of problems.

The foregoing range of values (from 5 to 10) may seem relatively small. However, some applications use even smaller ranges, but adaptively, increase and decrease the midpoint of the range for diversification and intensification purposes (see Chapter 4). Well-designed adaptive systems can significantly reduce or even eliminate the need to discover a best range of tenures by preliminary calibration. This is an important area of study.

These basic alternatives typically provide good starting tabu search implementations. In fact, most initial implementations apply only the simplest versions of these ideas.

2.6 Aspiration Criteria and Regional Dependencies

Aspiration criteria are introduced in tabu search to determine when tabu activation rules can be overridden, thus removing a tabu classification otherwise applied to a move. (The improved-best and aspiration-by-default criteria, as previously mentioned, are obvious simple instances.) The appropriate use of such criteria can be very important for enabling a TS method to achieve its best performance levels. Early applications employed only a simple type of aspiration criterion, consisting of removing a tabu classification from a trial move when the move yields a solution better than the best obtained so far. This criterion remains widely used. However, other aspiration criteria can prove effective for improving the search.

A basis for one of these criteria arises by introducing the concept of *influence,* which measures the degree of change induced in solution structure or feasibility. This notion can be illustrated for the Min k-Tree problem as follows. Suppose that the current solution includes edges (1,2), (1,4), (4,7) and (6,7), as illustrated in Figure 2.9, following. A high influence move, that significantly changes the structure of the current solution, is exemplified by dropping edge (1,2) and replacing it by edge (6,9). A low influence move, on the other hand, is exemplified by dropping edge (6,7) and adding edge (4,6). The weight difference of the edges in the high influence move is 15, while the difference is 9 for the low influence move. However, it is important to point out that differences on weight or cost are not the only — or even the primary — basis for distinguishing between moves of high and low influence. In the present example, the move we identify as a low influence move creates a solution that consists of the same set of nodes included in the current solution, while the move we identified as a high influence move includes a new node (number 9) from which new edges can be examined. (These moves correspond to those labeled static and dynamic in Figure 2.4.)

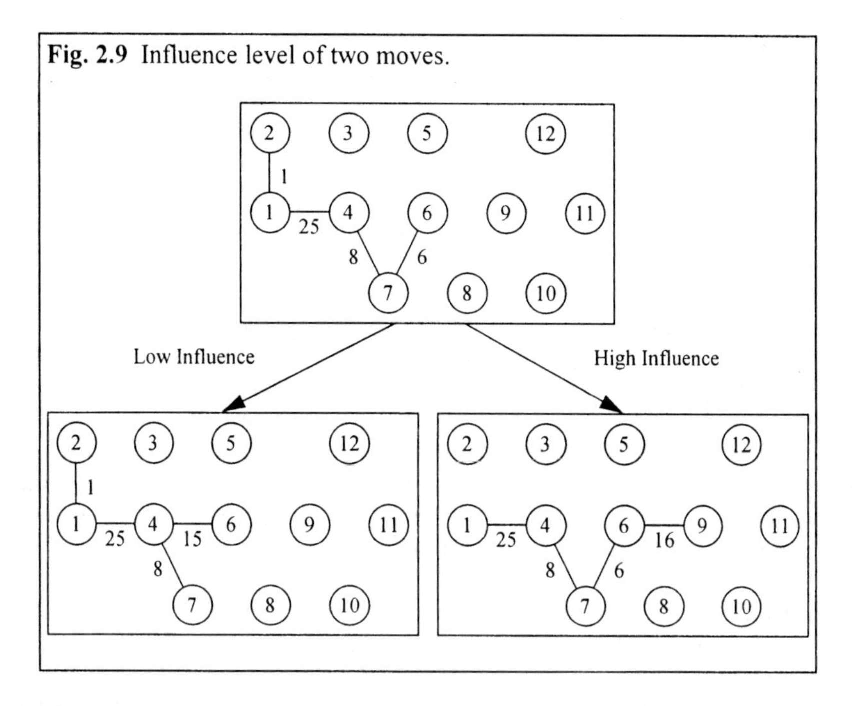

Fig. 2.9 Influence level of two moves.

As illustrated here, high influence moves may or may not improve the current solution, though they are less likely to yield an improvement when the current solution is relatively good. But high influence moves are important, especially during intervals of breaking away from local optimality, because a series of moves that is confined to making only small structural change is unlikely to uncover a chance for significant improvement. Executing the high influence move in Figure 2.9, for example, allows the search to reach the optimal edges (8,9) and (9,12) in subsequent iterations. (Of course, moves of much greater influence than those shown can be constructed by employing compound moves. Such considerations are treated in Chapter 5.)

Influence often is associated with the idea of move distance. The move distance concept has a straightforward interpretation in the context of job sequencing problems. Consider the sequencing problem that we have been using for illustrative purposes. The solution given by the sequence (1, 4, 2, 3, 6, 5) with a total tardiness of 19 is a local optimum with respect to a neighborhood defined by swap moves. Hence, no swap exists that can improve the current solution. In this context, a low influence move is given by the swap of jobs 1 and 4 which results in the same total tardiness of 19 (i.e., the move value associated with this move is zero). If the move distance of a swap (i, j) is defined as the number of positions that separate job i from job j, then the swap of jobs 1 and 4 has the minimum move distance of 1. Also, the absolute difference of the due date values for these jobs is one (see Table 2.7). The similarity of the jobs in the proposed swap classifies it as a low influence move.

Table 2.7 Comparison of two swap moves.		
Characteristics	Swap (1, 4)	Swap (4, 5)
Move value	0	36
Move distance	1	4
Due date difference	1	12
Influence	Low	High

A high influence move in this situation is the swap of jobs 4 and 5. The move distance of this swap is 4 (one unit less than the maximum of 5), the move value is 36, and the absolute due date difference of the corresponding jobs is 12. Clearly, this move has the potential of driving the search far from the current optimum in fewer steps than the swap of jobs 1 and 4. Although in this case the swap (4,5) has the largest move value in the neighborhood, this should not be interpreted as suggesting there is merit in selecting a move at a local optimum that creates a significant deterioration in solution quality. Generally, as in this example, reasonable measures of influence exist that do not depend simply on the magnitude of an objective function change, and generally it is preferable to choose moves with more attractive move values to achieve a given level of influence. The importance of establishing an appropriate balance between influence and quality is addressed more fully in Chapter 5. At present, we simply note that the nature of the balance depends on the phase of search. For example, moves of lower influence (including those with zero move values) may be tolerated until the opportunities for gain from them appear to be negligible. At such a point, and in the absence of improving moves, aspiration criteria should shift to give influential moves a higher rank. These considerations of move influence interact with considerations of regionality and search direction.

The information presented in Table 2.7 can be used to create evaluation functions to determine search directions in different regions of the solution space. One possible rule in this context could be to use the move value to choose the next point to be visited as long as there exists at least one neighbor that improves the current solution. In the absence of a neighbor with an improving move value (i.e., one that is negative in a minimization context, as considered here), the search may be directed by functions such as $\frac{\text{move value}}{\text{move distance}}$ or $\frac{\text{move value}}{\text{due date difference}}$, for all moves with strictly positive move values. These functions try to incorporate the influential aspect of a move without disregarding its value.

Although important, move influence is only one of several elements that commonly underlie the determination of aspiration criteria. We illustrate a few of these elements in Table 2.8. The realm of aspiration criteria is one that deserves much fuller study than it has so far received.

Table 2.8 Illustrative aspiration criteria.		
Aspiration by	*Description*	*Example*
Default	If all available moves are classified tabu, and are not render admissible by some other aspiration criteria, then a "least tabu" move is selected.	Revoke the tabu status of all moves with minimum *TabuEnd* value.
Objective	Global: A move aspiration is satisfied if the move yields a solution better than the best obtained so far.	Global: The best total tardiness found so far is 29. The current sequence is (4, 1, 5, 3, 6, 2) with T = 39. The move value of the tabu swap (5,2) is -20. Then, the tabu status of the swap is revoked and the search moves to the new best sequence (4, 1, 2, 3, 6, 5) with T = 19.
	Regional: A move aspiration is satisfied if the move yields a solution better than the best found in the region where the solution lies.	Regional: The best sequence found in the region defined by all sequences (1, 2, 3, *, *, *) is (1, 2, 3, 6, 4, 5) with $T = 31$. The current solution is (1, 4, 3, 2, 6, 5) with T = 23. The swap (4, 2) with move value of 6 is tabu. The tabu status is revoked because a new regional best (1, 2, 3, 4, 6, 5) with $T = 29$ can be found.
Search Direction	An attribute can be added and dropped from a solution (regardless of its tabu status), if the direction of the search (improving or nonimproving) has not changed.	For the Min *k*-Tree problem, the edge (11,12) has been recently dropped in the current improving phase making its addition a tabu-active attribute. The improving phase can continue if edge (11,12) is now added, therefore its tabu status may be revoked.
Influence	The tabu status of a low influence move may be revoked if a high influence move has been performed since establishing the tabu status for the low influence move.	If the low influence swap (1,4) described in Table 2.7 is classified tabu, its tabu status can be revoked after the high influence swap (4,5) is performed.

Aspirations such as those shown in Table 2.8 can be applied according to two implementation categories: aspiration by move and aspirations by attribute. A move aspiration, when satisfied, revokes the move's tabu classification. An attribute aspiration, when satisfied, revokes the attribute's tabu-active status. In the latter case the move may or may not change its tabu classification, depending on whether the tabu activation rule is triggered by more than one attribute. For example in our sequencing problem, if the swap of jobs 3 and 6 is forbidden because a tabu activation rule prevents job 3 from moving at all, then an attribute aspiration that revokes job 3's tabu-active status also revokes the move's tabu classification.

However, if the swap (3,6) is classified tabu because both job 3 and job 6 are not allowed to move, then revoking job 3's tabu-active status does not result in overriding the tabu status of the entire move.

Different variants of the aspiration criteria presented in Table 2.8 are possible. For example, the regional aspiration by objective can be defined in terms of bounds on the objective function value. These bounds determine the region being explored, and they are modified to reflect the discovery of better (or worse) regions. Another possibility is to define regions with respect to time. For example, one may record the best solution found during the recent past (defined as a number of iterations) and use this value as the aspiration level.

A very natural and useful type of aspiration criterion, which is often overlooked, involves reference to two levels of tabu restrictions, one strong and one weak — where the latter is capable of avoiding duplicate solutions, but allows "too much flexibility" in choosing moves under normal circumstances. (For example, the weak restriction may typically require a very large tabu tenure to avoid cycling.) Then any selected measure of attractiveness can be allowed to override the strong tabu restriction, but only a "safe" measure (such as identifying a new best solution) is allowed to override the weak restriction. Such an approach can be applied with more than two levels of restrictiveness (such as strong, moderate and weak), and is called the *Split Level Aspiration Criterion.* At each level, the strength of the associated restriction and the attractiveness of the associated aspiration can be made to vary throughout the search.

These fundamental short term memory considerations of tabu search are developed by additional examples in the exercises, and are elaborated to include other important short term elements in the next chapter.

2.7 Discussion Questions and Exercises

Short term memory tabu tenures and aspiration criteria (Exercises 1 - 9).

1. The tabu tenures used to illustrate the first level TS approach for the Min k-Tree problem in Section 2.3 are very small. To demonstrate the risks of using such tenures, show that if the weight of edge (3,6) in Figure 2.3 is changed from 20 to 17, then the optimal solution will not change but the illustrated TS approach with *TabuAddTenure* = 2 and *TabuDropTenure* = 1 will go into a cycle that will prevent the optimal solution from being found. (As in the illustrations of Sections 2.1 and 2.2, assume the tabu tenures are chosen to be the same for all edges, differentiated only relative to edges that are added and dropped. Also, we continue to suppose for the illustration that tenures are not varied on different iterations.)

2. Assign edge (3,6) of Figure 2.3 a weight of 17, as in Exercise 1. Show that changing the tenures in Exercise 1 so that *TabuAddTenure* = 3 and

TabuDropTenure = 1 will create a sequence of moves where the improved-best aspiration criterion will allow a tabu move to be selected. Also show that these tabu tenures will not avoid the cycling phenomenon identified in Exercise 1 when the weight of edge (3,6) is changed as indicated.

3. Use tenures of *TabuAddTenure* = 1 and *TabuDropTenure* = 2 for the changed example of Exercises 1 and 2. How does the behavior produced by this change compare to the behavior when the previous tabu tenures were used? Does the outcome reinforce the intuition that the *TabuDropTenure* has a stronger influence than the *TabuAddTenure* for this problem?

4. Practical experience with problems related to the Min *k*-Tree problem suggests that *TabuDropTenure*, which is stronger than *TabuAddTenure* in these settings, should be usually be kept relatively small even for somewhat larger problems. Can you argue that this is because such a stronger form of tabu tenure is likely to have a more chaotic effect — in the popular sense associated with chaos theory — on the solutions that are admitted for consideration?

5. Does the outcome of Exercise 3 suggest that a chaotic effect, as discussed in Exercise 4, can be valuable? Or, since a tenure of 2 is still small, would you anticipate that the outcome does not strongly relate to the occurrence of chaotic behavior? What experiments might be conducted to identify useful degrees of controlled chaos induced by different tenures?

6. What reasons may exist for assigning different edges different tabu tenures? Should the weights of the edges play a role? How about the change in the tree weight created when given edges are added and dropped?

7. Should past experience, such as the number of times an edge has previously appeared in solutions, play a role in determining tabu tenures? What other ways can such experience be used to directly influence evaluations and tabu status other than through tabu tenures? (This leads to a consideration of frequency-based memory, discussed in Chapter 4.)

8. Aspiration criteria often become increasingly important for regulating tabu restrictions as stronger or more elaborate forms of these restriction are used. Discuss the relevance of aspiration criteria illustrated in Section 2.6. What other types of aspiration criteria may be appropriate?

9. Care must be taken in creating aspiration criteria, because they can sometimes allow "unexpected loopholes" that encourage cycling. Can you identify criteria that are likely to be risky in this respect?

Alternative Neighborhoods (Exercises 10 - 12).

10. For any given set of $k+1$ nodes, an optimal (min weight) k-tree over these nodes can always be found by using the greedy constructive procedure illustrated in Table 2.2 to generate a starting solution (restricted to these nodes) or by beginning with an arbitrary tree on these nodes and performing a succession of static improving moves (which do not change the node set). The absence of a static improving move signals that no better solution can be found on this set.

    The preceding fact suggests that tabu search might advantageously be used to guide the search over a "node-swap" neighborhood instead of an "edge-swap" neighborhood, where each move consists of adding a non-tree node i and dropping a tree node j, followed by finding a min weight solution on the resulting node set. (Since the tree node j may not be a leaf node, and the reconnections may also not make node i a leaf node in the new tree, the possibilities are somewhat different than making a dynamic move in the edge-swap neighborhood.) The tabu tenures may reasonably be defined over nodes added and dropped, rather than over edges added and dropped.

    Devise a choice rule for a node-swap neighborhood to determine the node i to be added and the node j to be dropped at each step, and examine the application of a simple TS approach that chooses $TabuAddTenure = 2$ and $TabuDropTenure = 1$ for the example problem considered in Exercise 1.

11. Based on your findings from Exercise 10, discuss the relative strength of tabu tenures used in the node-swap neighborhood structure compared to those used in the edge-swap neighborhood structure. Would it also be possible (or convenient) to use tabu tenures based on both edges and nodes in these two different types of neighborhoods? Discuss the issues of choice rules and the added work of determining an optimal k-tree over each set of nodes selected.

12. Joining the perspectives of Exercises 1-3 and 10-11, what may be the value of a method that alternates between the two different neighborhoods considered? Consider a parallel processing application where different processors may use different neighborhoods (or different ways of combining neighborhoods) and different choice rules. What type of coordination would you imagine useful to establish between the different solution streams?

Critical event memory (Exercises 13 - 17).

13. Explain why the type of critical event memory used in the illustration of restarting the TS approach in Section 2.3.1 may not be best. Do you think it may be relevant to include edges of the worst solutions found (e.g., local maxima) as well as edges of the best solutions found?

14. It is reasonable to expect that the type of critical event memory used for restarting should normally be different from that used to continue the search from the current solution (when both are applied to drive the search into new regions). Nevertheless, a form that is popularly used in both situations consists of remembering *all* elements contained in solutions previously examined. One reason is that it is actually easier to maintain such memory than to keep track of elements that only occur in selected solutions. Explain why this is so.

15. Instead of keeping track only of which elements occur in past solution, critical event memory is more usually designed to monitor the frequency that elements have appeared in past solutions. Discuss why this information may be useful, and how you might take advantage of it. (These considerations are elaborated in Chapter 4.)

16. Can you see a difference between frequency-based critical event memory that monitors the number of times an element is added to (or dropped from) a solution, versus a memory than monitors the number of times an element was a member of (or not a member of) solutions previously examined? Should the information from these different types of memory be used differently? Is one easier to record than the other?

17. Would it make sense to select critical events of critical event memory differently if the search is in an intensification phase (seeking to reinforce or reinstate attributes of good solutions) as opposed to a diversification phase (seeking to drive the search into new regions)? What differences would you expect to be relevant? Would it be possible to pursue the goals of intensification and diversification simultaneously?

3 TABU SEARCH FOUNDATIONS: ADDITIONAL ASPECTS OF SHORT TERM MEMORY

We began the discussion of short term memory for tabu search in Chapter 2 by contrasting the TS designs with those of memoryless strategies such as simple or iterated descent, and by pointing out how candidate list strategies are especially important for applying TS in the most effective ways. We have deferred the discussion of such strategies in order to provide an understanding of some of the basic roles played by memory, and to illustrate the aggressive nature of decision criteria that typically are used to reinforce these roles. In this chapter we begin by describing types of candidate list strategies that often prove valuable in tabu search implementations. Then we examine the issues of logical restructuring, which provide important bridges to longer term considerations. Finally, we describe additional short term memory structures that build on the ideas used to apply TS memory to add/drop moves. We show how these ideas can be conveniently adapted to provide related structures for many different kinds of optimization problems, and associated neighborhoods for defining moves for these problems. These structures likewise provide a basis for incorporating longer term considerations.

3.1 Tabu Search and Candidate List Strategies

The aggressive aspect of TS is manifest in choice rules that seek the best available move that can be determined with an appropriate amount of effort. As addressed in Chapter 2, the meaning of best in TS applications is customarily not limited to an objective function evaluation. Even where the objective function evaluation may appear on the surface to be the only reasonable criterion to determine the best move, the non-tabu move that yields a maximum improvement or least deterioration is not always the one that should be chosen. Rather, as we have noted, the definition of best should consider factors such as move influence, determined by the search history and the problem context.

For situations where $\mathbf{N}^*(x)$ is large or its elements are expensive to evaluate, candidate list strategies are essential to restrict the number of solutions examined on a given iteration. In many practical settings, TS is used to control a search process that may involve the solution of relatively complex subproblems by way of linear programming or simulation. Because of the importance TS attaches to selecting elements judiciously, efficient rules for generating and evaluating good candidates are critical to the search process. The purpose of these values is to isolate regions of the neighborhood containing moves with desirable features and to put these moves on a list of candidates for current examination.

Before describing the kinds of candidate list strategies that are particularly useful in tabu search implementations, we note that the efficiency of implementing such strategies often can be enhanced by using relatively straightforward memory structures to give efficient updates of move evaluations from one iteration to another. Appropriately coordinated, such updates can appreciably reduce the effort of finding best or near best moves.

In sequencing, for example, the move values often can be calculated without a full evaluation of the objective function. We use the sequencing example introduced at the start of Chapter 2 as a basis for illustration. Consider the move value of the swap represented by the pair (5,6) as derived from Table 2.1 of Chapter 2. We show the information concerning this move value in Table 3.1.

Table 3.1 Swap move illustration.

Processing times:	(6, 4, 8, 2, 10 3)		
Due dates:	(9, 12, 15, 8, 20, 22)		
Tardiness penalties:	(1, 1, 1, 1, 1, 1)		
Current sequence	*Tardiness*	*Swap*	*Move value*
(1, 2, 3, 4, 5, 6)	36	(5,6)	-7
(1, 2, 3, 4, 6, 5)	29	—	—

The indicated move value of -7 can be found in a "direct" (but expensive) manner by temporarily exchanging the positions of jobs 5 and 6, calculating the new total tardiness value of 29, and subtracting the current tardiness value of 36 from the new one. After the evaluation, the jobs are returned to their original positions and the process continues. This is an $O(n)$ operation for each swap in the neighborhood (since on average the calculation of the new total tardiness in this design involves looking at roughly half the total number of job positions). Alternatively, the completion times for each job in the current solution can be stored and used to update the completion times of those jobs that are affected by a proposed swap. In this case, only jobs 5 and 6 are affected. In general, however, if a swap (i, j) is performed, the completion times for jobs i and j and all those jobs between the positions occupied by i and j must be updated. The new completion times C' for jobs 5 and 6 can be found by:

$$C'_5 = C_5 + p_6 = 30 + 3 = 33$$
$$C'_6 = C_6 - p_5 = 33 - 10 = 23$$

and the move value is given by $C'_5 + C'_6 - C_5 - C_6 = 33 + 23 - 30 - 33 = -7$.

Intelligent updating of this type can be useful even where candidate list strategies are not used. However, the inclusion of explicit candidate list strategies, for problems that are large, can significantly magnify the resulting benefits. Not only search speed but also solution quality can be influenced by the use of appropriate candidate list strategies. Perhaps surprisingly, the importance of such approaches is often overlooked.

3.2 Some General Classes of Candidate List Strategies

Candidate lists can be constructed from context related rules (such as the one illustrated for the sequencing problem) and from general strategies. In this section we focus on rules for constructing candidate lists that are context-independent. In considering such rules, we emphasize that the effectiveness of a candidate list strategy should not be measured in terms of the reduction of the computational effort in a single iteration. Instead, a preferable measure of performance for a given candidate list is the quality of the best solution found given a specified amount of computer time. For example, a candidate list strategy intended to replace an exhaustive neighborhood examination may result in more iterations per unit of time, but may require many more iterations to match the solution quality of the original method. If the quality of the best solution found within a desirable time limit (or across a graduated series of such limits) does not improve, we conclude that the candidate list strategy is not effective.

3.2.1 Aspiration Plus

The Aspiration Plus strategy establishes a threshold for the quality of a move, based on the history of the search pattern. The procedure operates by examining moves until finding one that satisfies this threshold. Upon reaching this point, additional moves are examined, equal in number to the selected value *Plus,* and the best move overall is selected.

To assure that neither too few nor too many moves are considered, this rule is qualified to require that at least *Min* moves and at most *Max* moves are examined, for chosen values of *Min* and *Max*. The interpretation of *Min* and *Max* is as follows. Let *First* denote the number of moves examined when the aspiration threshold is first satisfied. Then if *Min* and *Max* were not specified, the total number of moves examined would be *First* + *Plus*. However, if *First* + *Plus* < *Min*, then *Min* moves are examined while if *First* + *Plus* > *Max*, then *Max* moves are examined. (This conditions may be viewed as imposing limits on the move that is "effectively" treated as the *First* move. For example, if as many as *Max* - *Plus* moves are examined without finding one that satisfies the aspiration threshold, then *First* effectively becomes the same as *Max* - *Plus*.)

This strategy is graphically represented in Figure 3.1. In this illustration, the fourth move examined satisfies the aspiration threshold and qualifies as *First*. The value of *Plus* has been selected to be 5, and so 9 moves are examined in total, selecting the best over this interval. The value of *Min*, set at 7, indicates that at least 7 moves will be examined even if *First* is so small that *First* + *Plus* < 7. (In this case, *Min* is not very restrictive, because it only applies if *First* < 2.) Similarly, the value of *Max*, set at 11, indicates that at most 11 moves will be examined even if *First* is so large that *First* + *Plus* > 11. (Here, *Max* is strongly restrictive.) The sixth move examined is the best found in this illustration.

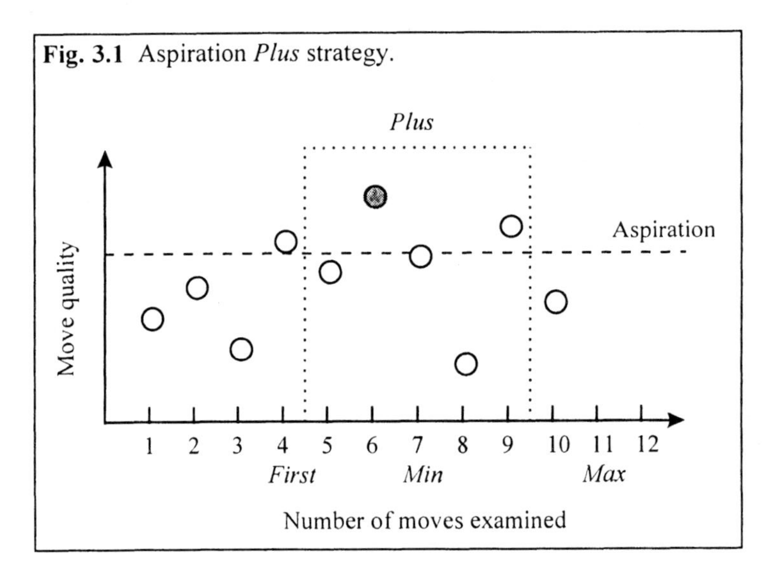

Fig. 3.1 Aspiration *Plus* strategy.

The "Aspiration" line in this approach is an established threshold that can be dynamically adjusted during the search. For example, during a sequence of improving moves, the aspiration may specify that the next move chosen should likewise be improving, at a level based on other recent moves and the current objective function value. Similarly, the values of *Min* and *Max* can be modified as a function of the number of moves required to meet the threshold.

During a nonimproving sequence the aspiration of the Aspiration Plus rule will typically be lower than during an improving phase, but rise toward the improving level as the sequence lengthens. The quality of currently examined moves can shift the threshold, as by encountering moves that significantly surpass or that uniformly fall below the threshold. As an elementary option, the threshold can simply be a function of the quality of the initial *Min* moves examined on the current iteration.

The Aspiration Plus strategy includes several other strategies as special cases. For example, a *first improving* strategy results by setting *Plus* = 0 and directing the aspiration threshold to accept moves that qualify as improving, while ignoring the values of *Min* and *Max*. Then *First* corresponds to the first move that improves the current value of the objective, if such a move can be found. A slightly more advanced strategy can allow *Plus* to be increased or decreased according to the variance in the quality of moves encountered from among some initial number examined. In general, in applying the Aspiration Plus strategy, it is important to assure on each iteration that new moves are examined which differ from those just reviewed. One way of achieving this is to create a circular list and start each new iteration where the previous examination left off.

3.2.2 Elite Candidate List

The Elite Candidate List approach first builds a Master List by examining all (or a relatively large number of) moves, selecting the k best moves encountered, where k is a parameter of the process. Then at each subsequent iteration, the current best move from the Master List is chosen to be executed, continuing until such a move falls below a given quality threshold, or until a given number of iterations have elapsed. Then a new Master List is constructed and the process repeats. This strategy is depicted in Figure 3.2, below.

This technique is motivated by the assumption that a good move, if not performed at the present iteration, will still be a good move for some number of iterations. More precisely, after an iteration is performed, the nature of a recorded move implicitly may be transformed. The assumption is that a useful proportion of these transformed moves will inherit attractive properties from their antecedents.

The evaluation and precise identity of a given move on the list must be appropriately monitored, since one or both may change as result of executing other moves from the list. For example, in the Min k-Tree problem the evaluations of many moves can remain unchanged from one iteration to the next. However, the identity and evaluation of specific moves will change as a result of deleting and adding particular edges, and these changes should be accounted for by appropriate updating (applied periodically if not at each iteration). An Elite Candidate List strategy can be advantageously extended by a variant of the Aspiration Plus strategy, allowing some additional number of moves outside the Master List to be examined at each iteration, where those of sufficiently high quality may replace elements of the Master List.

Fig. 3.2 Elite candidate list strategy.

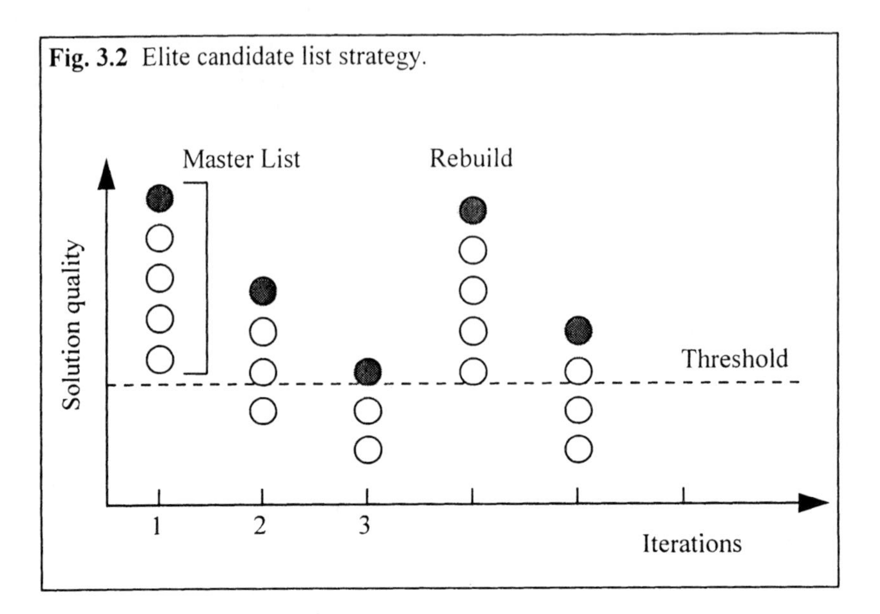

3.2.3 Successive Filter Strategy

Moves can often be broken into component operations, and the set of moves examined can be reduced by restricting consideration to those that yield high quality outcomes for each operation separately. For example, the choice of an exchange move that includes an "add component" and a "drop component" may restrict attention only to exchanges created from a relatively small subset of "best add" and "best drop" components. The gain in efficiency can be considerable. If there are 100 add possibilities and 100 drop possibilities, the number of add/drop combinations is 10,000. However, by restricting attention to the 8 best add and drop moves, considered independently, the number of combinations to examine is only 64. (Values of 8 and even smaller have been found effective in some practical applications.)

The evaluations of the separate components often will give only approximate information about their combined evaluation. Nevertheless, if this information is good enough to insure a significant number of the best complete moves will result by combining these apparently best components, then the approach can yield quite good outcomes. Improved information may be obtained by sequential evaluations, as where the evaluation of one component is conditional upon the prior (restricted) choices of another. Such strategies of subdividing compound moves into components, and then restricting consideration of complete compound moves only to those assembled from components that pass selected thresholds of quality, have proved quite effective in TS methods for partitioning problems and for telecommunication channel balancing problems.

Conditional uses of component evaluations are also relevant for sequencing problems, where a measure can be defined to identify preferred attributes using information such as due dates, processing times, and delay penalties. If swap moves are being used, then some jobs are generally better candidates than others to move early or later in the sequence. The candidate list considers those swaps whose composition includes at least one of these preferred attributes.

In the context of the traveling salesman problem, good solutions are often primarily composed of edges that are among the 20 to 40 shortest edges meeting one of their endpoints (depending on various factors). Some studies have attempted to limit consideration entirely to tours constructed from such a collection of edges. The successive filter strategy, by contrast, offers greater flexibility by organizing moves that do not have to be entirely composed of such special elements, provided one or more of these elements is incorporated as part of the move. This approach can be frequently controlled to require little more time than the more restricted standard approach, while affording a more desirable set of alternatives to consider.

3.2.4 Sequential Fan Candidate List

A type of candidate list that is highly exploitable by parallel processing is the sequential fan candidate list. The basic idea is to generate some p best alternative moves at a given step, and then to create a fan of solution streams, one for each alternative. The several best available moves for each stream are again examined, and only the p best moves overall (where many or no moves may be contributed by a given stream) provide the p new streams at the next step.

In the setting of tree search methods such a sequential fanning process is sometimes called *beam search.* For use in the tabu search framework, TS memory and activation rules can be carried forward with each stream and hence inherited in the selected continuations. Since a chosen solution can be assigned to more than one stream, different streams can embody different missions in TS. Alternatively, when two streams merge into the same solution other streams may be started by selecting a neighbor adjacent to one of the current streams.

The process is graphically represented in Figure 3.3. Iteration 0 constructs an initial solution or alternatively may be viewed as the starting point for constructing a solution. That is, the sequential fan approach can be applied using one type of move to create a set of initial solutions, and then can continue using another type of move to generate additional solutions. (We thus allow a "solution" to be a partial solution as well as a complete solution.) The best moves from this solution are used to generate p streams. Then at every subsequent iteration, the overall best moves are selected to lead the search to p different solutions. Note that since more than one move may lead the search to the same solution, more than p moves may be necessary to continue the exploration of p distinct streams.

Fig. 3.3 Sequential fan candidate list.

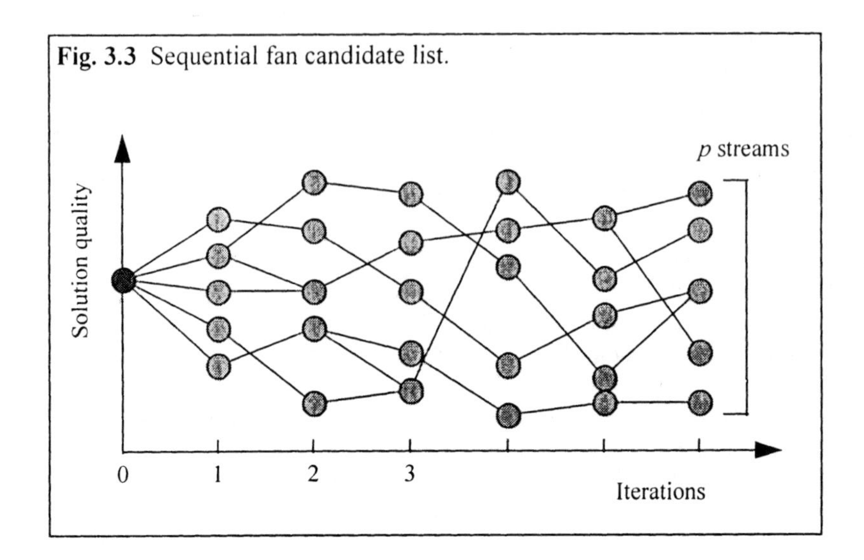

A more intensive form of the sequential fan candidate list approach, which is potentially more powerful but requires more work, is to use the process illustrated in Figure 3.3 as a "look ahead" strategy. In this case a limit is placed on the number of iterations that the streams are generated beyond iteration 0. Then the best outcome at this limiting iteration is used to identify a "best current move" (a single first branch) from iteration 0. Upon executing this move, the step shown as iteration 1 in Figure 3.3 becomes the new iteration 0, that is, iteration 0 always corresponds to the current iteration. Then this solution becomes the source of p new streams, and the process repeats.

There are a number of possible variants of this sequential fan strategy. For example, instead of selecting a single best branch at the limiting iteration, the method can select a small number of best branches, and thus give the method a handful of candidates from which to generate p streams at the new iteration 0.

The iteration limit that determines depth of the look ahead can be variable, and the value of p can change at various depths. Also the number of successors of a given solution that are examined to determine candidates for the p best continuations can be varied as by progressively reducing this number at greater depths.

The type of staging involved in successive solution runs of each stream may be viewed as a means of defining levels in the context of the Proximate Optimality Principle commonly associated with the strategic oscillation component of tabu search. Although we will study this principle in more detail later, we remark that the sequential fan candidate list has a form that is conveniently suited to exploit it.

3.2.5 Bounded Change Candidate List

A bounded change candidate list strategy is relevant in situations where an improved solution can be found by restricting the domain of choices so that no solution component changes by more than a limited degree on any step. A bound on this degree, expressed by a distance metric appropriate to the context, is selected large enough to encompass possibilities considered strategically relevant. The metric may allow large changes along one dimension, but limit the changes along another so that choices can be reduced and evaluated more quickly. Such an approach offers particular benefits as part of an intensification strategy based on decomposition, where the decomposition itself suggests the limits for bounding the changes considered.

3.3 Connections Between Candidate Lists, Tabu Status and Aspiration Criteria

It is useful to summarize the short term memory considerations embodied in the interaction between candidate lists, tabu status and aspiration criteria. The operations of these TS short term elements are shown in Figure 3.4. The representation of penalties in Figure 3.4 either as "large" or "very small" expresses a thresholding effect: either the tabu status yields a greatly deteriorated evaluation or else it chiefly serves to break ties among solutions with highest evaluations. Such an effect of course can be modulated to shift evaluations across levels other than these extremes. If all moves currently available lead to solutions that are tabu (with evaluations that normally would exclude them from being selected), the penalties result in choosing a "least tabu" solution.

The sequence of the *tabu test* and the *aspiration test* in Figure 3.4 evidently can be reversed (that is, by employing the tabu test only if the aspiration threshold is not satisfied). Also, the tabu evaluation can be modified by creating inducements based on the aspiration level, just as it is modified by creating penalties based on tabu status. In this sense, aspiration conditions and tabu conditions can be conceived roughly as "mirror images" of each other.

For convenience Figure 3.4 expresses tabu restrictions solely in terms of penalized evaluations, although we have seen that tabu status is often permitted to serve as an all-or-none threshold, without explicit reference to penalties and inducements (by directly excluding tabu options from being selected, subject to the outcome of aspiration tests). Whether or not modified evaluations are explicitly used, the selected move may not be the one with the best objective function value, and consequently the solution with the best objective function value encountered throughout the search history is recorded separately.

Fig. 3.4 Short term memory operation.

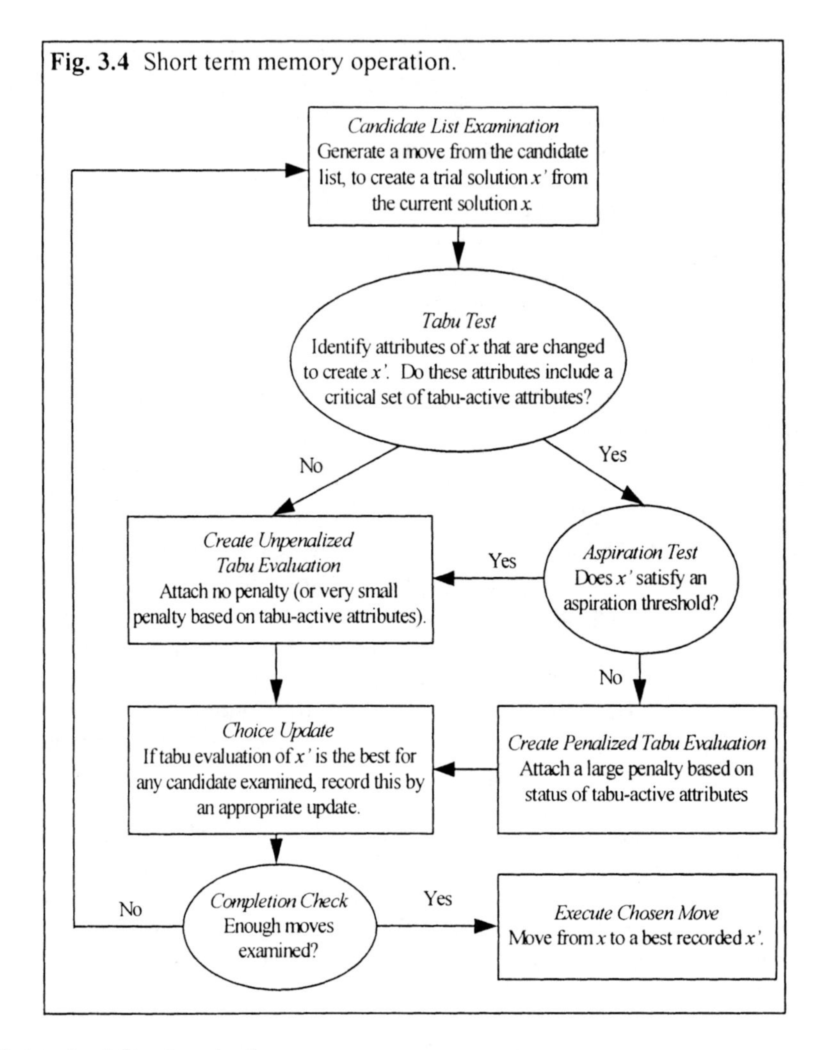

3.4 Logical Restructuring

Logical restructuring is an important element of adaptive memory solution approaches, which gives a connection between short and long term strategies. Logical restructuring is implicit in strategic oscillation, path relinking and ejection chain constructions, which we examine in subsequent chapters, but its role and significance in these strategies is often overlooked. By extension, the general usefulness of logical restructuring is also often not clearly understood. We examine some of its principal features before delving into longer term considerations, and show how it can also be relevant for improving the designs of short term strategies.

Logical restructuring emerges as a way to meet the combined concerns of quality and influence. Its goal is to exploit the ways in which influence (structural, local and global) can uncover improved routes to high quality solutions. For this purpose, a critical step is to re-design standard strategies to endow them with the power to ferret out opportunities otherwise missed. This step particularly relies on integrating two elements: (1) the identification of changes that satisfy properties that are essential (and limiting) in order to achieve improvement, in contrast to changes that simply depart from what has previously been seen; (2) the use of anticipatory ("means-ends") analysis to bring about such essential changes. Within the context of anticipatory analysis, logical restructuring seeks to answer the following questions: "What conditions assure the existence of a trajectory that will lead to an improved solution?" and "What intermediate moves can create such conditions?" The "intermediate moves" of the second question may be generated either by modifying the evaluations used to select transitions between solutions or by modifying the neighborhood structure that determines these transitions.

To illustrate the relevant considerations, we return again to the example of the Min k-Tree problem discussed in Chapter 2. We replace the previous graph by the one shown in Figure 3.5, but continue to consider the case of $k = 4$.

Fig. 3.5 Illustrative Min k-Tree Problem.

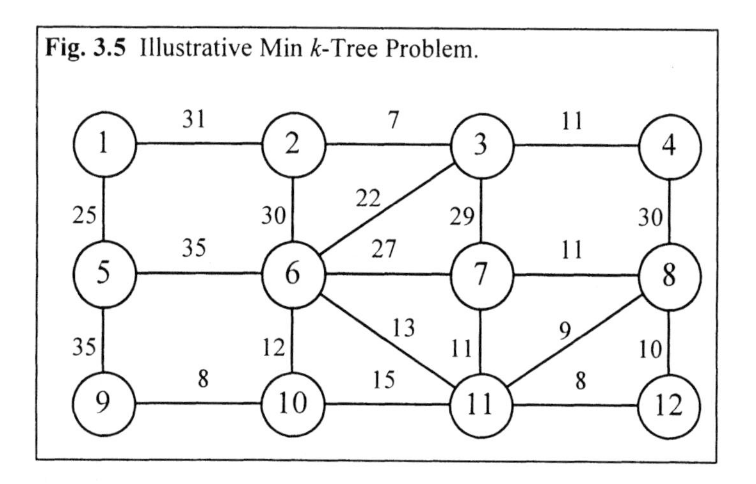

Using the same rules to execute a first-level tabu search approach as in our earlier illustrations of Chapter 2 (including the rules for generating a starting solution), we obtain the sequence of solutions shown in Table 3.2 and Figure 3.6. As before, asterisks identify local optima, subject to tabu conditions, and the "Tabu Edges" are those that are tabu for the iteration that follows.

It is not necessary to trace the steps depicted in the table and diagram in detail. Instead we note certain features of these steps and highlight their implications by focusing on two key iterations (4 and 5). The solution shown at iteration 11 is optimal for this problem, and we have terminated the illustration upon reaching this solution.

Table 3.2 TS iterations for Min *k*-Tree Problem in Fig. 3.5.

Iteration	Tabu-active net tenure		Add	Delete	Move Value	Weight
	1	2				
0						69
1			(6,10)	(3,7)	-17	52
2	(6,10)	(3,7)	(9,10)	(3,4)	-3	49*
3	(3,7), (9,10)	(3,4)	(6,11)	(2,3)	6	55
4	(3,4), (6,11)	(2,3)	(11,12)	(3,6)	-14	41*
5	(2,3), (11,12)	(3,6)	(8,11)	(9,10)	1	42
6	(3,6), (8,11)	(9,10)	(7,8)	(6,10)	-1	41
7	(9,10), (7,8)	(6,10)	(8,12)	(8,11)	1	42
8	(6,10), (8,12)	(8,11)	(7,11)	(7,8)	0	42
9	(8,11), (7,11)	(7,8)	(10,11)	(6,11)	2	44
10	(7,11), (10,11)	(6,11)	(9,10)	(7,11)	-3	41
11	(7,8), (9,10)	(7,11)	(8,11)	(8,12)	-1	40*

Examination of the solution sequence shows that the method quickly reached the vicinity of the optimal solution, but required some effort to actually find this solution. In fact, all edges of the optimal solution except one, edge (10,11), were contained in the union of the two solutions in iterations 4 and 5. Yet it was not until iteration 9 that this edge became incorporated into the current solution, and 2 further iterations were required to locate the optimum. This delayed process of finding a route to an optimal solution (which can be greatly magnified for larger and more complex problems) can be substantially accelerated by means of logical restructuring. More generally, such restructuring can make it possible to uncover fertile options that can otherwise be missed entirely.

We use the present example, as depicted in Figure 3.6 following, to illustrate two different forms of logical restructuring, which are instances of more advanced strategies that will be examined later. (Exercises at the end of the chapter also develop these forms of logical restructuring in greater detail.)

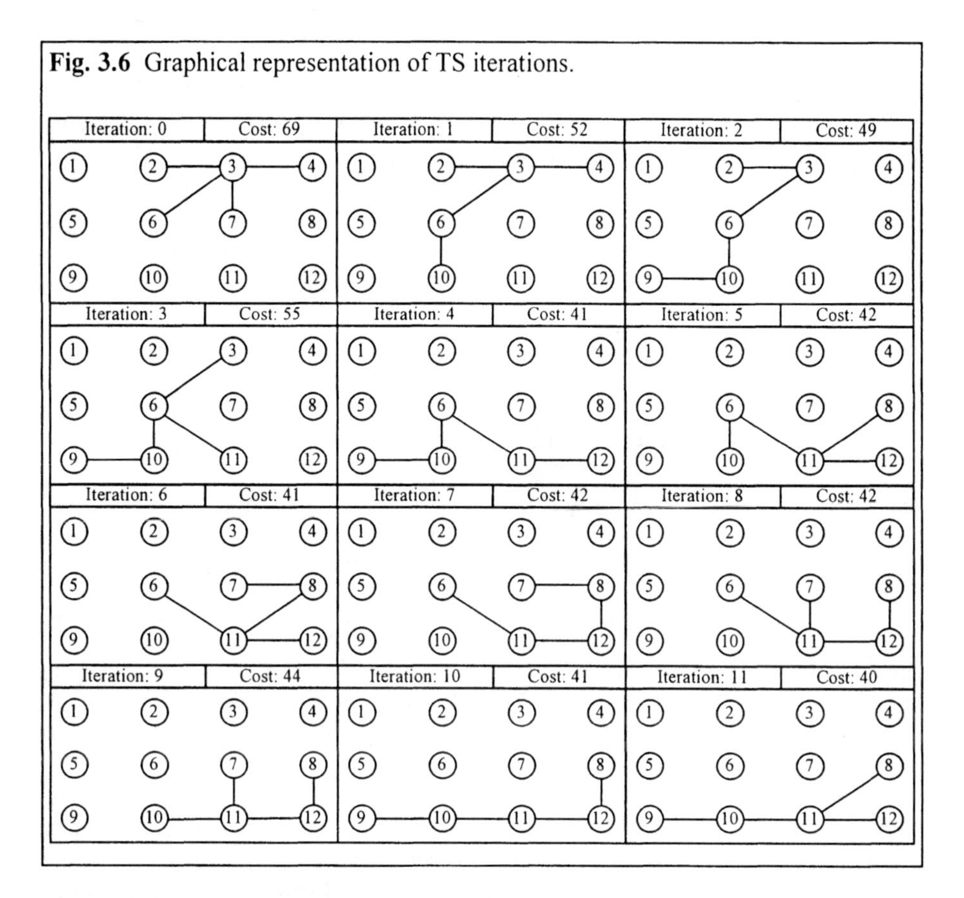

Fig. 3.6 Graphical representation of TS iterations.

3.4.1 Restructuring by Changing Evaluations and Neighborhoods

The first type of logical restructuring we illustrate makes use both of modified evaluations and an amended neighborhood structure. As pointed out in Figure 2.4 of Chapter 2, the swap moves we have employed for the Min *k*-Tree problem may be subdivided into two types: *static* swaps, which leave the nodes of the current tree unchanged, and *dynamic* swaps, which replace one of the nodes currently in the tree with another that is not in the tree. This terminology was chosen to reflect the effect that each swap type has on the nodes of the tree. Since dynamic swaps in a sense are more influential, we give them special consideration. We observe that a dynamic swap can select an edge to be dropped only if it is a *terminal edge* — i.e., one that meets a *leaf node* of the tree, which is a node that is met by only a single tree edge (the terminal edge). For example, at iteration 5, as depicted in Figure 3.6, the terminal edge (9,10) that is dropped from the tree of iteration 4 is the unique edge that meets the leaf node 9, and node 9 also is dropped when (9,10) is dropped.

Although it is usually advantageous to drop an edge with a relatively large weight, this may not be possible — as in the dynamic swap of iteration 5, where the only terminal edges are low weight edges. Thus, we are prompted to consider an "anticipatory goal" of making moves that cause more heavily weighted edges to become terminal edges, and hence eligible to be dropped. By this means, static swaps can be used to set up desirable conditions for dynamic swaps.

The solution obtained at iteration 4 (before executing iteration 5) gives a basis for showing what is involved. We clarify the situation by showing the current solution as heavy edges and the candidate edges to add as light edges in Figure 3.7.

Fig. 3.7 Solution and candidate edges to add to iteration 4 tree.

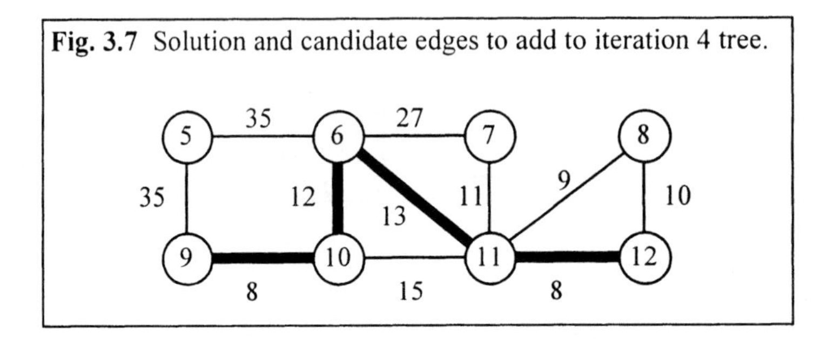

The move that changes the tree at iteration 4 to that of iteration 5 is a dynamic swap that adds edge (8,11) with a weight of 9 and drops edge (9,10) with a weight of 8. We make use of the information contained in this choice to construct a more powerful move using logical restructuring as follows.

Having identified (8,11) as a candidate to be added, the associated anticipatory goal is to identify a static swap that will change a larger weight edge into a terminal edge. Specifically, the static swap that adds edge (10,11) and drops edge (6,10), with a move value of 3, produces a terminal edge from the relatively high weight edge (6,11) (which has a weight of 13). Since the candidate edge (8,11) to be added has a weight of 9, the result of joining the indicated static swap with the subsequent dynamic swap (that respectively adds and drops (8,11) and (6,11)) will be a net gain. (The static move value of 3 is joined with the dynamic move value of -4, yielding a result of -1.)

Effectively, such anticipatory analysis leads to a way to extract a fruitful outcome from a relatively complex set of options by focusing on a simple set of features. It would be possible to find the same outcome by a more ponderous approach that checks all sequences in which a dynamic move follows a static move. This requires a great deal of computational effort — in fact, considerably more than involved in the approach without logical restructuring that succeeded in finding an optimal solution at iteration 11 (considering the trade-off between number of iterations and work per iteration).

By contrast, the use of logical restructuring allows the anticipatory analysis to achieve the benefits of a more massive exploration of alternatives, but without incurring the burden of undue computational effort. In this example, the restructuring is accomplished directly as follows. First, it is only necessary to identify the two best edges to add for a dynamic swap (independent of matching them with an edge to drop), subject to requiring that these edges meet different nodes of the tree. (In the tree of iteration 4, seen in Figure 3.7, these two edges are (8,11) and (8,12).) Then at the next step, during the process of looking at candidate static swaps, a modified "anticipatory move value" is created for each swap that creates a terminal edge, by subtracting the weight of this edge from the standard move value.

This gives all that is needed to find (and evaluate) a best "combined move sequence" of the type we are looking for. In particular, every static move that generates a terminal edge can be combined with a dynamic move that drops this edge and then adds one of the two "best edges" identified in first of the two preceding steps. Hence, the restructuring is completed by adding the anticipatory move value to the weight of one of these two edges (appropriately identified) thereby determining a best combined move. The illustrated process therefore achieves restructuring in two ways — by modifying customary move values and by fusing certain sequences of moves into a single compound move.

Although this example appears on the surface to be highly problem specific, its basic features are shared by applications that arise in a variety of problem settings. Later the reader will see how variants of logical restructuring embodied in this illustration are natural components of the strategies of path relinking and ejection chain constructions.

3.4.2 Threshold Based Restructuring and Induced Decomposition

The second mode of logical restructuring that we illustrate by reference to the Min *k*-Tree problem example is more complex (in the sense of inducing a more radical restructuring), but relatively easy to sketch and also potentially more powerful.

Consider again the solution produced at iteration 4. This is a local optimum and also the best solution found up to the current stage of search. We seek to identify a property that will be satisfied by at least one solution that has a smaller weight than the weight of this solution (41), and which will impose useful limits on the composition of such a solution. A property that in fact must be shared by all "better" solutions can be expressed as a threshold involving the average weight of the tree edges. This average weight must be less than the threshold value of 41/4 (i.e., 10 1/4). Since some of the edges in any improved solution must have weights less than this threshold, we are motivated to identify such "preferred" edges as a foundation for a restructured form of the solution approach. In this type of restructuring, we no longer confine attention to swap moves, but look for ways to link the preferred edges to produce an improved solution. (Such a restructuring can be based on threshold values derived from multiple criteria.)

When the indicated strategy is applied to the present example, a large part of the graph is eliminated, leaving only 3 separate connected components: (a) the edge (2,3), (b) the edge (9,10), and (c) the three edges (8,11), (8,12) and (11,12). The graph that highlights these components is shown in Figure 3.8. At this point a natural approach is to link such components by shortest paths, and then shave off terminal edges if the trees are too large, before returning to the swapping process. Such an approach will immediately find the optimal solution that previously was not found until iteration 11.

Fig. 3.8 Threshold generated components.

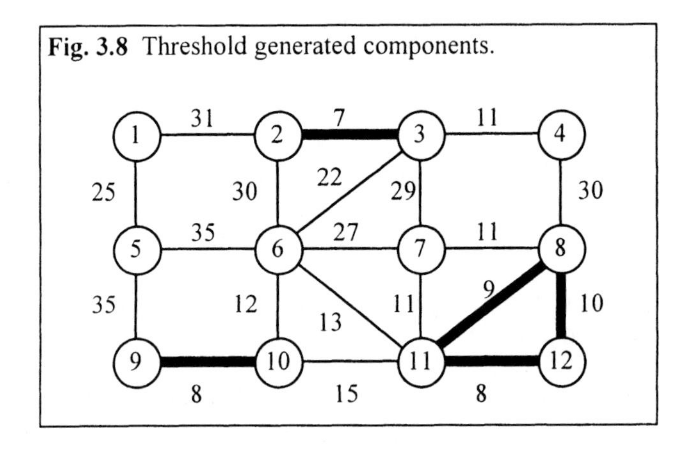

This second illustrated form of restructuring is a fundamental component of the strategic oscillation approach which we have alluded to in Chapter 1 and describe in more detail in the next chapter. A salient feature of this type of restructuring is its ability to create an *induced decomposition* of either the solution space or the problem space. This outcome, coupled with the goal of effectively joining the decomposed components to generate additional solution alternatives, is also a basic characteristic of path relinking, which is also examined in the next chapter. More particularly, the special instance of path relinking known as vocabulary building, which focuses on assembling fragments of solutions into larger units, offers a direct model for generalizing the "threshold decomposition" strategy illustrated here.

In some applications, specific theorems can be developed about the nature of optimal solutions and can be used to provide relevant designs for restructuring. The Min k-Tree problem is one for which such a theorem is available (Glover and Laguna, 1997a). Interestingly, the second form of restructuring we have illustrated, which is quite basic, exploits several aspects of this theorem — although without "knowing" what the theorem is. In general, logical restructuring and the TS strategies such as path relinking and strategic oscillation which embody it, appear to behave as if they similarly have a capacity to exploit underlying properties of optimal solutions in broader contexts — contexts whose features are not sufficiently uniform or easily characterized to permit the nature of optimal solutions to be expressed in the form of a theorem.

3.5 Special Cases and Extensions of Recency-Based Implementations

We conclude the chapter by showing how the basic ideas for implementing recency-based memory described in Section 2.4 of Chapter 2 lead to several useful alternatives that are relevant to other settings.

The ideas and data structures of this section can be deferred by readers who prefer to gain a broad overall understanding of TS before focusing on details at this level. However, those who undertake serious TS implementations will find this section valuable, as a basis for handling specific types of problems and neighborhoods. The material that follows is designed to provide fuller understanding of how to establish short term memory structures and how to take advantage of strategic options they offer.

We have noted earlier in this chapter that tabu-active status can be handled by the use of penalties (and, in conjunction with aspiration levels, by inducements or "negative penalties"). We first show how the ideas of Section 2.4 can easily be applied to implement penalty-based strategies. By reference to the *TabuEnd* array as introduced in Section 2.4.2, if we define

$$NetAddTenure = TabuEnd(TestAdd) - Iter$$
$$NetDropTenure = TabuEnd(TestDrop) - Iter$$

then we may consider differentiating among different degrees of being tabu-active, as a function of the values of these net tenures (restricting attention to those that are positive). For example, we can simply penalize a move value by multiplying the net tenures by associated penalty factors, which therefore will make a given move less attractive as the values of these component net tenures increase. Other functions that cause penalties to change other than proportionately with changes in net tenures are also possible.

Net tenures of course can also be defined relative to the *TabuStart* array by noting that, for any element E,

$$TabuEnd(E) = TabuStart(E) + TabuTenure(E),$$

where $TabuTenure(E)$ represents an add tenure or drop tenure according to the current state of E. The use of net tenures and penalties to generate graduated levels of tabu-active status was proposed and implemented in some of the early tabu search applications (e.g., Glover, McMillan and Novick, 1985). However, little experimentation has been done to determine whether such penalty based approaches are more effective than the commonly used methods that treat tabu-status as an all-or-none condition, subject to the moderating effect of aspiration criteria. On the other hand, as will be seen, penalties (and inducements) are quite often used in conjunction with frequency-based memory, and thus are particularly natural to be used when frequency-based memory and recency-based memory are combined. As in a variety of other areas that will be pointed out in this book, this represents an area that invites additional study. Fuller appreciation for the possibilities will be gained once the foundations are laid for incorporating frequency-based memory (and combining this memory with recency-based memory).

3.5.1 Tabu Restrictions: Transforming Attribute Status into Move Status

By the foregoing development, we can identify various conditions that show how the tabu-active status of attributes (treated as elements with implicit states, in the add/drop setting) can give rise to various conditions for classifying moves tabu — or more precisely, for classifying solutions tabu that will visited by various associated moves.

We first examine the conditions that were applied in the Min *k*-Tree Problem example, where a move was classified tabu if either its added or its dropped element (edge) was tabu-active. Denote the provisional move of adding *TestAdd* and dropping *TestDrop* by the ordered pair (*TestAdd,TestDrop*). Then the classification used in the Min *k*-Tree Problem example can be expressed in the form:

Example Tabu Classification 1.
The candidate move (*TestAdd,TestDrop*) is tabu if:

$$Iter \leq TabuStart(TestAdd) + TabuAddTenure$$

or

$$Iter \leq TabuStart(TestDrop) + TabuDropTenure.$$

(As in Section 2.4.2, we can alternately replace the right hand sides of the two preceding inequalities by *TabuEnd*(*TestAdd*) and *TabuEnd*(*TestDrop*).) This type of classification is a common one, because it is easy to implement and because it is generally easy to identify values of *TabuAddTenure* and *TabuDropTenure* that prove effective. In many settings, the impact of preventing an element from being dropped, as in preventing an edge from being dropped from a tree, is so strong that *TabuDropTenure* is chosen quite small. Often *TabuAddTenure* is a simple multiple of the square root of a relevant problem dimension, such as the number of nodes or edges of a graph. (A number of applications select this multiple to be between 0.5 and 2).

Clearly, however, the tabu classification illustrated above is not the only one possible. Even simpler is to base the tabu status of a move solely on the tabu-active status of a single element, as in the following alternative, where $X = Add$ or $X = Drop$:

Example Tabu Classification 2.
The candidate move (TestAdd,TestDrop) is tabu if:

$$Iter \leq TabuStart(TestX) + TabuXTenure.$$

When such a determination of tabu status is used, normally the choice is to let $X = Add$, due to the greater restrictiveness of letting $X = Drop$, as previously discussed. Of course, the restrictiveness is offset to a degree by choosing the tabu tenure to be smaller for $X = Drop$. However, in applications where the difference between the impact of dropping and adding elements is not great, the preceding "single element determination" of tabu status is somewhat arbitrary (and usually inappropriate).

The Example Tabu Classification 2 is relevant in settings that are slightly different from those so far discussed. Useful neighborhoods for some kinds of problems consist simply of making either an add move or a drop move at each step, rather than making both types of moves simultaneously. For example, in solving a linear or nonlinear pure 0-1 integer programming problem (where each variable receives only the value 0 or 1), it can be convenient to employ moves that simply set a single chosen variable to 1 or 0 at each iteration. By treating these options as corresponding to "add" and "drop" moves we therefore find it useful to employ Classification 2, where $X = Add$ or $Drop$ according to the type of move currently considered. (Both add and drop moves may be considered on the same step, or may be considered in alternate waves as in strategic oscillation approaches.) When special structures such as generalized upper bound (GUB) constraints are present, then combined add/drop moves are often preferable.

In addition, there are settings where it is appropriate to use independent add and drop moves on some iterations, and to use combined add/drop moves on other iterations. Since combined add/drop moves are typically far more numerous than independent add and drop moves, and may also be more expensive to evaluate, a natural approach is to implement them only periodically and more sparingly. When both independent and combined add/drop options are employed, then both Classification 1 and Classification 2 are relevant. In this case, the choice of tabu tenures depends on the type of move (and hence the Tabu Classification) used. Applications where such mixes of moves have proved effective include digital line network design problems in telecommunications, capital budgeting problems, multidimensional knapsack problems and 0-1 quadratic optimization problems.

Another evident classification possibility for add/drop moves in combination is given by

Example Tabu Classification 3.
The candidate move (*TestAdd,TestDrop*) is tabu if:

Iter ≤ *TabuStart*(*TestAdd*) + *TabuAddTenure*

and

Iter ≤ *TabuStart*(*TestDrop*) + *TabuDropTenure*.

This classification results by replacing "or" with "and" in Classification 1, to require that both of the elements *TestAdd* and *TestDrop* are tabu-active. The condition of Classification 3 will clearly be satisfied less often than the condition of Classification 1, and hence will less frequently result in classifying a move to be tabu. We say that Classification 1 gives rise to a *stronger* tabu restriction. (It may be noted that Classification 2 in turn gives a stronger tabu restriction than Classification 1, since it will be more frequently satisfied, for the same choice of tenure values.)

When tabu status is mediated by penalties, instead of constituting an all-or-none designation, the penalty function associated with Classification 3 is a non-decreasing function of the minimum of the two associated net tenure values. That is, the function is 0 at nonpositive arguments, and otherwise is a positive non-decreasing function of Min(*NetAddTenure,NetDropTenure*), where

NetAddTenure = *TabuStart*(*TestAdd*) + *TabuAddTenure* - *Iter*
NetDropTenure = *TabuStart*(*TestDrop*) + *TabuDropTenure* - *Iter*.

By contrast, the penalty function associated with Classification 1 is a non-decreasing function of Max(*NetAddTenure,NetDropTenure*).

Although Tabu Classification 3 is less often used in practice than Tabu Classification 1, a preliminary implementation can easily test them both (without noticeable added coding effort) to see if one is more suitable for a given context. Appropriate tabu tenure values for Classification 3 will evidently be larger than those for Classification 1, to compensate for the fact that Classification 3 embodies a weaker restriction.

The fact that Classifications 1 and 3 affect the solution path in somewhat different ways suggests the potential merit of combining the two classifications. (Classification 2 can be considered a special case of Classification 1 by choosing one of the two tenures large enough to make it redundant.) Such an option may be represented by choosing values *LargerTabuAddTenure* and *LargerTabuDropTenure* (respectively larger than *TabuAddTenure* and *TabuDropTenure*), and introducing the following classification:

Example Tabu Classification 4.
The candidate move (*TestAdd,TestDrop*) is tabu if:

Iter ≤ *TabuStart*(*TestAdd*) + *TabuAddTenure*

or

Iter ≤ *TabuStart*(*TestDrop*) + *TabuDropTenure*.

or both

$$Iter \leq TabuStart(TestAdd) + LargerTabuAddTenure$$

and

$$Iter \leq TabuStart(TestDrop) + LargerTabuDropTenure.$$

Tabu Classification 4 is a little less convenient to apply than the other classifications, since it involves a bit more calibration effort, but presumably may offer some benefits — as by providing a larger or more stable range of tenure parameters that yield good results. (This involves another area, at a relatively simple level, that invites exploration.) An appropriate penalty function for Classification 4 combines the stipulations for Classification 1 and 3 penalty functions — that is, it is a non-decreasing function of the larger of the minimum of the net tenures for the Classification 1 portion and the maximum of the net tenures for the Classification 3 portion.

Now we will see how these ideas for add/drop moves translate into related ideas for other kinds of moves.

3.5.2 Applications with Increase/Decrease Moves

Increase/Decrease moves arise in settings where a solution vector is modified to generate neighbor solutions by increasing the value of one component variable and decreasing the value of another component variable. Such moves closely resemble Add/Drop moves, and can be handled quite similarly. (As already noted, they can be interpreted to be the same as Add/Drop moves in the case of pure 0-1 integer programming problems.)

Increase/Decrease moves are relevant for applications involving disjoint GUB constraints with right hand sides that may differ from 1, i.e., which take the form

$$\sum_{j \in J} x_j = k$$

where k is a positive constant. (The value of k can differ for different sets J. As often pointed out in GUB applications, such constraints include those of the form

$$\sum_{j \in J} x_j \leq k$$

by adding a non-negative slack variable to change the inequality into an equality.) For such problems, the amount of increase in one variable must the matched by the amount of the decrease in another. Problems of this type show up in a wide range of applications, notably including network distribution problems, such as arise in routing and resource allocation.

Increase/Decrease moves are also relevant for applications involving "distribution of effort" (or "proportional allocation") constraints, which take the form

$$\sum_{j \in J} a_j x_j = k$$

where the coefficients a_j are positive (again dependent on J). Here an increase in one variable is matched by a decrease in another, where the ratio of those changes is determined by the a_j coefficient values. When variables are not required to take discrete values, as usually occurs in problems of this sort, they may be scaled so that the constraints become ordinary GUB constraints.

In the case of GUB problems that involve integer-valued variables, sometimes Adjacent Increase/Decrease moves are employed, which increase and decrease variables only by a single unit (on any given iteration) to an adjacent integer value. Such adjacent moves may be treated exactly in the same way as Add/Drop moves. For notational convenience one can simply replace the terms "Add" and "Drop" (in *TestAdd*, *TestDrop*, *TabuAddTenure*, *TabuDropTenure*, etc.) by the terms "Increase" and "Decrease." (Depending on the context, the tabu tenures for increase and decrease moves may be more nearly symmetric than those for add and drop moves.) This correspondence applies also to situations where adjacent increase and decrease moves are applied independently, rather than in combination. The observations about the use of Tabu Classification 2 in the preceding section are relevant in this case.

Adjacent Increase/Decrease moves can also be implemented for non-discrete variables, by means of *variable scaling* strategies. Such approaches consist of applying adjacent increase / decrease moves that change values of variables by discrete units but where the size of those units varies by submitting them to a scaling operation. For example, in nonlinear optimization problems in finance, an effective approach has resulted by scaling these units so that they differ on different blocks of iterations. The approach starts with moderate to large units and then reduces their size in successive phases to create a more refined search. When the lowest level of refined search fails to bring improvement, the unit sizes are scaled upward again and the process repeats. Clearly other variable scaling strategies are possible. The tabu search memory structures are superimposed as indicated to provide appropriate guidance.

3.5.3 General Case Increase/Decrease Moves

This section and the next (concluding) section of this chapter examine more elaborate variations and uses of short term memory which are useful for neighborhoods that take more complex forms. We first consider how the memory structure previously described can be conveniently extended when the moves employed are not restricted to be Adjacent Increase/Decrease moves, and variables are allowed to change by amounts that may "jump over" adjacent values.

Assume that each variable x_j is constrained to lie within upper and lower bounds, given by

$$U_j \geq x_j \geq L_j .$$

Further suppose that the range from L_j to U_j is subdivided into H segments, numbered h = 1, ..., H in order of increasing x_j values, for each variable. The interval for segment h consists of values $U_j(h) \geq x_j \geq L_j(h)$, except for h = 1, where $x_j \geq L_j(1)$. (The sizes of the intervals need not be the same for all h, and may differ from variable to variable. Also, by variable scaling, the value of H may change in different phases.)

For an n variable problem, we create an array *TabuStart*(h,j) which constitutes a matrix of H rows and n columns. The interpretation of this matrix corresponds closely to the interpretation of the *TabuStart* array used for Add/Drop moves. Given a current solution x' that satisfies the lower and upper bound requirements, the value x'_j lies in some interval h for x_j and hence can be associated with an entry *TabuStart*(h,j). (In the simplest case, which requires the least bookkeeping, each interval for x_j contains a single "central" value, and moves are restricted to assigning variables such central values. Such a situation is natural when the variables are discrete.)

Each entry of TabuStart is initialized to equal a large negative number, and current information begins to be recorded after generating a starting solution (or after reaching a first local optimum). An Increase/Decrease move from a current solution x' to a new solution x'' takes the form

$$\begin{array}{l} \text{increase } x'_p \text{ to } x''_p \\ \text{decrease } x'_q \text{ to } x''_q \end{array}$$

while all other variables retain their current values. (An exception occurs where variables are divided into independent and dependent classes, in which the dependent variables may change their values as a result of changes in the independent variables. This is commonly encountered in basis exchange methods.)

Let *From*(j) denote the current interval h (for x_j) that contains the value x'_j (which we are "moving from"), and let *To*(j) denote the interval h that contains the value $x''(j)$ (which we are "moving to"). When an Increase/Decrease move is executed, the current iteration *Iter* is recorded in the *TabuStart* array by setting

$TabuStart(h,j) = Iter$ for $j = p$ and q, and
for $h = From(j)$ and $To(j)$.

That is, we are concerned with four attributes for each move, summarized by the following pairs, which represent cells in the $TabuStart(h,j)$ array:

(1) $(From(p),p)$
(2) $(To(p),p)$
(3) $(From(q),q)$
(4) $(To(q),q)$

The connection between the foregoing update of *TabuStart,* for the indicated attribute cells, and the related update for Add/Drop moves is as follows. If there were only two states (intervals) for each variable, as in the case of Add/Drop moves, then it would be unnecessary to refer to the "From" attributes, and the foregoing record would be exactly equivalent to the record used for Add/Drop moves. Specifically, the assignment $TabuStart(To(p),p) = Iter$ then corresponds to $TabuStart(Added) = Iter$, and the assignment $TabuStart(To(q),q) = Iter$ corresponds to $TabuStart(Dropped) = Iter$. (We have associated the increase move summarized by the attribute $(To(p),p)$ with "Added" and associated the decrease move summarized by the attribute $(To(q),q)$ with "Dropped.")

For the general case, there typically is no particular asymmetry between increase and decrease moves (as there is, by contrast, between add and drop moves in many settings). Consequently, we create just two classes of tabu tenure, *TabuFromTenure* associated with the *From* attributes (1) and (3), and *TabuToTenure* associated with the *To* attributes (2) and (4). (Individual variations in these two classes of tabu tenure, which may depend both on the variables and the intervals considered, are still allowed. For simplicity, however, we use a single symbol for each tenure class, as we have done for add and drop moves, without bothering to introduce subscripts for additional variations.)

Then, the attributes for a candidate move from x' to x'' may be characterized as tabu-active under the following circumstances:

The attribute $(From(j),j)$, for $j = p$ or q, is tabu-active if:
$$Iter \le TabuStart(From(j),j) + TabuFromTenure$$
The attribute $(To(j),j)$, for $j = p$ or q, is tabu-active if:
$$Iter \le TabuStart(To(j),j) + TabuToTenure.$$

We emphasize the two classes of *From* and *To* attributes because they embody a natural asymmetry for general Increase/Decrease moves. While there is a single *From* attribute for each variable x_j in the current solution x' (corresponding to the interval $From(j)$ in which x'_j lies), there may be many *To* attributes (the remaining

$H-1$ intervals to which x_j might move). Thus, forbidding a variable to move from its current value is much more restrictive than forbidding it to move to a new value. (In Add/Drop settings, as with 0-1 variables, there is no asymmetry in these cases because $H-1 = 1$.) Thus, normally the value of *TabuFromTenure* will appropriately be somewhat smaller than the value of *TabuToTenure*. (If H itself is small or only moderate, then *TabuFromTenure* is usually chosen to be somewhere in the range from 1 to 3.)

Because there are four attributes, a number of ways exist to translate tabu-active conditions for attributes into tabu classifications for moves. Rather than itemize the possibilities, we identify one whose rationale for classifying a move tabu differs slightly from those considered previously. Define a *From* attribute (*From*(j),j) to be *strongly-tabu-active* if

$$Iter \leq TabuStart(From(j),j) + 1.$$

This definition corresponds to creating a special *TabuFromTenure* that equals 1. The motive for this special tenure is the disposition to classify an Increase/Decrease move tabu if either one of its variables had just been assigned its current value on the preceding iteration. (This assumes that other values for both of the variables, and for other variables they might be partnered with, were also considered on the preceding iteration, and passed over in preference for the move executed. If this assumption does not hold, the indicated tabu classification may still be reasonable if aspiration criteria are appropriately included to overrule the classification when conditions warrant.)

The following tabu classification then represents one that may be applied where the influence of remaining factors is roughly symmetric (balanced):

> Example Increase/Decrease Tabu Classification
> The candidate move of increasing x_p and decreasing x_q (to move from x' to x'') is tabu when
>
> Either one of the attributes (*From*(p),p) or (*From*(q),q) is strongly-tabu-active
>
> or
>
> Any two of the four attributes (1) to (4) are tabu-active.

The *TabuStart* array and the associated move attributes also can be used to provide short term memory guidance for increase and decrease moves that are executed independently, rather than in combination. In this case, conceiving the asymmetry between *From* and *To* attributes to be analogous to the asymmetry between "Add" and "Drop" in Add/Drop moves, each of the Example Tabu Classifications described for Add/Drop moves can be translated into a corresponding rule applicable to the independent increase and decrease moves.

3.5.4 Moves for Permutation Problems Related to Add/Drop Moves

The basic procedures indicated for Add/Drop moves can also be extended to handle the types of insert and swap moves that are often embedded in heuristics for permutation problems, such as those which arise in scheduling. We illustrate this by reference to the sequencing problem introduced at the start of this Chapter. It suffices to consider a swap move that exchanges two jobs i and j, since a simple variant of the design also applies to insert moves.

Let *Before*(h) and *After*(h) identify the jobs that immediately precede and follow a given job h in the permutation. We make use of a directed graph representation in which each job h corresponds to a node, and hence the ordered pairs (*Before*(h),h) and (*After*(h),h) identify the two arcs of the graph that meet node h. (If h is the first or last job in the sequence, *Before*(h) and *After*(h) respectively correspond to initial and terminal dummy nodes.) By means of this graph representation, we may view the swap move as performing the following operations (see Figure 3.9):

Add the arcs:	(*Before*(i),j) and (j,*After*(i))
	(*Before*(j),i) and (i,*After*(j))
Drop the arcs:	(*Before*(i),i) and (i,*After*(i))
	(*Before*(j),j) and (j,*After*(j)).

(A special case occurs where (i,j) or (j,i) is an arc of the permutation graph. In this instance the swap degenerates to become an insert move, to which a simple adaptation of the observations of our discussion can be applied.)

These component add and drop operations identify eight different solution attributes that change as a result of the move. The attributes correspond to the four arcs added and the four arcs dropped, as shown in Figure 3.9. (To be precise, each such attribute takes the form (arc,present) and (arc,absent), where the added arcs identify arcs currently absent and the dropped arcs identify those currently present.)

Fig. 3.9 Graphical representation of a swap move.

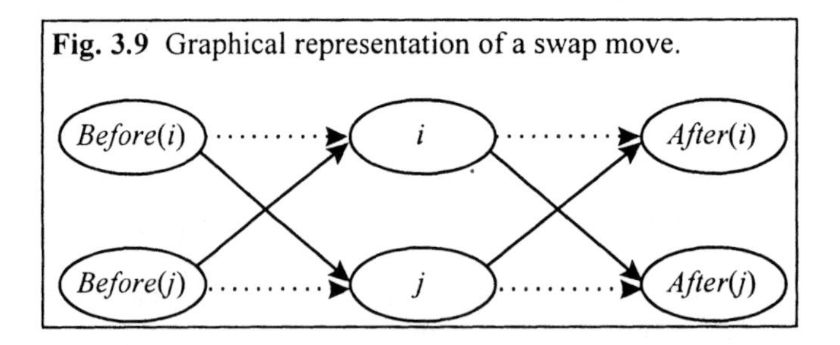

As before, we speak of such solution attributes as move attributes. However, we emphasize that moves can have certain defining features that may invite the designation of "move attributes," but which do not correspond to solution attributes. The reader is to be warned that such features give a risky basis for defining tabu status. (An example is the pair (i,j), which is a feature of the move of exchanging jobs i and j, yet which does not correspond to an attribute of a solution. If we create tabu restrictions that forbid the re-appearance of such pairs for a specified number of iterations, we nevertheless can generate a path that repeatedly cycles among the same solutions during the period that these restrictions are in force.

For the attributes associated with these eight operations, it is natural to create two classes of tenures, *TabuAddTenure* and *TabuDropTenure*, exactly as in the case of the simple Add/Drop moves examined at the beginning of Section 2.4. Thus, for the swap move that is chosen and executed at iteration *Iter*, we set

$$TabuStart(u,v) = Iter$$

for each of the eight arcs (u,v) identified to be added and dropped in the move of swapping nodes i and j. Then arcs that later become candidates to be added and dropped are tabu-active by the customary designation

A candidate arc (r,s) to be added is tabu-active when

$$Iter \leq TabuStart(r,s) + TabuAddTenure$$

A candidate arc (r,s) to be dropped is tabu-active when

$$Iter \leq TabuStart(r,s) + TabuDropTenure$$

There are generally far fewer arcs available to be dropped than added (since there are many fewer arcs "in solution" than "out of solution"), and hence *TabuDropTenure* will characteristically be considerably smaller than *TabuAddTenure,* as in the case of simple Add/Drop moves. (Likewise, *TabuAddTenure* and *TabuDropTenure* may vary according to characteristics of the arc (r,s) considered, and according to patterns specifically introduced to avoid uniformity from iteration to iteration.)

The challenge is to determine the composition of tabu-active attributes that will result in classifying a candidate move tabu. Options range from stipulating that at least one of the eight candidate arcs (r,s) is tabu-active to stipulating that all of these arcs must be tabu-active. Although tabu search has been applied to a wide range of permutation problems in scheduling and routing, virtually no study has sought to determine the comparative merit of more than one or two simple alternatives from the collection referred to here. Consequently, in spite of the successes reported by many of these studies, considerably more remains to be learned about the options that work best. (Our earlier observations about Add/Drop moves and Increase/Decrease moves may be a starting point for determining alternatives worth examining. Moreover, move constructions such as those based on ejection chains may benefit from similar analysis.)

Before leaving the subject of permutation problems, we point out that the graph representation is not the only way to define attributes. A commonly used representation is based on identifying the positions occupied by different elements, e.g., in our illustrated sequencing problem, by identifying the position $\pi(i)$ to which each job i is assigned. Using this job/position representation, an effective rule for classifying a swap move tabu distinguishes between the job that is being moved to an earlier position in the sequence (*Expedite*) and the job that is being moved to a later position in the sequence (*Delay*). When such distinction is made, the swap of jobs i and j, for $\pi(i) < \pi(j)$, in iteration *Iter* results in the following update:

$$TabuExpediteStart(i) = Iter$$
$$TabuDelayStart(j) = Iter.$$

The arrays *TabuExpediteStart* and *TabuDelayStart* record the iteration number where jobs i and j become tabu-active, to respectively prevent job i from later being expedited and prevent job j from later being delayed. To create a tabu classification, we also define *TabuExpediteTenure* as the number of iterations to forbid a job from being expedited (moved to an earlier position in the sequence) and *TabuDelayTenure* as the number of iterations to forbid a job from being delayed (moved to a later position in the sequence). A tabu classification using these definitions is illustrated as follows.

Example Sequencing Tabu Classification
The candidate move of swapping the positions of jobs i and j, where $\pi(i) < \pi(j)$, is tabu when

$$Iter \leq TabuDelayStart(i) + TabuDelayTenure$$
or
$$Iter \leq TabuExpediteStart(j) + TabuExpediteTenure$$

A variety of alternative classifications are clearly possible. Comparisons among alternative means for determining tabu classifications have been more extensive relative to the job/position representation, and it is likely that the rule indicated above is one of the better ones.

3.6 Discussion Questions and Exercises

General candidate list strategies (Exercises 1 - 3).

1. For each of the candidate list strategies described in Section 3.2, identify one or more kinds of optimization problems (and/or neighborhoods applied to these problems) where such a strategy may prove useful.

2. Identify situations in the examples for Exercise 1 where two or more types of candidate lists may prove useful in combination. How would you create such a combination?

3. What types of experiments might be designed to determine appropriate parameter values for candidate list strategies given in Section 3.2? Does your

intuition lead you to favor looking first at some particular (limited) range of values? (What form of preliminary experimentation might sharpen this intuition?)

Problem Specific Logical Restructuring (Exercises 4 - 8).

4. The logical restructuring described in Section 3.4.1. shows how to apply anticipatory analysis to efficiently identify a best outcome of applying a static move and then a dynamic move in combination (where the static move "sets the stage" for the dynamic move). Review the nature of this restructuring and verify as in the discussion of Section 3.4.1 that it immediately identifies a move to apply to the tree of iteration 4 (shown in the illustration of Figure 3.7) that finds the optimal solution.

5. Show how a form of anticipatory analysis related to that of Exercise 4 can be used to efficiently identify a best outcome of applying two dynamic moves in succession. Restrict consideration to the case where both added edges meet the current tree, but possibly one of the dropped edges may not be a terminal edge until the other edge is dropped first. (An expanded discussion of the types of analysis underlying Exercises 4 and 5, is provided in a note at the end of the exercises of this section.)

6. Show how the restructuring of Exercise 5 can be applied to break out of cycling and find an optimal solution in the example of Exercise 1 of Chapter 2.

7. The logical restructuring of Exercises 4 and 5 can be extended to provide guidance for combining more than 2 moves in sequence. Can you create an example that shows the relevance of this? Since the possibilities to examine grow with the number of moves to be combined, how may an embedded TS approach be used in a subroutine to allow a succession of such possibilities to be examined without cycling? (Note this gives an option of setting up TS at two levels, a lower level that guides the generation of each combined move sequence, and a higher level that guides the transition from one combined move sequence to another.)

8. Describe how the ideas of the problem specific design illustrated in Exercises 4-7 can be adapted and applied to problems in other settings. (Specifically, consider the utility of integrating the perspectives provided by different neighborhood structures, using anticipatory analysis to identify features that are essential and limiting in the choice of good moves, designing restructured evaluations and linked neighborhoods to achieve these features.) Select a particular type of problem you are familiar with to illustrate how you would introduce such a restructuring. Can you determine characteristics that would make some problems more difficult to be treated by such restructuring than others?

Threshold based logical restructuring (Exercises 9 - 13).

9. In place of the problem specific approach used to illustrate logical restructuring in Exercises 4-8, we examine the use of threshold based forms of logical restructuring. The instance in Figure 3.10 of the Min k-Tree problem provides a foundation.

Fig. 3.10 Min k-Tree problem instance.

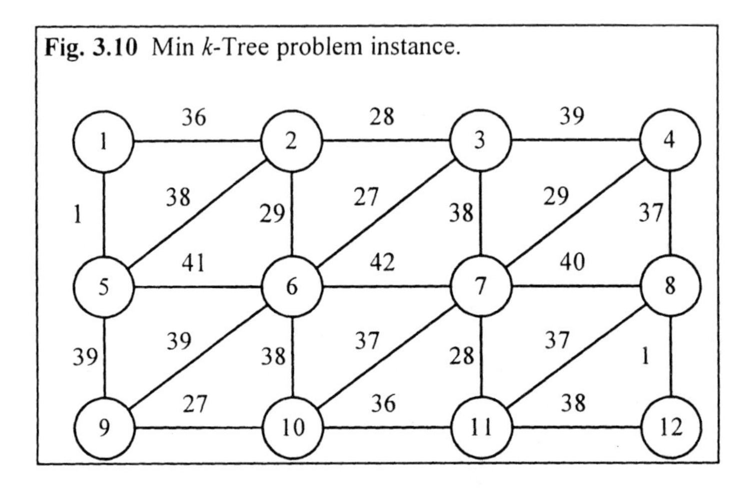

Choose k = 5, for the goal of finding a minimum weight tree with 5 edges. Then demonstrate each of the following:

(a) The first level TS approach illustrated in Chapter 2 will not succeed in finding an optimal solution to this problem — even when using the modified tabu tenures of exercises at the end of Chapter 2, or other somewhat larger tenures.

(b) The types of logical restructuring introduced in Exercises 4 and 5 will also not succeed in finding an optimal solution. Only a more advanced (and complex) application based on Exercise 7 will succeed.

(c) By contrast, a straightforward application of threshold based restructuring will immediately identify the optimal solution (by creating a shortest path between the solution components generated).

10. Show that the simple shortest path approach for joining solution components will no longer work in the example of Exercise 9, if the example is modified by adding an edge (1,4) between nodes 1 and 4 with a weight of 60. Can you think of another way to find the optimal solution using the information provided by the threshold based restructuring? (The next exercise gives one way of addressing this issue.)

11. Show that the threshold based restructuring applied in Exercise 9 not only gives candidates to incorporate into the solution, but can also identify elements that may be excluded. Specifically, show in the present example that:

    (a) Given the assumption that relevant components obtained from the threshold based restructuring will be incorporated into solution, the largest weight of any edge that can be in a 5-tree better than the best found (in Exercise 9) must be less than 43.

    (b) The edge (1,4) of weight 60 introduced in Exercise 10 cannot be given a weight less than 58 or else it will become part of a new optimal solution, which likewise can be found by the approach already considered.

    (c) Whether edge (1,4) retains its weight of 60, or is given any smaller weight, the threshold based restructuring identifies an optimal solution.

12. Describe how the ideas of Exercises 9 and 10 can be applied more generally when there are several components to consider, as a basis for generating additional solutions which may be starting points for repeated implementation of tabu search processes.

13. The concept of a "shortest path" has a numerous analogs in other problem settings, including settings that have no reference to graph constructions. For example, a simple heuristic variant involves a restricted search in a constructive neighborhood space to find a least costly succession of steps to transform partial solutions into feasible complete solutions. (See the discussion of path relinking in Chapter 4.)

    (a) What other constructions can you envision to take the role of a shortest path construction used in the preceding exercises? (Select one or two example problem classes as a basis for explaining your answer.)

    (b) What types of thresholds other than weight (or cost) thresholds may be relevant for generating solution components that are candidates to be joined or expanded to find better solutions? (Again, make reference to one or two specific example problem classes.)

Candidate list strategies and logical restructuring (Exercises 14 - 16).

14. Describe how the issue of candidate list strategies overlaps with the issue of logical restructuring, using the example of Exercises 4-11 as a basis for consideration.

15. Discuss the relation between threshold based forms of logical restructuring (as considered in Exercise 9) and candidate list strategies.

16. From a general point of view, all search methods can be characterized as a collection of candidate list strategies (where memory and other aids may be used to determine the candidate lists), in which some rule ultimately filters the list to a size of 1 to yield a specific choice. Explain why it is useful to focus on a narrower and more highly specific set of designs for candidate list strategies, following models such as those discussed in this chapter.

Note on Problem Specific Logical Restructuring

The logical restructuring illustrations considered in Exercises 4 and 5 are developed more thoroughly in this note to give a fuller sense of how such restructuring may be applied. Although problem specific, the analysis discloses features of logical restructuring in general. In the following discussion, scenario (1) corresponds to the approach of Exercise 4 and scenario (2) corresponds to the approach of Exercise 5.

First we observe that exercises of Chapter 2 motivate a form of restructuring that joins aspects of both edge-swap and node-swap neighborhoods. We start from the edge-swap perspective, which offers faster iterations and updates, and then extend it to exploit the broader possibilities of the node-swap perspective. Our goal is to detect situations where a better k-tree exists by adding a new node and dropping a current node, but where the dynamic moves of an edge-swap neighborhood offer no improvement. The analysis proceeds by first seeking to identify changes that are essential and limiting in order to obtain improvement — hence that will allow us to "set up" a dynamic improving edge-swap move where none currently exists.

Accordingly, we undertake to create a situation where a non-tree edge (i,j) can be added (where i is a non-tree node and j is a tree node) and an associated edge (p,q) can be dropped (where q is a leaf node different from j), so that

$$\text{weight}(i,j) < \text{weight}(p,q).$$

Designate the move that creates this situation to consist of adding an edge (i',j') and dropping an edge (p',q'), where the combination of the two moves is improving, to yield

$$\text{weight}(i',j') + \text{weight}(i,j) < \text{weight}(p',q') + \text{weight}(p,q).$$

To assure q becomes a leaf node we can stipulate $q' = q$. Moreover, q must have degree 2 (i.e., must be met by exactly 2 edges) in the tree before making the swap move that drops the edge (p',q). Then two scenarios can create the desired situation:

(1) The edge-swap that adds (i',j') and drops (p',q) is static: The best candidate edge for (i,j) is the least weight non-tree edge such that j is a tree node different from q. We can limit attention to two edges (i_1,j_1) and (i_2,j_2), where (i_1,j_1) is the least weight non-tree edge such that j_1 is in the tree, and where (i_2,j_2) is the least weight non-tree edge such that j_2 is in the tree and $j_2 \neq j_1$.

(2) The edge-swap that adds (i',j') and drops (p',q) is dynamic: The node p' must be a leaf node, and we can limit attention to the following possibility. Given that (p',q) is a tree edge such that p' and q respectively have degrees 1 and 2 in the tree, we may choose (i,j) to be the least weight non-tree edge such that j is a tree node different from p' and q, and choose (i',j') to be the least weight non-tree edge such that $i' \neq i$, and $j' \neq p', q$. (These choices can be affected by tabu status, but are unaffected in considering aspiration criteria.)

The advantage of this logical restructuring is clear. In scenario (1), since (i,j) is just one of two edges known in advance, we can immediately determine if an improving possibility exists (from the combination of the indicated static and dynamic moves) by examining only the current static moves. (This would be done anyway in the simpler setting where anticipatory analysis is not used to consider effects of moves in combination.) If any static move causes a node q to become a leaf node, then at most 2 checks will determine whether an associated dynamic move exists that creates a net improvement. The amount of work is almost no different than looking for a best static move by itself, as opposed to looking for all two move combinations where a static move is followed by a dynamic move.

Similarly, scenario (2) avoids examining all combined moves. Only a very limited set of edges can qualify to be the edge (p',q), and for each of these choices of (p',q), the edges (p,q), (i,j) and (i',j') can be identified immediately. (Extended analysis can still further limit the possibilities for these latter edges in advance.)

4 TABU SEARCH FOUNDATIONS: LONGER TERM MEMORY

In some applications, the short term TS memory components are sufficient to produce very high quality solutions. However, in general, TS becomes significantly stronger by including longer term memory and its associated strategies. In the longer term TS strategies, the modified neighborhood produced by tabu search may contain solutions not in the original one, generally consisting of selected elite solutions (high quality local optima) encountered at various points in the solution process. Such elite solutions typically are identified as elements of a regional cluster in intensification strategies, and as elements of different clusters in diversification strategies. In addition, elite solution components, in contrast to the solutions themselves, are included among the elements that can be retained and integrated to provide inputs to the search process.

Perhaps surprisingly, the use of longer term memory does not require long solution runs before its benefits become visible. Often its improvements begin to be manifest in a relatively modest length of time, and can allow solution efforts to be terminated somewhat earlier than otherwise possible, due to finding very high quality solutions within an economical time span. The fastest methods for some types of routing and scheduling problems, for example, are based on including longer term TS memory. On the other hand, it is also true that the chance of finding still better solutions as time grows — in the case where an optimal solution is not already found — is enhanced by using longer term TS memory in addition to short term memory.

4.1 Frequency-Based Approach

Frequency-based memory provides a type of information that complements the information provided by recency-based memory, broadening the foundation for selecting preferred moves. Like recency, frequency often is weighted or decomposed into subclasses by taking account of the dimensions of solution quality and move influence. Also, frequency can be integrated with recency to build a composite structure for creating penalties and inducements that modify move evaluations. (Although recency-based memory is often used in the context of short term memory, it can also be a foundation for longer term forms of memory.)

For our present purposes, we conceive frequencies to consist of ratios, whose numerators represent counts expressed in two different measures: a *transition measure* — the number of iterations where an attribute changes (enters or leaves) the solutions visited on a particular trajectory, and a *residence measure* — the number of iterations where an attribute belongs to solutions visited on a particular trajectory, or the number of instances where an attribute belongs to solutions from a particular subset. The denominators generally represent one of three types of quantities: (1) the total number of occurrences of all events represented by the numerators (such as the total number of associated iterations), (2) the sum (or average) of the numerators, and (3) the maximum numerator value. In cases where the numerators represent weighted counts, some of which may be negative, denominator (3) is expressed as an absolute value and denominator (2) is expressed as a sum of absolute values (possibly shifted by a small constant to avoid a zero denominator). The ratios produce *transition frequencies* that keep track of how often attributes change, and *residence frequencies* that keep track of how often attributes are members of solutions generated. In addition to referring to such frequencies, thresholds based on the numerators alone can be useful for indicating when phases of greater diversification are appropriate. (The thresholds for particular attributes can shift after a diversification phase is executed.)

Residence frequencies and transition frequencies sometimes convey related information, but in general carry different implications. They are sometimes confused (or treated identically) in the literature. A noteworthy distinction is that residence measures, by contrast to transition measures, are not concerned with the characteristics of a particular solution attribute or whether it is an attribute that changes in moving from one solution to another. For example in the Min k-Tree problem, a residence measure may count the number of times edge (i,j) was part of the solution, while a transition measure may count the number of times edge (i,j) was added to the solution. (More complex joint measures, such as the number of times edge (i,j) was accompanied in the solution by edge (k,l), or was deleted from the solution in favor of edge (k,l), can also selectively be generated. Such frequencies relate to the issues of creating more complex attributes out of simpler ones, and to the strategies of vocabulary building.)

A high residence frequency may indicate that an attribute is highly attractive if the domain consists of high quality solutions, or may indicate the opposite, if the domain consists of low quality solutions. On the other hand, a residence frequency that is high (or low) when the domain is chosen to include both high and low quality solutions may point to an entrenched (or excluded) attribute that causes the search space to be restricted, and that needs to be jettisoned (or incorporated) to allow increased diversity. For example, an entrenched attribute may be a job that is scheduled in the same position during a sequence of iterations that include both low and high quality objective function evaluations.

As a further useful distinction, a high transition frequency, in contrast to a high residence frequency, may indicate an associated attribute is a "crack filler," that shifts in and out of solutions to perform a fine tuning function. In this context, a transition frequency may be interpreted as a measure of volatility. For example, the Min k-Tree problem instance in Figure 2.3 of Chapter 2 contains a number of edges whose weight may give them the role of crack fillers. Specifically, edges (3,5) and (6,7) both have a weight of 6, which makes them attractive relative to other edges in the graph. Since these edges are not contained in an optimal solution, there is some likelihood that they may repeatedly enter and leave the current solution in a manner to lure the search away from the optimal region. (Exercises at the end of Chapter 2 show this is true for the edge (6,7).) In general, crack fillers are determined not simply by cost or quality but by structure, as in certain forms of connectivity. (Hence, for example, the edge (3,5) of Figure 2.3 does not repeatedly enter and leave solutions in spite of its cost.) Some subset of such elements is also likely to be a part of an optimal solution. This subset can typically be identified with much less difficulty once other elements are in place. On the other hand, a solution (full or partial) may contain the "right" crack fillers but offer little clue as to the identity of the other attributes that will transform the solution into one that is optimal.

Table 4.1. Example of frequency measures.

Problem	*Residence Measure*	*Transition Measure*
Sequencing	Number of times job j has occupied position $\pi(j)$.	Number of times job i has exchanged positions with job j.
	Sum of tardiness of job j when this job occupies position $\pi(j)$.	Number of times job j has been moved to an earlier position in the sequence.
Min k-Tree Problem	Number of times edge (i, j) has been part of the current solution.	Number of times edge (i, j) has been deleted from the current solution when edge (k, l) has been added.
	Sum of total solution weight when edge (i, j) is part of the solution.	Number of times edge (i, j) has been added during improving moves.

We use the sequencing problem and the Min k-Tree problem introduced in Chapter 2 to further illustrate both residence and transition frequencies. Only numerators are indicated, understanding that denominators are provided by the conditions (1) to (3) previously defined. The measures are given in Table 4.1.

Attributes that have greater frequency measures, just as those that have greater recency measures (i.e., that occur in solutions or moves closer to the present), can trigger a tabu activation rule if they are based on consecutive solutions that end with the current solution. However, frequency-based memory often finds its most productive use as part of a longer term strategy, which employs incentives as well as restrictions to determine which moves are selected. In such a strategy, tabu activation rules are translated into evaluation penalties, and incentives become evaluation enhancements, to alter the basis for qualifying moves as attractive or unattractive.

To illustrate, an attribute such as a job j with a high residence frequency in position $\pi(j)$ may be assigned a strong incentive ("profit") to serve as a *swap attribute,* thus resulting in the choice of a move that yields a new sequence Π' with $\pi'(j) \neq \pi(j)$. Such an incentive is particularly relevant in the case where the *TabuEnd* value of job j is small compared to the current iteration, since this value (minus the corresponding tabu tenure) identifies the latest iteration that job j was a *swap attribute*, and hence discloses that job j has occupied position $\pi(j)$ in every solution since.

Frequency-based memory therefore is usually applied by introducing graduated tabu states, as a foundation for defining penalty and incentive values to modify the evaluation of moves. A natural connection exists between this approach and the recency-based memory approach that creates tabu status as an all-or-none condition. If the tenure of an attribute in recency-based memory is conceived as a conditional threshold for applying a very large penalty, then the tabu classifications produced by such memory can be interpreted as the result of an evaluation that becomes strongly inferior when the penalties are activated. Conditional thresholds are also relevant to determining the values of penalties and incentives in longer term strategies. (See Exercises 4 to 6 at the end of the chapter.) Most applications at present, however, use a simple linear multiple of a frequency measure to create a penalty or incentive term. The multiplier is adjusted to create the right balance between the incentive or penalty and the cost (or profit) coefficients of the objective function.

4.2 Intensification Strategies

Intensification strategies are based on modifying choice rules to encourage move combinations and solution features historically found good. They may also initiate a return to attractive regions to search them more thoroughly. A simple instance of this second type of intensification strategy is shown in Figure 4.1. The strategy for selecting elite solutions is italicized in Figure 4.1 due to its importance. Two variants have proved quite successful. One introduces a diversification measure to assure the solutions recorded differ from each other by a desired degree, and then erases all short term memory before resuming from the best of the recorded solutions. A diversification measure may be related to the number of moves that are necessary to transform one solution into another. Or the measure may be defined independently from the move mechanism. For example, in sequencing, two solutions may be considered diverse if the number of swaps needed to move from one to the other is "large." On the other hand, the diversification measure may be the number of jobs that occupy a different position in the two sequences being compared. (This shows that intensification and diversification often work together, as elaborated in the next section.)

Fig. 4.1 Simple TS intensification approach.

```
Apply TS short term memory.
Apply an elite selection strategy.
do {
        Choose one of the elite solutions.
        Resume short term memory TS from chosen solution.
        Add new solutions to elite list when applicable.
} while (iterations < limit and list not empty)
```

The second variant that has also proved successful, keeps a bounded length sequential list that adds a new solution at the end only if it is better than any previously seen. The current last member of the list is always the one chosen (and removed) as a basis for resuming search. However, TS short term memory that accompanied this solution is also saved, and the first move also forbids the move previously taken from this solution, so that a new solution path will be launched.

A third variant of the approach of Figure 4.1 is related to a strategy that resumes the search from unvisited neighbors of solutions previously generated. Such a strategy keeps track of the quality of these neighbors to select an elite set, and restricts attention to specific types of solutions, such as neighbors of local optima or neighbors of solutions visited on steps immediately before reaching such local optima. This type of "unvisited neighbor" strategy has been little examined. It is noteworthy, however, that the two variants previously indicated have provided solutions of remarkably high quality.

Another type of intensification approach is *intensification by decomposition*, where restrictions may be imposed on parts of the problem or solution structure in order to generate a form of decomposition that allows a more concentrated focus on other parts of the structure. A classical example is provided by the traveling salesman problem, where edges that belong to the intersection of elite tours may be "locked into" the solution, in order to focus on manipulating other parts of the tour. The use of intersections is an extreme instance of a more general strategy for exploiting frequency information, by a process that seeks to identify and constrain the values of *strongly determined* and *consistent variables*. We discuss the identification and use of such variables in Section 4.4.1.

Intensification by decomposition also encompasses other types of strategic considerations, basing the decomposition not only on indicators of strength and consistency, but also on opportunities for particular elements to interact productively. Within the context of a permutation problem as in scheduling or routing, for example, where solutions may be depicted as selecting one or more sequences of edges in a graph, a decomposition may be based on identifying subchains of elite solution, where two or more subchains may be assigned to a common set if they contain nodes that are "strongly attracted" to be linked with nodes of other subchains in the set. An edge disjoint collection of subchains can be treated by an intensification process that operates in parallel on each set, subject to the restriction that the identity of the endpoints of the subchains will not be altered. As a result of the decomposition, the best new sets of subchains can be reassembled to create new solutions. Such a process can be applied to multiple alternative decompositions in broader forms of intensification by decomposition.

These ideas are lately finding favor in other procedures, and may provide a bridge for interesting components of tabu search with components of other methodologies. We address the connections with these methodologies in Chapter 9.

4.3 Diversification Strategies

Search methods based on local optimization often rely on diversification strategies to increase their effectiveness in exploring the solution space defined by a combinatorial optimization problem. Some of these strategies are designed with the chief purpose of preventing searching processes from *cycling,* i.e., from endlessly executing the same sequence of moves (or more generally, from endlessly and exclusively revisiting the same set of solutions). Others are introduced to impart additional robustness or vigor to the search. Genetic algorithms use randomization in component processes such as combining population elements and applying crossover (as well as occasional mutation), thus providing an appropriate diversifying effect. Simulated annealing likewise incorporates randomization to make diversification a function of temperature, whose gradual reduction correspondingly diminishes the directional variation in the objective function trajectory of solutions generated. Diversification in GRASP (Greedy Randomized Adaptive Search Procedures) is achieved (in a certain sense) within repeated construction phases by means of a random sampling over elements that pass a threshold of attractiveness by a greedy criterion.

In tabu search, diversification is created to some extent by short term memory functions, but is particularly reinforced by certain forms of longer term memory. TS diversification strategies, as their name suggests, are designed to drive the search into new regions. Often they are based on modifying choice rules to bring attributes into the solution that are infrequently used. Alternatively, they may introduce such attributes by periodically applying methods that assemble subsets of these attributes into candidate solutions for continuing the search, or by partially or fully restarting the solution process. Diversification strategies are particularly helpful when better solutions can be reached only by crossing barriers or "humps" in the solution space topology.

4.3.1 Modifying Choice Rules

Consider a TS method designed to solve a graph partitioning problem which uses full and partial swap moves to explore the local neighborhood. The goal of this problem is to partition the nodes of the graph into two equal subsets so that the sum of the weights of the edges that join nodes in one subset to nodes in the other subset is minimized. Full swaps exchange two nodes that lie in two different sets of the partition. Partial swaps transfer a single node from one set to the other set. Since full swaps do not modify the number of nodes in the two sets of the partition, they maintain feasibility, while partial swaps do not. Therefore, under appropriate guidance, one approach to generate diversity is to periodically disallow the use of non-improving full swaps for a chosen duration (after an initial period where the search "settles down"). The partial swaps must of course be coordinated to allow feasibility to be recovered after achieving various degrees of infeasibility. (This relates to the approach of strategic oscillation, described in Section 4.4.) Implemented appropriately, this strategy has the effect of intelligently perturbing the current solution, while escaping from a local optimum, to an extent that the search is directed to a region that is different than the one being currently explored. The implementation of this strategy as applied to experimental problems has resulted in significant improvements in problem-solving efficacy.

The incorporation of partial swaps in place of full swaps in the previous example can be moderated by using the following penalty function:

$$MoveValue' = MoveValue + d * Penalty.$$

This type of penalty approach is commonly used in TS, where the *Penalty* value is often a function of frequency measures such as those indicated in Table 4.1, and d is an adjustable diversification parameter. Larger d values correspond to a desire for more diversification. (E.g., nodes that change sets more frequently are penalized more heavily to encourage the choice of moves that incorporate other nodes. Negative penalties, or "inducements," may also be used to encourage low frequency elements.) The penalty can be applied to classes of moves as well as to attributes of moves. Thus, during a phase where full swaps moves are excluded, all such moves receive a large penalty (with a value of d that is effectively infinite).

In some applications where d is used to inhibit the selection of "feasibility preserving" moves, the parameter can be viewed as the reciprocal of a Lagrangean multiplier in that "low" values result in nearly infinite costs for constraint violation, while "high" values allow searching through infeasible regions. The adjustment of such a parameter can be done in a way to provide a strategic oscillation around the feasibility boundary, again as discussed in Section 4.4. The parameter can also be used to control the amount of randomization in probabilistic versions of tabu search.

In TS methods that incorporate the simplex method of linear programming, as in "adjacent extreme point approaches" for solving certain nonlinear and mixed-integer programming problems, a diversification phase can be designed based on the number of times variables become basic. For example, a diversification step can give preference to bringing a nonbasic variable into the basis that has remained out of the basis for a relatively long period (cumulatively, or since its most recent inclusion, or a combination of the two). The number of successive iterations such steps are performed, and the frequency with which they are initiated, are design considerations of the type that can be addressed, for example, by the approach of target analysis (Chapter 9).

4.3.2 Restarting

Frequency information can be used in different ways to design restarting mechanisms within tabu search. In our sequencing problem, for example, the overall frequency of jobs occupying certain positions can be used to bias a construction procedure and generate new restarting points. Suppose that the earliest due date rule (EDD) is used to construct an initial solution to the example problem introduced at the beginning of Chapter 2. This rule assigns the highest priority to the job with the earliest due date. Since the due dates in our example are (9, 12, 15, 8, 20, 22), then the initial sequence is (4, 1, 2, 3, 5, 6). A restarting sequence can be found by using frequency information to modify the priority index of each job. In this case, the priority indexes are simply given by the due date values, which can be biased using an appropriate frequency measure. We may represent the modified index by:

$$PriorityIndex' = PriorityIndex + d * FrequencyMeasure.$$

The construction process is then changed to consider the value of *PriorityIndex'*. For our sequencing problem, the *FrequencyMeasure* for a job at given iteration *Iter* may be the percentage of time that the job was completed on time, considering all the sequences from the last restarting point to *Iter*. For diversification purposes, we might want to decrease the priority of those jobs whose *FrequencyMeasure* is high and increase the priority of those jobs with low *FrequencyMeasure* value. Suppose that the *FrequencyMeasure* values for the jobs in our example during a restarting phase are those given in the third column of Table 4.2. If in addition we assume that the value of d is set to 10, then the modified *PriorityIndex'* results in the restarting sequence (1, 4, 2, 5, 3, 6).

Table 4.2 Illustration of a restarting mechanism.			
Job	*PriorityIndex*	*FrequencyMeasure*	*PriorityIndex'*
1	9	0.01	9.1
2	12	0.23	14.3
3	15	0.83	23.3
4	8	0.13	9.3
5	20	0.31	23.1
6	22	0.93	31.3
Assume $d = 10$.			

This type of procedure can be incorporated in many different applications, and finds its greatest usefulness where frequency information is easily recorded and a fairly large number of iterations is performed.

In a TS method for a location/allocation problem, a diversification phase can be developed using frequency counts on the number of times a depot has changed its status (from open to closed or vice versa). The diversification phase can be started from the best solution found during the search. Based on the frequency information, d depots with the lowest counts are selected and their status is changed. The search starts from the new solution which differs from the best by exactly d components. To prevent a quick return to the best solution, the status of the d depots is also recorded in short term memory. (This is another case where residence frequency measures may provide useful alternatives or supplements to transition frequency measures.)

Additional forms of memory functions are possible when a restarting mechanism is implemented. For example, in the location/allocation problem, it is possible to keep track of recent sets of depots that were selected for diversification and avoid the same selection in the next diversification phase. Similarly, in a sequencing problem, the positions occupied by jobs in recent starting points can be recorded to avoid future repetition. This may be viewed as a very simple forms of the critical event memory discussed in Chapter 2, and more elaborate forms will often yield greater benefits. The exploitation of such memory is very important in TS designs that are completely deterministic, since in these cases a given starting point will always produce the same search path. Experience also shows, however, that uses of TS memory to guide probabilistic forms of restarting can likewise yield benefits (Rochat and Taillard, 1995; Fleurent and Glover, 1996; Lokketangen and Glover, 1996).

Before concluding this section, it is appropriate to provide a word of background about the orientation underlying diversification strategies within the tabu search framework. Often there appears to be a hidden assumption that diversification is somehow tantamount to randomization. Certainly the introduction of a random element to achieve a diversifying effect is a widespread theme among search procedures, and is fundamental to the operation of simulated annealing and genetic algorithms. From an abstract standpoint, there is clearly nothing wrong with equating randomization and diversification, but to the extent that diversity connotes differences among elements of a set, and to the extent that establishing such differences is relevant to an effective search strategy, then the popular use of randomization is at best a convenient proxy (and at worst a haphazard substitute) for something quite different.

When randomization is used as part of a restarting mechanism, for example, frequency information can be employed to approximate probability distributions that bias the construction process. In this way, randomization is not a "blind" mechanism, but instead it is guided by search history. We examine inappropriate roles of randomization in Section 4.6, where we also explore the intensification / diversification distinction more thoroughly.

4.4 Strategic Oscillation

Strategic oscillation is closely linked to the origins of tabu search, and provides a means to achieve an effective interplay between intensification and diversification over the intermediate to long term. We have briefly sketched some of the features of the approach in Chapter 1. The recurring usefulness of this approach documented in a variety of studies (such as a number of those reported in Chapter 8) warrants a more detailed examination of its characteristics.

Strategic oscillation operates by orienting moves in relation to a critical level, as identified by a stage of construction or a chosen interval of functional values. Such a critical level or *oscillation boundary* often represents a point where the method would normally stop. Instead of stopping when this boundary is reached, however, the rules for selecting moves are modified, to permit the region defined by the critical level to be crossed. The approach then proceeds for a specified depth beyond the oscillation boundary, and turns around. The oscillation boundary again is approached and crossed, this time from the opposite direction, and the method proceeds to a new turning point (see Figure 4.2).

Fig. 4.2 Strategic oscillation.

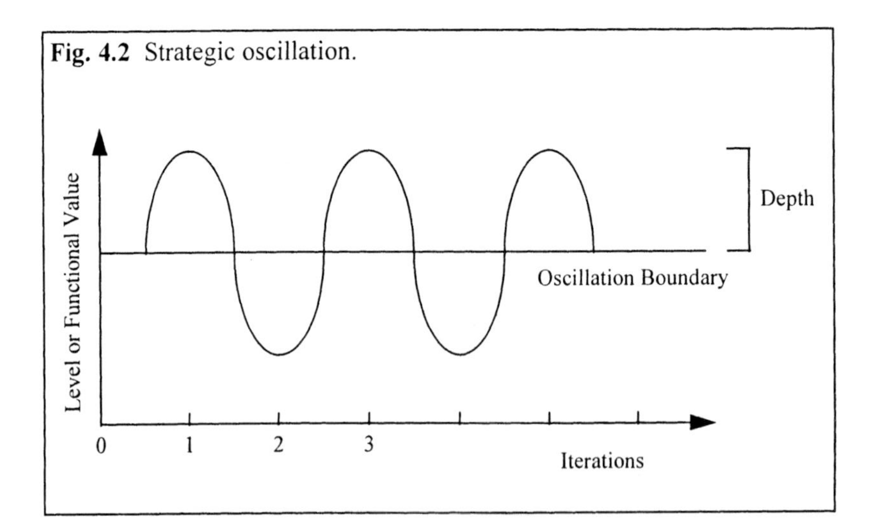

The process of repeatedly approaching and crossing the critical level from different directions creates an oscillatory behavior, which gives the method its name. Control over this behavior is established by generating modified evaluations and rules of movement, depending on the region navigated and the direction of search. The possibility of retracing a prior trajectory is avoided by standard tabu search mechanisms, like those established by the recency-based and frequency-based memory functions.

A simple example of this approach occurs for the multidimensional knapsack problem, where values of zero-one variables are changed from 0 to 1 until reaching the boundary of feasibility. The method then continues into the infeasible region using the same type of changes, but with a modified evaluator. After a selected number of steps, the direction is reversed by choosing moves that change variables from 1 to 0. Evaluation criteria to drive toward improvement vary according to whether the movement occurs inside or outside the feasible region (and whether it is directed toward or away from the boundary), accompanied by associated restrictions on admissible changes to values of variables. The turnaround towards feasibility can also be triggered by a maximum infeasibility value, which defines the depth of the oscillation beyond the critical level (i.e., the feasibility boundary).

A somewhat different type of application occurs for graph theory problems where the critical level represents a desired form of graph structure, capable of being generated by progressive additions (or insertions) of basic elements such as nodes, edges, or subgraphs. One type of strategic oscillation approach for this problem results by a constructive process of introducing elements until the critical level is reached, and then introducing further elements to cross the boundary defined by the critical level. The current solution may change its structure once this boundary is crossed (as where a forest becomes transformed into a graph that contains loops), and hence a different neighborhood may be required, yielding modified rules for selecting moves. The rules again change in order to proceed in the opposite direction, removing elements until again recovering the structure that defines the critical level.

In the Min k-Tree problem, for example, edges can be added beyond the critical level defined by k. Then a rule to delete edges must be applied. The rule to delete edges will typically be different in character from the one used for adding (i.e., will not simply be its "inverse"). In this case, all feasible solutions lie on the oscillation boundary, since any deviation from this level results in solutions with more or less than k edges.

Such rule changes are typical features of strategic oscillation, and provide an enhanced heuristic vitality. The application of different rules may be accompanied by crossing a boundary to different depths on different sides. An option is to approach and retreat from the boundary while remaining on a single side, without crossing (i.e., electing a crossing of "zero depth").

These examples constitute a constructive/destructive type of strategic oscillation, where constructive steps "add" elements (or set variables to 1) and destructive steps "drop" elements (or set variables to 0). (Types of TS memory structures for add / drop moves discussed in Chapter 2 and 3 are relevant for such procedures.) One-sided oscillations (that remain on a single side of a critical boundary) are appropriate in a variety of scheduling and graph theory applications where constructive solution approaches are often applied. The alternation with destructive processes, which strategically dismantle and then rebuild successive trial solutions, affords an enhancement of such traditional constructive procedures. In both one-sided and two-sided oscillation approaches it is frequently important to spend additional search time in regions close to the critical level, and especially to spend time at the critical level itself. This may be done by inducing a sequence of tight oscillations about the critical level, as a prelude to each larger oscillation that proceeds to a greater depth. Alternately, if greater effort is permitted for evaluating and executing each move, the method may use "exchange moves" (broadly interpreted) to stay at the critical level for longer periods. In the case of the Min *k*-Tree problem, for example, once the oscillation boundary has been reached, the search can stay on it by performing swap moves (either of nodes or edges). An option is to use such exchange moves to proceed to a local optimum each time the critical level is reached.

When the level or functional values in Figure 4.2 refer to degrees of feasibility and infeasibility, a vector-valued function associated with a set of problem constraints can be used to control the oscillation. In this case, controlling the search by bounding this function can be viewed as manipulating a parameterization of the selected constraint set. A preferred alternative is often to make the function a Lagrangean or surrogate constraint penalty function, avoiding vector-valued functions and allowing tradeoffs between degrees of violation of different component constraints.

4.4.1 Strategic Oscillation Patterns and Decisions

Strategic oscillation may be viewed as composed of two interacting sets of decisions, one at a macro level and one at a micro level. These decisions are depicted in Figure 4.3. We discuss these two types of decisions in sequence.

Macro Level Decisions. The oscillation guidance function of the first macro level decision corresponds to the element controlled, e.g., tabu restrictions, infeasibility penalties, or the objective function. (More precisely, the function provides a measure of this element that permits control to be established.) In the second decision at the macro level, a variety of target levels are possible, and their form is often determined by the problem setting. For example, in an alternating process of adding or deleting edges in a graph, the target can be the stage at which the current set of edges creates a spanning tree. Similarly in a process of assigning (and "unassigning") jobs to machines, the target can be the stage at which a complete assignment is achieved. Each of these can be interpreted as special cases of the situation where the target constitutes a boundary of feasibility, approached from inside or outside. (For a variety of scheduling problems, it is natural to approach the boundary — a complete schedule — only from one side, progressively building a schedule and then choosing moves to dismantle it, in alternating waves.) In some instances, the target may appropriately be adaptive, as by representing the average number of employees on duty in a set of best workforce schedules. Such a target gives a baseline for inducing variations in the search, and does not necessarily represent an ideal to be achieved.

Fig. 4.3 Strategic oscillation decisions.

Macro level decisions

1. Select an oscillation guidance function.
2. Choose a target level for the function.
3. Choose a pattern of oscillation.

Micro level decisions

1. Choose a target rate of change (for moving toward or away from the target level).
2. Choose a target band of change.
3. Identify aspiration criteria to override target restrictions.

The pattern of oscillation at a macro level deals with features such as the depth by which the search goes beyond the target in a given direction, and more generally the way in which the depth varies over time. An example of a simple type of oscillation is shown in Figure 4.4, where the search begins by approaching the target "from above" and then oscillates with a constant amplitude thereafter.

Fig. 4.4 Simple uniform oscillation.

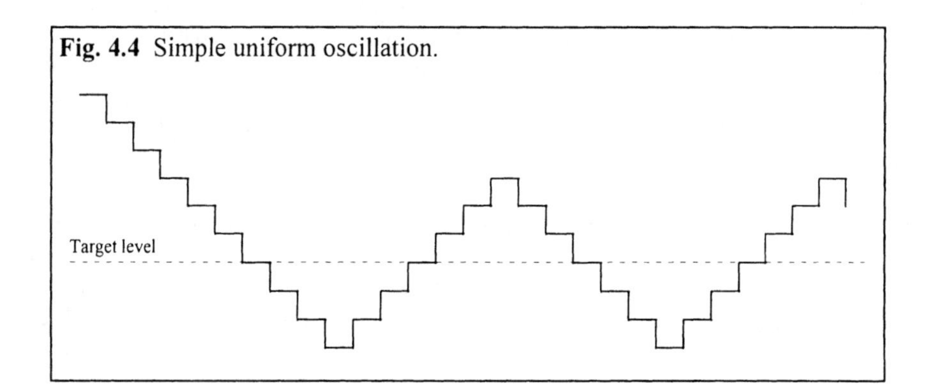

The pattern in Figure 3.4 is shown as a broken line rather than a smooth curve, to indicate that the search may not flow continuously from one level to the next, but may remain for a period at a given level. The diagram is suggestive rather than precise, since in reality there are no vertical lines, i.e., the guidance function does not change its value in zero time. Also, the dashed line for the target level should be interpreted as spanning an interval (as conveyed by using the term *target level* rather than *target value*). Similarly, each of the segments of the oscillation curve may be conceived as having a "thickness" or "breadth" that spans a region within which particular values lie.

A type of oscillation pattern often employed in short term tabu search strategies adopts an aggressive approach to the target, in some cases slowing the rate of approach and spending additional time at levels that lie in the near vicinity of the target. When accompanied by a policy of closely hugging the target level once it is attained, the pattern is an instance of an *intensification strategy*. Such a pattern is shown in Figure 4.5.

Fig. 4.5 Patterned oscillation (intensification).

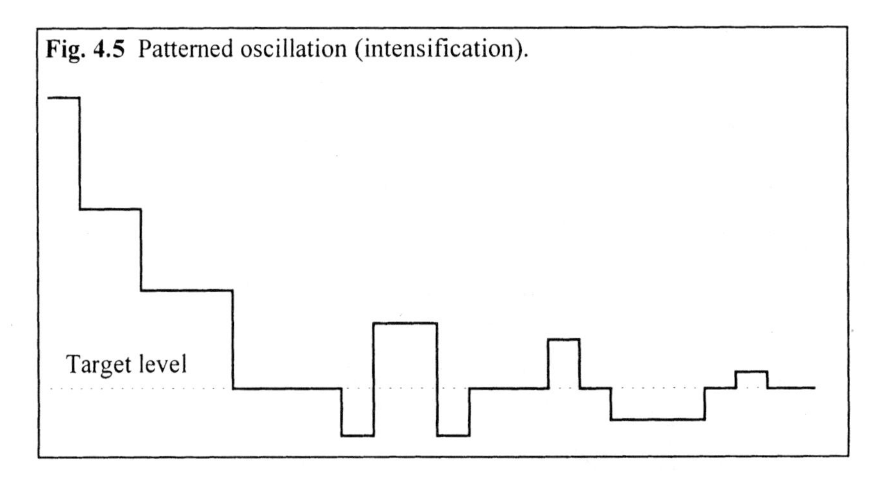

The relative larger times at or near the target, in contrast to the periodic steps that drive the search somewhat farther away, are often more pronounced than shown in this diagram. On the other hand, the variable distance to which the search is propelled away from the target is generally made larger when frequency measures indicate that specific subsets of attributes are unproductively "hogging the action" at the target level.

In both Figures 4.4 and 4.5 the oscillations that occur above and below the target may be replaced by oscillations on a single side of the target. Over a longer duration, for example, the pattern may predominantly focus on one side and then the other, or may alternate periods of such a one-sided focus with periods of a more balanced focus.

Patterns that represent intensification strategies generally benefit by introducing some variation over the near to intermediate term. Figure 4.5 contains a modest degree of variation in its pattern, and other simple types of variation, still predominantly hugging the target level (and hence qualifying as intensification strategies), are shown in Figures 4.6 and 4.7. Evidently, intensification patterns with greater degrees of variability can easily be created (as by using pseudo randomization to sequence their components).

Fig. 4.6 Patterned oscillation (varied intensification).

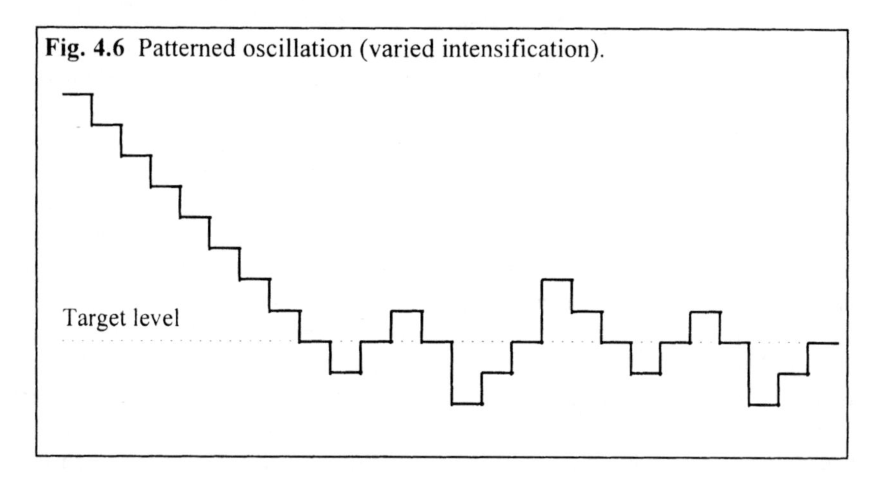

The types of intensification illustrated in the preceding diagrams are not required to make recourse to memory, except to keep track of the current phase of a pattern being executed. Nevertheless, the use of TS memory structures enhances the ability to ferret out and focus on regions judged to be strategically important. (The use of a special form of critical event memory for strategic oscillation is discussed in Section 4.7.3.)

Fig. 4.7 Patterned oscillation (varied intensification).

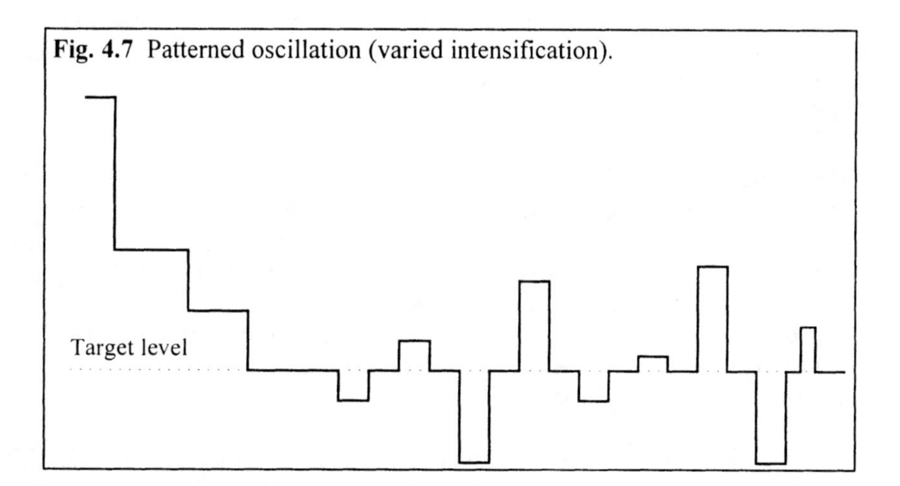

Intensification processes can readily be embedded in strategic oscillation by altering choice rules to encourage the incorporation of particular attributes — or at the extreme, by locking such attributes into the solution for a period. Such processes can be viewed as designs for exploiting *strongly determined* and *consistent* variables. A strongly determined variable is one that cannot change its value in a given high quality solution without seriously degrading quality or feasibility, while a consistent variable is one that frequently takes on a specific value (or a highly restricted range of values) in good solutions. The development of useful measures of "strength" and "consistency" is critical to exploiting these notions, particularly by accounting for tradeoffs determined by context. However, straightforward uses of frequency-based memory for keeping track of consistency, sometimes weighted by elements of quality and influence, have produced methods with very good performance outcomes.

An example of where these kinds of approaches are also beginning to find favor in other settings occurs in recently developed variants of genetic algorithms for sequencing problems. The more venturesome of these approaches are coming to use special forms of "crossover" to assure offspring will receive attributes shared by good parents, thus incorporating a type of intensification based on consistency. Extensions of such procedures using TS ideas of identifying elements that qualify as consistent and strongly determined according to broader criteria, and making direct use of memory functions to establish this identification, provide an interesting area for investigation. (Additional links to GA methods, and ways to go beyond current explorations of such methods, are discussed in Chapter 9.)

Longer term processes, following the type of progression customarily found beneficial in tabu search, may explicitly introduce supplemental diversification strategies into the oscillation pattern. When oscillation is based on constructive and destructive processes, the repeated application of constructive phases (rather than moving to intermediate levels using destructive moves) embodies an extreme type of oscillation that is analogous to a restart method. In this instance the restart point is always the same (i.e., a null state) instead of consisting of different initial solutions, and hence it is important to use choice rule variations to assure appropriate diversification.

A connection can also be observed between an extreme version of strategic oscillation — in this case a relaxed version — and the class of procedures known as *perturbation* approaches. An example is the subclass known as "large-step simulated annealing" or "large-step Markov chain" methods (Martin, Otto and Felten, 1991 and 1992; Johnson, 1990; Lourenco and Zwijnenburg, 1996; Hong, Kahng and Moon, 1997). Such methods try to drive an SA procedure (or an iterated descent procedure) out of local optimality by propelling the solution a greater distance than usual from its current location.

Perturbation methods may be viewed as loosely structured procedures for inducing oscillation, without reference to intensification and diversification and their associated implementation strategies. Similarly, perturbation methods are not designed to exploit tradeoffs created by parametric variations in elements such as different types of infeasibility, measures of displacement from different sides of boundaries, etc. Nevertheless, at a first level of approximation, perturbation methods seek goals similar to those pursued by strategic oscillation.

The trajectory followed by strategic oscillation is imperfectly depicted in the figures of the preceding diagrams, insofar as the search path will not generally conform to precisely staged levels of a functional, but more usually will lie in regions with partial overlaps. (This feature is taken into account in the micro level decisions, examined in the next subsection.) Further, in customary approaches where diversification phases are linked with phases of intensification, the illustrated patterns of hovering about the target level are sometimes accompanied by hovering as well at other levels, in order to exploit a notion called the Proximate Optimality Principle or POP (see Chapter 5). According to this notion, good solutions at one level can often be found *close to* good solutions at an adjacent level. (E.g., only a modest number of steps will be required to reach good solutions at one level from those at another.) This condition of course depends on defining levels — and ways for moving within them — appropriately for given problem structures.

The challenge is to identify oscillation parameters and levels that will cause this potential relationship to become manifest. In strategies for applying the POP notion, the transition from one level to another is normally launched from a chosen high quality solution, rather than from the last solution generated before the transition is made (representing another feature difficult to capture in the preceding diagrams).

The POP notion motivates the tabu search strategy called *path relinking* (examined in Section 4.5, following), which provides an important means for enhancing the outcomes of strategic oscillation. Path relinking gives a way to refine strategic oscillation by choosing elite solutions from different levels, and joining them by new trajectories that pass through the target level. Elite solutions from the target level itself likewise provide a basis for path relinking processes to launch new searches at this level.

The POP notion implies a form of connectivity for the search space that may be usefully exploited by this approach. That is, path relinking trajectories which guided by elite solutions, whether deterministically or probabilistically, are likely to go through regions where new elite solutions reside, provided appropriate neighborhood definitions are used. The result is a type of focused diversification that is more effective than “sampling.”

Micro Level Decisions. The decision of selecting a target rate of change for strategic oscillation is placed at a micro level because it evidently involves variability of a particularly local form. Figure 4.8 illustrates three alternative rates of change, ranging from mildly to moderately aggressive, where the current direction is one of “descent.” Again the changes are shown as broken lines, to suggest they do not always proceed smoothly or uniformly. The issue of using target rates as a component of search strategy may seem evident, but its significance is often overlooked in other types of search frameworks. (For example, in simulated annealing, all improving moves are considered of equal status.)

Fig. 4.8 Target rates of change (micro level).

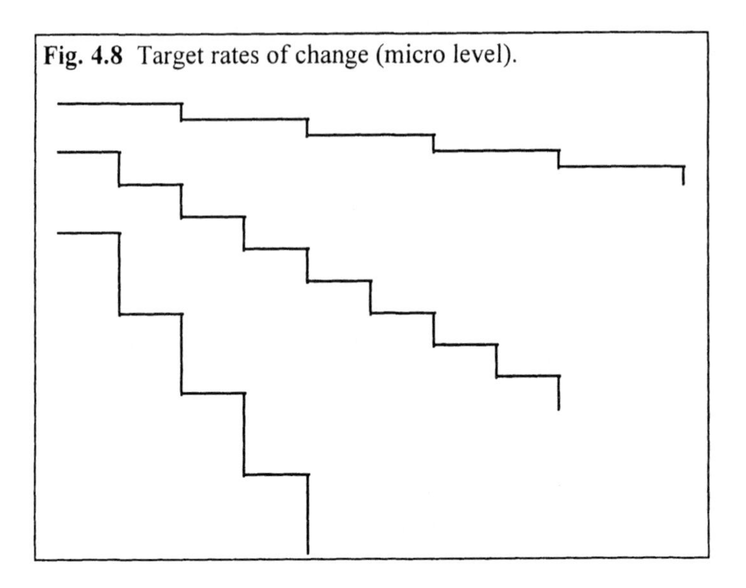

When the parameter of oscillation is related to values of the objective function, tabu search normally prescribes aggressive changes, seeking the greatest improvement or least disimprovement possible (in a steepest descent, mildest ascent orientation). Such an approach applies chiefly to intensification phases, and may be tempered or even reversed in diversification phases. More precisely, as previously emphasized, the notion of seeking aggressive (best or near best) changes is qualified in tabu search by specifying that the meaning of *best* varies in different settings and search phases, where rates of change constitute one of the components of this varying specification. This may be viewed as constituting another level of oscillation, which likewise is applied adaptively rather than subjected to monotonic control.

Accompanying the target rate of change is the micro level decision of choosing a target band of change, which sets boundaries on deviations from the target rate. The lines of the preceding diagrams, as noted earlier, should be interpreted as having a certain breadth (so that the lower limit of one segment may overlap with the upper limit of the next), and the target band of change is introduced to control this breadth.

Targeted rates and bands are not determined independently of knowledge about accessible solutions, but are based on exploring the current neighborhood to determine the possibilities available (hence bending the curves of Figure 4.8 by a factor determined from these possibilities).

Finally, aspiration criteria at the micro level permit the controls previously indicated to be abandoned if a sufficiently attractive alternative emerges, such as a move that leads to a new best solution, or to a solution that is the best one encountered at the current oscillation level.

4.5 Path Relinking

A useful integration of intensification and diversification strategies occurs in the approach called *path relinking*. This approach generates new solutions by exploring trajectories that connect elite solutions — by starting from one of these solutions, called an *initiating solution*, and generating a path in the neighborhood space that leads toward the other solutions, called *guiding solutions*. This is accomplished by selecting moves that introduce attributes contained in the guiding solutions.

The approach may be viewed as an extreme (highly focused) instance of a strategy that seeks to incorporate attributes of high quality solutions, by creating inducements to favor these attributes in the moves selected. However, instead of using an inducement that merely encourages the inclusion of such attributes, the path relinking approach subordinates all other considerations to the goal of choosing moves that introduce the attributes of the guiding solutions, in order to create a "good attribute composition" in the current solution. The composition at each step is determined by choosing the best move, using customary choice criteria, from the restricted set of moves that incorporate a maximum number (or a maximum weighted value) of the attributes of the guiding solutions. As in other applications of TS, aspiration criteria can override this restriction to allow other moves of particularly high quality to be considered.

Specifically, upon identifying a collection of one or more elite solutions to guide the path of a given solution, the attributes of these guiding solutions are assigned preemptive weights as inducements to be selected. Larger weights are assigned to attributes that occur in greater numbers of the guiding solutions, allowing bias to give increased emphasis to solutions with higher quality or with special features (e.g., complementing those of the solution that initiated the new trajectory).

More generally, it is not necessary for an attribute to occur in a guiding solution in order to have a favored status. In some settings attributes can share degrees of similarity, and in this case it can be useful to view a solution vector as providing "votes" to favor or discourage particular attributes. Usually the strongest forms of aspiration criteria are relied upon to overcome this type of choice rule.

In a given collection of elite solutions, the role of initiating solution and guiding solutions can be alternated. The distinction between initiating solutions and guiding solutions effectively vanishes in such cases. For example, a set of current solutions may be generated simultaneously, extending different paths, and allowing an initiating solution to be replaced (as a guiding solution for others) whenever its associated current solution satisfies a sufficiently strong aspiration criterion.

Because their roles are interchangeable, the initiating and guiding solutions are collectively called *reference solutions*. These reference solutions can have different interpretations depending on the solution framework under consideration. Reference points can be created by any of a number of different heuristics that result in high quality solutions.

An idealized form of such a process is shown in Figure 4.9. The chosen collection of reference solutions consists of the three members, A, B, and C. Paths are generated by allowing each to serve as initiating solution, and by allowing either one or both of the other two solutions to operate as guiding solutions. Intermediate solutions encountered along the paths are not shown. The representation of the paths as straight lines of course is oversimplified, since choosing among available moves in a current neighborhood will generally produce a considerably more complex trajectory. Intensification can be achieved by generating paths from similar solutions, while diversification is obtained creating paths from dissimilar solutions. Appropriate aspiration criteria allow deviation from the paths at attractive neighbors.

Fig. 4.9 Paths relinking in neighborhood space.

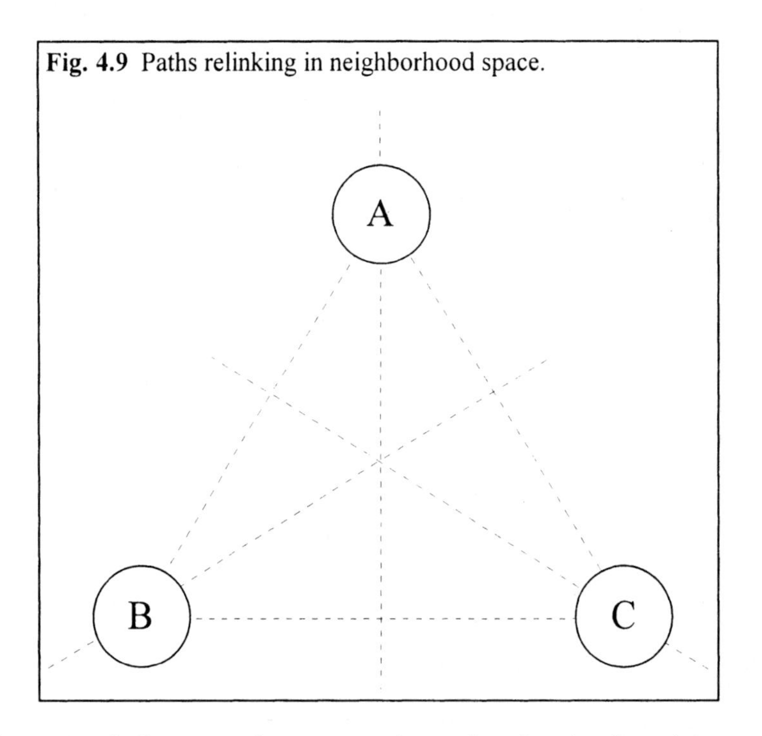

As Figure 4.9 indicates, at least one path continuation is allowed beyond each initiating/guiding solution. Such a continuation can be accomplished by penalizing the inclusion of attributes dropped during a trajectory, including attributes of guiding solutions that may be compelled to be dropped in order to continue the path. (An initiating solution may also be repelled from the guiding solutions by penalizing the inclusion of their attributes from the outset.) Probabilistic TS variants operate in the path relinking setting, as they do in others, by translating evaluations for deterministic rules into probabilities of selection, strongly biased to favor higher evaluations.

Promising regions are searched more thoroughly in path relinking by modifying the weights attached to attributes of the guiding solutions, and by altering the bias associated with solution quality and selected solution features. Figure 4.10 depicts the type of variation that can result, where the point X represents an initiating solution, the points A, B and C represent guiding solutions, and the dashed, dotted and solid lines are different searching paths. Variations of this type within a promising domain are motivated by the proximate optimality principle discussed in connection with strategic oscillation and elaborated in Chapter 5. For appropriate choices of the reference points (and neighborhoods for generating paths from them), this principle suggests that additional elite points are likely to be found in the regions traversed by the paths, upon launching new searches from high quality points on these paths.

Fig. 4.10 Path relinking by attribute bias.

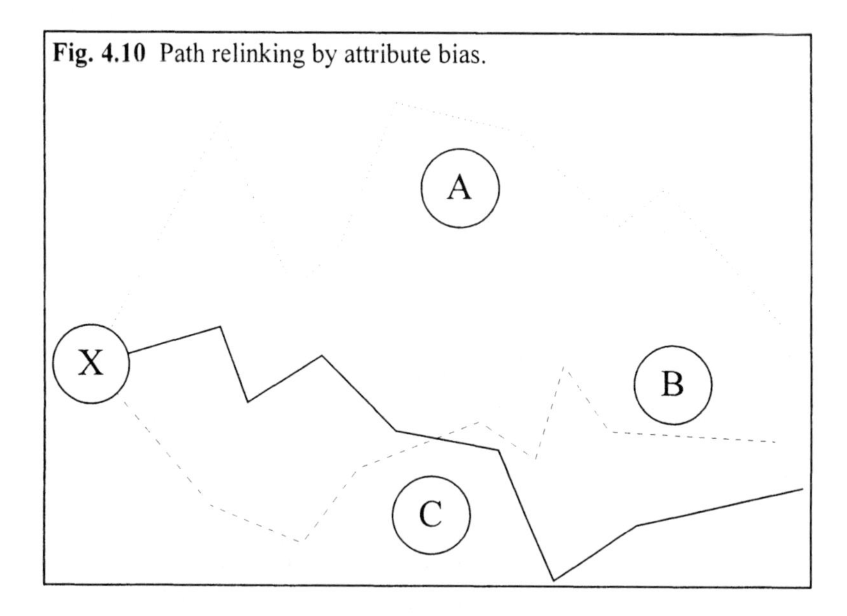

To illustrate some of the concepts associated with path relinking, we again consider the sequencing problem introduced in Chapter 2. The initiating solution with total weighted tardiness of 37 in Figure 4.11 following, is used to start a path relinking process towards the guiding solution (2, 1, 3, 4, 6, 5). The swaps are chosen in order to create a short path from one solution to the other. In this case, we can define a simple "score" for the current solution by counting the number of jobs that are in the same absolute position in the guiding solution. The score of the initiating solution is therefore zero. After the first move, jobs 5 and 6 are placed in their target absolute positions, hence the score of the second solution is 2. The scores of the following two solutions are 3 and 4, respectively. At that point, as indicated in Figure 4.11, a solution with an objective function value better than any other seen before can be found. Thus the use of an improved-best aspiration criterion can allow the path to deviate from its main goal (of ultimately reaching the guiding solution). Upon evaluating and recording the solution, the procedure alternatively may continue to look for other good solutions on or adjacent to its path to the guiding solution, before returning to explore more intensively in the region of attractive solutions found along the way. (Such regions may also be implicitly explored by including these attractive solutions among future reference points.) We indicate this evaluation, and the associated step of recording without deviating, by a dotted line in Figure 4.11.

Fig. 4.11 Path relinking example.

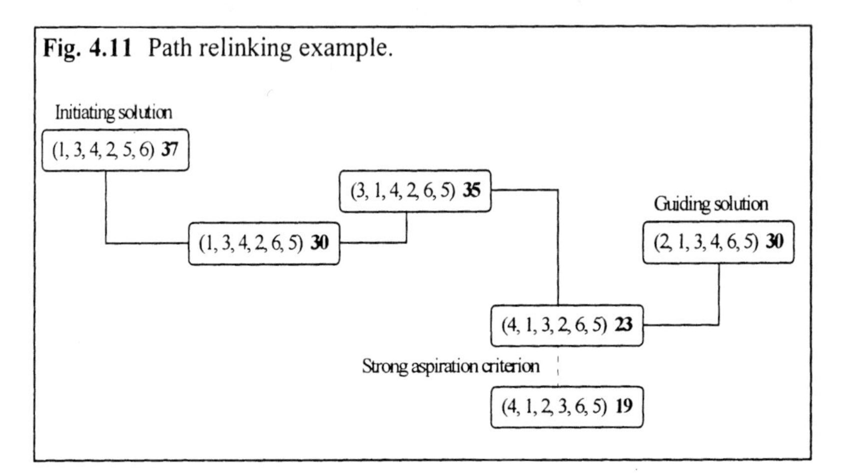

Similar path relinking mechanisms can be developed for other applications by creating an appropriate guiding (or scoring) procedure. Note that in our sequencing problem the score could have been defined in terms of relative job positions as opposed to absolute positions.

4.5.1 Roles in Intensification and Diversification

Path relinking, in common with strategic oscillation, gives a natural foundation for developing intensification and diversification strategies. Intensification strategies in this setting typically choose reference solutions to be elite solutions that lie in a common region or that share common features. Similarly, diversification strategies based on path relinking characteristically select reference solutions that come from different regions or that exhibit contrasting features. Diversification strategies may also place more emphasis on paths that go beyond the reference points. Collections of reference points that embody such conditions can be usefully determined by clustering and conditional analysis methods (Chapter 10).

These alternative forms of path relinking also offer a convenient basis for parallel processing, contributing to the approaches for incorporating intensification and diversification tradeoffs into the design of parallel solution processes generally.

4.5.2 Incorporating Alternative Neighborhoods

Path relinking strategies in tabu search can occasionally profit by employing different neighborhoods and attribute definitions than those used by the heuristics for generating the reference solutions. For example, it is sometimes convenient to use a constructive neighborhood for path relinking, i.e., one that permits a solution to be built in a sequence of constructive steps (as in generating a sequence of jobs to be processed on specified machines using dispatching rules). In this case the initiating solution can be used to give a beginning partial construction, by specifying particular attributes (such as jobs in particular relative or absolute sequence positions) as a basis for remaining constructive steps. Similarly, path relinking can make use of destructive neighborhoods, where an initial solution is "overloaded" with attributes donated by the guiding solutions, and such attributes are progressively stripped away or modified until reaching a set with an appropriate composition.

When path relinking is based on constructive neighborhoods, the guiding solution(s) provide the attribute relationships that give options for subsequent stages of construction. At an extreme, a full construction can be produced, by making the initiating solution a *null solution*. The destructive extreme starts from a "complete set" of solution elements. Constructive and destructive approaches differ from transition approaches by typically producing only a single new solution, rather than a sequence of solutions, on each path that leads from the initiating solution toward the others. In this case the path will never reach the additional solutions unless a transition neighborhood is used to extend the constructive neighborhood.

Constructive neighborhoods can often be viewed as a special case of feasibility restoring neighborhoods, since a null or partially constructed solution does not satisfy all conditions to qualify as feasible. Similarly, destructive neighborhoods can also represent an instance of a feasibility restoring function, as where an excess of elements may violate explicit problem constraints. A variety of methods have been devised to restore infeasible solutions to feasibility, as exemplified by flow augmentation methods in network problems, subtour elimination methods in traveling salesman and vehicle routing problems, alternating chain processes in degree-constrained subgraph problems, and value incrementing and decrementing methods in covering and multidimensional knapsack problems. Using neighborhoods that permit restricted forms of infeasibilities to be generated, and then using associated neighborhoods to remove these infeasibilities, provides a form of path relinking with useful diversification features. Upon further introducing transition neighborhoods, with the ability to generate successive solutions with changed attribute mixes, the mechanism of path relinking also gives a way to tunnel through infeasible regions. The following is a summary of the components of path relinking:

Step 1. Identify the neighborhood structure and associated solution attributes for path relinking (possibly different from those of other TS strategies applied to the problem).

Step 2. Select a collection of two or more reference solutions, and identify which members will serve as the initiating solution and the guiding solution(s). (Reference solutions can be infeasible, such as "incomplete" or "overloaded" solution components treated by constructive or destructive neighborhoods.)

Step 3. Move from the initiating solution toward (or beyond) the guiding solution(s), generating one or more intermediate solutions as candidates to initiate subsequent problem solving efforts. (If the first phase of this step creates an infeasible solution, apply an associated second phase with a feasibility restoring neighborhood.)

In Chapter 9 we will see how the path relinking strategy relates to a strategy called scatter search, which provides additional insights into the nature of both approaches.

4.6 The Intensification / Diversification Distinction

The relevance of the intensification/diversification distinction is supported by the usefulness of TS strategies that embody these notions. Although both operate in the short term as well as the long term, we have seen that longer term strategies are generally the ones where these notions find their greatest application.

In some instances we may conceive of intensification as having the function of an intermediate term strategy, while diversification applies to considerations that emerge in the longer run. This view comes from the observation that in human problem solving, once a short term strategy has exhausted its efficacy, the first (intermediate term) response is often to focus on the events where the short term approach produced the best outcomes, and to try to capitalize on elements that may be common to those events. When this intensified focus on such events likewise begins to lose its power to uncover further improvement, more dramatic departures from a short term strategy are undertaken. (Psychologists do not usually differentiate between intermediate and longer term memory, but the fact that memory for intensification and diversification can benefit from such differentiation suggests that there may be analogous physical or functional differences in human memory structures.) Over the truly long term, however, intensification and diversification repeatedly come into play in ways where each depends on the other, not merely sequentially, but also simultaneously.

There has been some confusion between the terms intensification and diversification, as applied in tabu search, and the terms *exploitation* and *exploration,* as popularized in the literature of genetic algorithms. The differences between these two sets of notions deserve to be clarified, because they have substantially different consequences for problem solving.

The exploitation/exploration distinction comes from control theory, where exploitation refers to following a particular recipe (traditionally memoryless) until it fails to be effective, and exploration then refers to instituting a series of random changes — typically via multi-armed bandit schemes — before reverting to the tactical recipe. (The issue of exploitation versus exploration concerns how often and under what circumstances the randomized departures are launched.)

By contrast, intensification and diversification in tabu search are both processes that take place when simpler exploitation designs play out and lose their effectiveness — although as we have noted, the incorporation of memory into search causes intensification and diversification also to be manifest in varying degrees even in the short range. (Similarly, as we have noted, intensification and diversification are not opposed notions, for the best form of each contains aspects of the other, along a spectrum of alternatives.)

Intensification and diversification are likewise different from the control theory notion of exploration. Diversification, which is sometimes confused with exploration, is not a recourse to a Game of Chance for shaking up the options invoked, but is a collection of strategies — again taking advantage of memory — designed to move purposefully rather than randomly into uncharted territory.

The source of these differences is not hard to understand. Researchers and practitioners in the area of search methods have had an enduring love affair with randomization, perhaps influenced by the much publicized Heisenberg Uncertainty Principle in Quantum Mechanics. Einstein's belief that God does not roll dice is out of favor, and many find a special enchantment in miraculous events where blind purposelessness creates useful order. (We are less often disposed to notice that this way of producing order requires an extravagant use of time, and that order, once created, is considerably more effective than randomization in creating still higher order.)

Our "scientific" reports of experiments with nature reflect our fascination with the role of chance. When apparently chaotic fluctuations are brought under control by random perturbations, we seize upon the random element as the key, while downplaying the importance of attendant restrictions on the setting in which randomization operates. The diligently concealed message is that under appropriate controls, *perturbation* is effective for creating desired patterned outcomes — and in fact, if the system and attendant controls are sufficiently constrained, perturbation works even when random. (Instead of accentuating differences between workable and unworkable kinds of perturbation, in our quest to mold the universe to match our mystique we portray the central consideration to be randomization versus nonrandomization.)

The tabu search orientation evidently contrasts with this perspective. As manifest in the probabilistic TS variant (Chapter 7), elements subjected to random influence are preferably to be strongly confined, and uses of randomization are preferably to be modulated through well differentiated probabilities. In short, the situations where randomization finds a place are very highly structured. From this point of view God may play with dice, but beyond any question the dice are *loaded.*

4.7 Some Basic Memory Structures for Longer Term Strategies

To give a foundation for describing fundamental types of memory structures for longer term strategies, we first briefly review the form of the recency-based memory structure introduced in Chapter 2 for handling add/drop moves. However, we slightly change the notation, to provide a convenient way to refer to a variety of other types of moves such as those treated in the latter part of Chapter 3.

4.7.1 Conventions

Let $S = \{1, 2,..., s\}$ denote an index set for a collection of solution attributes. For example, the indexes $i \in S$ may correspond to indexes of zero-one variables x_i, or they may be indexes of edges that may be added to or deleted from a graph, or the job indexes in a production scheduling problem. More precisely, by the attribute/element distinction discussed in Section 2.4.1 of Chapter 2, the attributes referenced by S in these cases consist of the specific values assigned to the variables, the specific add/drop states adopted by the edges, or positions occupied by the jobs. In general, to give a correspondence with developments of Section 3.5 of Chapter 3, an index $i \in S$ can summarize more detailed information; e.g., by referring to an ordered pair (j, k) that summarizes a value assignment $x_j = k$ or the assignment of job j to position k, etc. Hence, broadly speaking, the index i may be viewed as a notational convenience for representing a pair or a vector.

To keep our description at the simplest level, suppose that each $i \in S$ corresponds to a 0-1 variable x_i. As before, we let *Iter* denote the counter that identifies the current iteration, which starts at 0 and increases by 1 each time a move is made.

For recency-based memory, following the approach indicated in Section 2.4.1 of Chapter 2, when a move is executed that causes a variable x_i to change its value, we record $TabuStart(i) = Iter$ immediately after updating the iteration counter. (This means that if the move has resulted in $x_i = 1$, then the attribute $x_i = 0$ becomes tabu-active at the iteration $TabuStart(i)$.) Further, we let $TabuTenure(i)$ denote the number of iterations this attribute will remain tabu-active. Thus, by our previous design, the recency-based tabu criterion says that the previous value of x_i is tabu-active throughout all iterations such that

$$TabuStart(i) + TabuTenure(i) \leq Iter.$$

Similarly, in correspondence with earlier remarks, the value *TabuStart(i)* can be set to 0 before initiating the method, as a convention to indicate no prior history exists. Then we automatically avoid assigning a tabu-active status to any variable with *TabuStart(i)* = 0 (since the starting value for variable x_i has not yet been changed).

4.7.2 Frequency-Based Memory

By our foregoing conventions, allowing the set $S = \{1, \ldots, s\}$ for illustration purposes to refer to indexes of 0-1 variables x_i, we may indicate structures to handle frequency-based memory as follows.

Transition frequency-based memory is by far the simplest to handle. A transition memory, *Transition(i)*, to record the number of times x_i changes its value, can be maintained simply in the form of a counter for x_i that is incremented at each move where such a change occurs. Since x_i is a zero-one variable, *Transition(i)* also discloses the number of times x_i changes to and from each of its possible assigned values. In more complex situations, by the conventions already noted, a matrix memory *Transition(j,k)* can be used to determine numbers of transitions involving assignments such as $x_j = k$. Similarly, a matrix memory may be used in the case of the sequencing problem where both the index of job j and position k may be of interest. In the context of the Min k-Tree problem, an array dimensioned by the number of edges can maintain a transition memory to keep track of the number of times that specific edges have been brought in and out of the solution. A matrix based on the edges can also identify conditional frequencies. For example, the matrix *Transition(j,k)* can be used to count the number of times edge j replaced edge k. It should be kept in mind in using transition frequency memory that penalties and inducements are often based on *relative* numbers (rather than absolute numbers) of transitions, hence requiring that recorded transition values are divided by the total number of iterations (or the total number of transitions). As noted earlier, other options include dividing by the current maximum transition value. Raising transition values to a power, as by squaring, is often useful to accentuate the differences in relative frequencies.

Residence memory requires only slightly more effort to maintain than transition memory, by taking advantage of the recency-based memory stored in *TabuStart(i)*. The following approach can be used to track the number of solutions in which $x_i = 1$, thereby allowing the number of solutions in which $x_i = 0$ to be inferred from this. Start with *Residence(i)* = 0 for all i. Then, whenever x_i changes from 1 to 0, after updating *Iter* but before updating *TabuStart(i)*, set

$$Residence(i) = Residence(i) + Iter - TabuStart(i).$$

Then, during iterations when $x_i = 0$, *Residence(i)* correctly stores the number of earlier solutions in which $x_i = 1$. During iterations when $x_i = 1$, the true value of

Residence(i) is the right hand side of the preceding assignment, however the update only has to be made at the indicated points when x_i changes from 1 to 0. Table 4.3 illustrates how this memory structure works when used to track the assignments of a variable x during 100 iterations. The variable is originally assigned to a value of zero by a construction procedure that generates an initial solution. In iteration 10 a move is made that changes the assignment of x from zero to one, however the *Residence* value remains at zero. *Residence* is updated at iterations 22 and 73, when moves are made that change the assignment of x from 1 to 0. At iteration 65, for example, x has received a value of 1 for 27 iterations (i.e., *Residence* + *Iter* - *TabuStart* = 12 + 65 - 50 = 27), while at iteration 90 the count is 35 (i.e., the value of *Residence*).

Table 4.3 Illustrative residence memory.

Iter	*Assignment*	*Residence*
0	$x = 0$	0
10	$x = 1$	0
22	$x = 0$	22 - 10 = 12
50	$x = 1$	12
73	$x = 0$	12 + 73 -50 = 35

As with transition memory, residence memory should be translated into a relative as a basis for creating penalties and inducements.

The indicated memory structures can readily be applied to multivalued variables (or multistate attributes) by the extended designs illustrated in Chapter 3. In addition, the 0-1 format can be adapted to reference the number of times (and last time) a more general variable changed its value, which leads to more restrictive tabu conditions and more limiting ("stronger") uses of frequency-based memory than by referring separately to each value the variable receives. As in the case of recency-based memory, the ability to affect larger numbers of alternative moves by these more aggregated forms of memory can be useful for larger problems, not only for conserving memory space but also for providing additional control over solutions generated.

4.7.3 Critical Event Memory

Strategic oscillation offers an opportunity to make particular use of both short term and long term frequency-based memory. To illustrate, let $A(Iter)$ denote a zero-one vector whose jth component has the value 1 if attribute j is present in the current solution and has the value 0 otherwise. The vector A can be treated "as if" it is the same as the solution vector for zero-one problems, though implicitly it is twice as large, since $x_j = 0$ is a different attribute from $x_j = 1$. This means that rules for operating on the full A must be reinterpreted for operating on the condensed form of A. The sum of the A vectors over the most recent t critical events provides a simple memory that combines recency and frequency considerations. To maintain the sum requires remembering $A(k)$, for k ranging over the last t iterations. Then the sum vector A^* can be updated quite easily by the incremental calculation

$$A^* = A^* + A(Iter) - A(Iter - t + 1).$$

Associated frequency measures, as noted earlier, may be normalized, in this case for example by dividing A^* by the value of t. A long term form of A^* does not require storing the $A(k)$ vectors, but simply keeps a running sum. A^* can also be maintained by exponential smoothing.

Such frequency-based memory is useful in strategic oscillation where critical events are chosen to be those of generating a complete (feasible) construction, or in general of reaching the targeted boundary (or a best point within a boundary region). Instead of using a customary recency-based TS memory at each step of an oscillating pattern, greater flexibility results by disregarding tabu restrictions until reaching the turning point, where the oscillation process alters its course to follow a path toward the boundary. At this point, assume a choice rule is applied to introduce an attribute that was not contained in any recent solution at the critical level. If this attribute is maintained in the solution by making it tabu to be dropped, then upon eventually reaching the critical level the solution will be different from any seen over the horizon of the last t critical events. Thus, instead of updating A^* at each step, the updating is done only for critical level solutions, while simultaneously enhancing the flexibility of making choices.

In general, the possibility occurs that no attribute exists that allows this process to be implemented in the form stated. That is, every attribute may already have a positive associated entry in A^*. Thus, at the turn around point, the rule instead is to choose a move that introduces attributes which are least frequently used. (Note, "infrequently used" can mean either "infrequently present" or "infrequently absent," depending upon the current direction of oscillation.) This again can be managed conveniently by using penalties and inducements. Such an approach has been found very effective for multidimensional knapsack problems and 0-1 quadratic optimization problems in Glover and Kochenberger (1996) and Glover, Kochenberger and Alidaee (1997).

For greater diversification, this rule can be applied for r steps after reaching the turn around point. Normally r should be a small number, e.g., with a baseline value of 1 or 2, which is periodically increased in a standard diversification pattern. Shifting from a short term A^* to a long term A^* creates a global diversification effect. A template for this approach is given in Figure 4.12.

The approach of Figure 4.12 is not symmetric. An alternative form of control is to seek immediately to introduce a low frequency attribute upon leaving the critical level, to increase the likelihood that the solution at the next turn around will not duplicate a solution previously visited at that point. Such a control enhances diversity, though duplication at the turn around will already be inhibited by starting from different solutions at the critical level.

Fig. 4.12 Strategic oscillation illustrative memory.

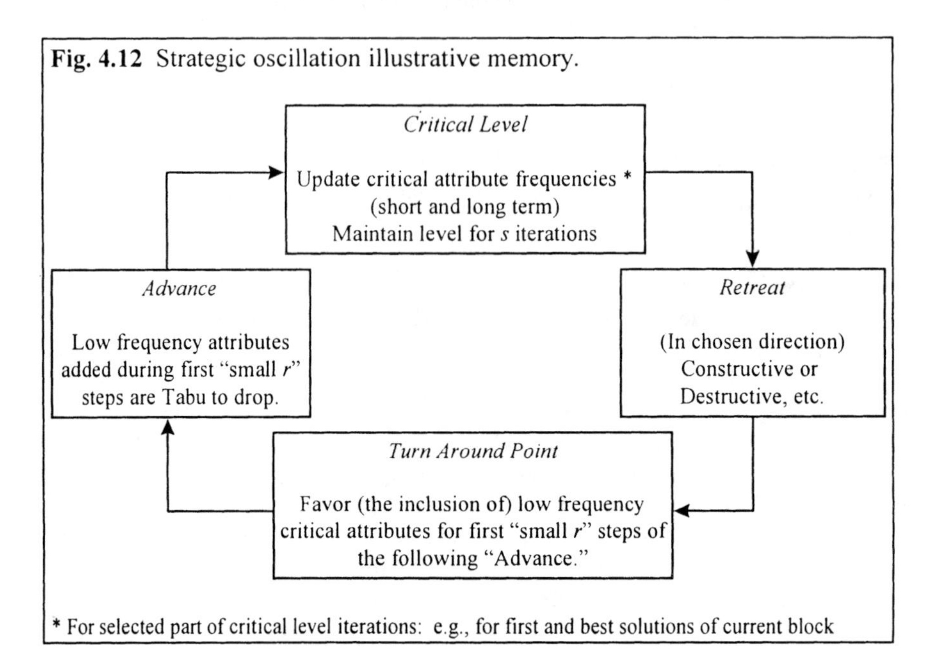

* For selected part of critical level iterations: e.g., for first and best solutions of current block

4.8 Discussion Questions and Exercises

1. Add at least three additional examples to Table 4.1 of Residence Measures and Transition Measures for frequency. Indicate the pros and cons of each measure.

2. Specify whether the measures you have provided in Exercise 1 are primarily useful for intensification or diversification, or whether they can be used for both. Can you identify different subsets of solutions for defining these measures that will change the type of strategy they are best design to support? Is it easy or hard to maintain frequency measures over these subsets? (Explain.)

3. Identify a different class of problems than the sequencing and Min k-Tree problems. Can you specify types of frequency memory for these problems analogous to those specified in Table 4.1 and the examples developed for Exercises 1 and 2?

4. Discuss the types of penalties and inducements that may be used in conjunction with the memories of Exercises 1 to 3 to pursue goals of intensification and diversification. What value can thresholds have in determining such penalties and inducements?

5. Indicate the general form of a strategy for the Min *k*-Tree problem that uses thresholds on the numerator of transition frequencies and residence frequencies as a basis for executing a diversification phase. Why should these thresholds be changed differently for different attributes after the diversification phase? Discuss the pros and cons of changing the recorded numerators instead of changing the thresholds.

6. Describe how strategic oscillation may be used in setting the magnitudes of penalties and inducements, and in setting threshold levels. Can you think of an adaptive way to determine appropriate values to assign to penalties, inducements and thresholds? Is there any way to learn what good values may be, other than by trail and error? (See the discussion of the target analysis learning approach in Chapter 9.)

7. Develop an intensification by decomposition for the sequencing problem, the Min *k*-Tree problem, and the 0-1 knapsack problem.

8. Design at least two different restarting mechanisms for the Min *k*-Tree problem based on frequency information.

9. Discuss the implementation of a strategic oscillation procedure in the context of a sequencing problem.

10. How can the use of critical event memory as considered in Chapter 2 be expanded by introducing frequency-based forms of memory as examined in Exercises 1 to 6? To what extent does the specification of different subsets of solutions for defining frequency memory already qualify this as a form of critical event memory?

11. Describe how critical event memory might be used with strategic oscillation, where critical events are defined to be those that occur at the oscillation boundary. Making reference to the fact that critical event memory can be recency-based as well as frequency-based, describe how critical event memory may be used to integrate recency and frequency concerns. How might exponential smoothing be used to provide a simple way of blending recency and frequency? What advantages and disadvantages may come from alternative "decay rates" for exponential smoothing?

12. Using the metaphor of "voting" as a way to integrate the influence of different choice criteria or neighborhood structures, consider how different voting schemes may be devised to integrate recency and frequency memory. (For example, suppose that three different exponential smoothing approaches are used to combine recency and frequency. How might these approaches be used to vote on decisions of attributes to favor or disfavor for inclusion in current solutions?) Can you envision a way to learn good voting schemes?

13. How may parallel processing be used to take advantage of alternative strategies for integrating recency and frequency?

14. The threshold based approach for logical restructuring described in Chapter 3, generates a form of path relinking that uses constructive and/or destructive neighborhoods. Explain how this approach fits within the path relinking framework. (Elaborate your answer by additionally considering the example treated in Exercises 9 to 11 of Chapter 3.)

15. Exercise 14 demonstrates that initiating and guiding solutions for path relinking can be interchangeable (or indistinguishable), and also that they can be infeasible (or incomplete). The exercise further demonstrates that the type of neighborhood used by path relinking does not have to be the same one used in other phases of the search. Would you expect that threshold based forms of logical restructuring will almost invariably constitute a form of path relinking — and that it also will typically be a type of path relinking that embraces these three features (i.e., of initiating and guiding solutions that are interchangeable, infeasible (or incomplete), and suitably joined by neighborhoods that differ from those used in other search phases)?

16. Discuss the creation of a "multilevel" tabu search that exploits the POP notion in the context of a sequencing problem.

17. Consider the p-median problem, which consists of locating p facilities on a graph such that the distances from each vertex to its nearest facility is minimized. If the graph contains n nodes and the distance from node i to j is given by d_{ij}, formulate this problem as a 0-1 integer program. Design a TS procedure for the solution of this problem, including at least one candidate list strategy, and a strategic oscillation or a path relinking component.

5 TABU SEARCH PRINCIPLES

This chapter describes concepts and issues that are important for an effective application of the TS methodology and that merit fuller investigation. We begin by re-examining the notion of influence, followed by considering the generation of compound moves. We also expand the description of the proximate optimality principle and introduce a series of additional principles that are relevant for designing better solution procedures

5.1 Influence and Measures of Distance and Diversity.

The notion of influence, and of influential moves, has several dimensions in tabu search. This notion is particularly relevant upon encountering an *entrenched regionality* phenomenon, where local optima — or regions encompassing a particular collection of local optima — are "mini black holes" that can be left behind, once visited, only by particularly strong effort. Viewed from a minimization perspective, these regions are marked by the presence of humps which can only be crossed by choosing moves with significantly inferior evaluations, or alternately by the presence of long valleys, where the path to a better solution can only be found by a long (and possibly erratic) climb. In such cases, a faster and more direct withdrawal may be desirable.

A strategy of seeking influential moves, or an *influential series* of moves, becomes important in such situations. The notion of influence does not simply refer to anything that creates a "large change," however, but rather integrates the two key aspects of diversification and intensification in tabu search by seeking change that holds indication of promise. This requires reference to memory and/or strategic uses of probabilities while paying careful attention to evaluations.

Diversification in its "pure" form, which solely strives to reach a destination that is markedly different from all others encountered, is incomplete as a basis for an effective search strategy. It is nevertheless important to characterize how such a pure form would operate, in order to overlay it with balancing considerations of intensification. The essential elements of pure diversification, and their differences from randomization, can be characterized as follows.

Sequential Diversification

In general, we are interested not just in diversified collections of moves but also in diversified sequences, since often the order of examining elements is important in tabu search. This can apply, for example, where we seek to identify a sequence of new solutions (not seen before) so that each successive solution is *maximally diverse* relative to all solutions previously generated. This includes possible reference to a baseline set of solutions, often of the form $x \in S$, which takes priority in establishing the diversification objective (i.e., where the first level goal is to establish diversification relative to S, and then in turn relative to other solutions generated). The diversification concept applies as well to generating a diverse sequence of numbers or a diverse set of points from the vertices of a unit hypercube.

Let $Z(k) = (z(1), z(2), \ldots, z(k))$ represent a sequence of points drawn from a set Z. For example, Z may be a line interval if the points are scalars. We take $z(1)$ to be a *seed point* of the sequence. The seed point need not belong to Z.) Then we define $Z(k)$ to be a *diversified sequence* (or simply a *diverse sequence*), relative to a chosen distance metric d over Z by requiring each point $z = z(h+1)$ for $h = 1$ to k-1, to satisfy the following hierarchy of conditions relative to the subsequence $Z(h) = (z(1), \ldots, z(h))$:

(A) z maximizes the minimum distance $d(z, z(i))$ for $i \leq h$;
(B) subject to (A), z maximizes the minimum distance $d(z, z(i))$ for $1 < i \leq h$, then for $2 < i \leq h$, ..., etc. (in strict priority order).
(C) subject to (A) and (B), z maximizes the distance $d(z, z(i))$ for $i = h$, then for $i = h - 1$, ..., and finally for $i = 1$. (Additional ties may be broken arbitrarily.)

To handle diversification relative to an initial baseline set Z^* (such as a set of solutions $x \in S$), the preceding hierarchy of conditions is preceded by a condition stipulating that z first maximizes the minimum distance $d(z, z^*)$ for $z^* \in Z^*$. A useful (weaker) variant of this condition simply treats points of Z^* as if they constitute the last elements of the sequence $Z(h)$.

Variations on (A), (B), and (C), including going deeper in the hierarchy before arbitrary tie breaking, are evidently possible. Such conditions make it clear that a diverse sequence is considerably different from a random sequence. Further, they are computationally very demanding to satisfy. Even by omitting condition (B), and retaining only (A) and (C), if the elements $z(i)$ refer to points on a unit hypercube, then by our present state of knowledge the only way to generate a diverse sequence of more than a few points is to perform comparative enumeration. However, a diverse sequence of points on a line interval, particularly if $z(1)$ is an endpoint or midpoint of the interval, can be generated with much less difficulty. Because of this, it can sometimes be useful to generate sequences by approximating the foregoing conditions.

Taking a broader view, an extensive effort to generate diverse sequences can be performed in advance, independent of problem solving efforts, so that such sequences are precomputed and available as needed. Further, a diverse sequence for elements of a high dimensional unit hypercube may be derived by reverse projection techniques ("lifting" operations) from a sequence for a lower dimensional hypercube, ultimately making reference to sequences from a line interval.

Observations about generating diverse subsets, as opposed to diverse subsequences, are given in Chapter 9.

5.1.1 Influential Diversity

The notion of influence enters into this by conceiving *influential diversity* to result when a new solution is not only different from (or far from) others seen, but also has a notably attractive structure or objective function value. A variant of this notion has also surfaced more recently in "large-step" optimization approaches, though without reference to memory, as noted in Chapter 4.

From a probability standpoint, solutions that satisfy such requirements of attractiveness are much rarer than those that meet the conditions of pure diversification, and hence in this sense involve a stronger form of diversity (Kelly, Laguna and Glover, 1994). Search spaces commonly have the property that solutions with progressively better objective function values are distributed with a "diminishing tail," which produces a relatively small likelihood of encountering the better representatives of such solutions. Where this is not the case, the problems are generally somewhat easier to solve. A strategy of treating high quality solutions as improbable is not a liability in any event.

One way to capture the relevant concerns is to create a measure of distance that identifies the magnitude of change in structure or location of a solution. Distance can refer to change induced by a single move or by a collection of moves viewed as a compound move. Natural measures of distance in different contexts, for example, may refer to weights of elements displaced by a move, costs of elements added or deleted, degrees of smoothness or irregularity created in a pattern, shifts in levels of aggregation or disaggregation, variation in step sizes, alterations in levels of a hierarchy, degrees of satisfying or violating critical constraints, and so forth.

5.1.2 Influence/Quality Tradeoffs

Given a particular distance measure, the tradeoffs between change in distance and change in quality embodied in the notion of influence can be addressed by partitioning distances into different classes. The word "class" is employed to reflect the fact that a measure may encompass more than one of the elements illustrated above, and different combinations invite categorical distinctions. Even where a measure is unidimensional, the effects of different levels of distance may not be proportional to their magnitudes, which again suggests the relevance of differentiation by class.

Under conditions of entrenched regionality, where moves that involve greater distances are likely to involve greater deterioration in solution quality, the goal is to determine when an evaluation for a given distance should be regarded attractive, although superficially such an evaluation may appear less attractive than an evaluation for a smaller distance. Such a determination of relative attractiveness depends on the extent to which the current solution is affected by the entrenched regionality phenomenon — hence, for example, by the distance it has already moved away from a local optimum. The importance of accounting for the quality of solutions produced when retreating from a local optimum is illustrated by the study of market niche clusters by Kelly (1995). The study used a form of strategic oscillation to periodically induce steps that degrade the objective function. It was found that selecting moves of least degradation was more effective than selecting moves of greater degradation. Thus, while the notion of influence suggests that moves that create greater changes are to be favored, provided they represent alternatives of comparable quality, it remains important not to be lured by change for the sake of change alone.

The graph of Figure 5.1 illustrates some of the relevant considerations. The horizontal axis identifies an influence measure (as reflected by a "move distance," for example) and the vertical axis identifies a quality measure. We refer to this measure as *apparent quality*, to indicate that it is only an estimate of how good the quality of move may be, as a basis for leading the search to a better solution. (For instance, apparent quality may simply refer to the change in the objective function value produced by a move.) This measure is to be differentiated from *effective quality*, which represents the progress made by the move toward obtaining an optimal solution, or less restrictively, a solution better than the best solution currently known.

The effective quality of a move is of course unknown during the search process (or else a move with high effective quality would automatically be selected), but Figure 5.1 illustrates the type of relationship that may exist between effective quality and the two observable measures given by apparent quality and influence. Each "stairstep curve" in the figure corresponds to an isovalue contour for a given level of effective quality; that is, all points on the curve have the same effective quality. Contours that are higher and farther to the right represent higher levels of effective quality.

Fig. 5.1 Effective quality isovalue contours.

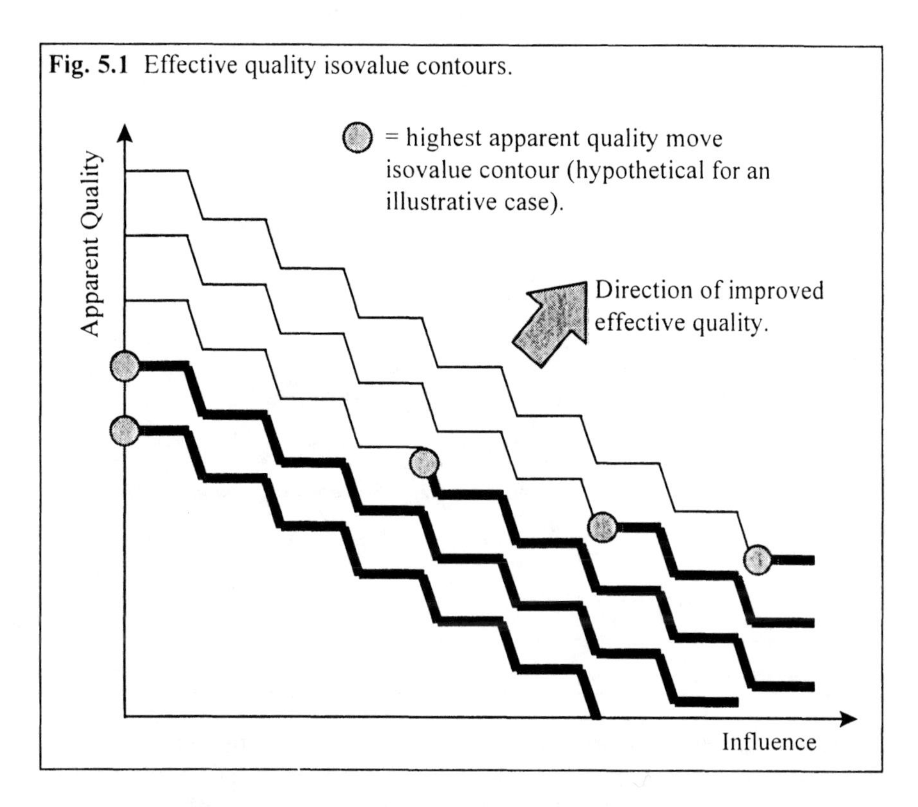

The graph illustrates the situation where moves of higher influence may have lower apparent quality and yet still produce an outcome of greater effective quality. The graph also illustrates that the higher influence moves may tend to be those that have lower levels of apparent quality. In particular, the darkened circle on each isovalue contour identifies the highest apparent quality move that exists on that contour, for a hypothetical current iteration. Thus, only the darkened portion of each contour contains moves that currently exist. There are no moves on the lighter portion, which identifies regions of higher apparent quality (and typically lower influence) for the given level of effective quality represented by the contour. In the case of the highest level of effective quality (the "highest contour"), the graph shows that the move with highest apparent quality (given by the darkened circle) lies below the moves of highest apparent quality on other contours. Nevertheless, due to the tradeoffs between influence and quality, this move is "better than" the moves represented by the other darkened circles, even though its apparent quality is less.

The structure of the graph shown in Figure 5.1 is intended to illustrate a situation where no improvements in the best solution have been found by a tabu search method for a fairly long period and none appear imminent. (Consequently the best move only occurs at a relatively high level of influence and a relatively low level of apparent quality.) Once moves of higher influence have been made, however, the structure of the graph is likely to shift to become more like that of Figure 5.2.

Fig. 5.2 Effective quality isovalue contours (after high influence moves have been made.

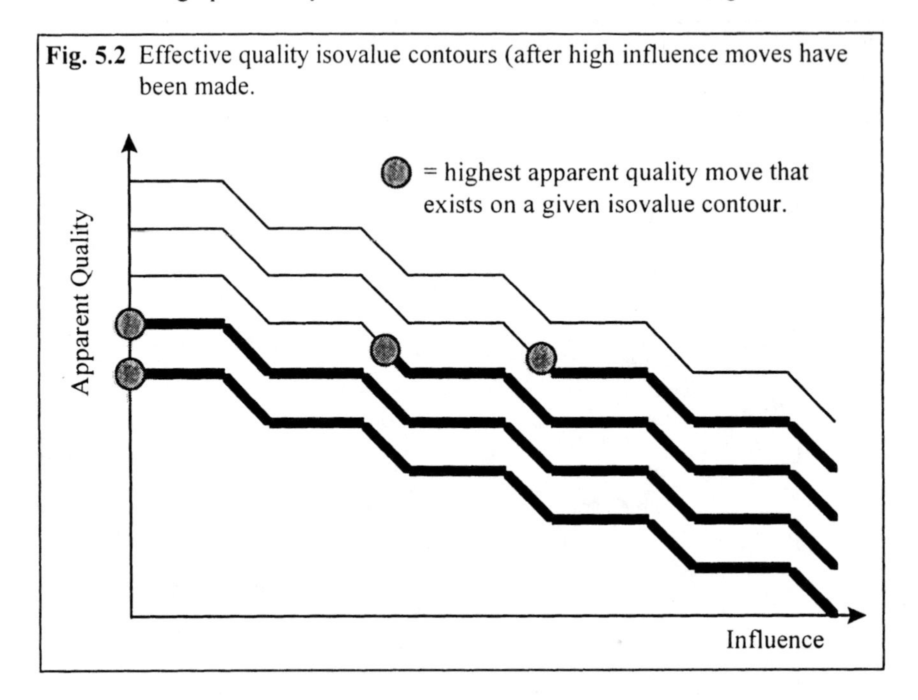

The effective quality isocontours of Figure 5.2 are more nearly horizontal than in Figure 5.1, which shows that the level of influence now plays a reduced role in determining the "true" (effective) quality of a move. In this situation, apparent quality is a more reliable guide. (Indeed, while horizontal at the left, such curves may drop off steeply at the right, indicating that there may not be many high influence moves that lead to better solutions. This is increasingly likely to occur as the search gets "close to" an optimal solution.) However, in both figures it is clear that there are tradeoffs between apparent quality and influence. Choices that account for only one of the two factors are not likely to be as good as those that seek to balance these factors.

The graphs of Figure 5.1 and 5.2 of course oversimplify the conditions that may be expected to hold. Some of the more realistic complexities that accompany the tradeoffs between quality and influence are depicted in Table 5.1. Within this table, apparent quality and influence have each been subdivided into 6 different levels (ranges of values), numbered from 1 to 6 across the top and side of the table. Cell (i,j) of the table refers to an apparent quality of level i and an influence of level j. Each such cell contains an entry of the form $x : y$, where x corresponds to the number of currently available moves (for the indicated levels of quality and influence), and y corresponds to the number of "good" moves; e.g., moves that have a sufficient degree of effective quality to move the search closer to an improved solution. Thus, in cell (2,5) which applies to the case where apparent quality is at level 2 and the influence is at level 5, the entry 4 : 3 shows that there are 4 available moves, of which 3 are good. (This is the most favorable ratio of available moves to good moves in the table.) Blank cells of the table refer to those cases where no moves are currently available that have the indicated levels of quality and influence.

Table 5.1 Available moves versus good moves at different levels of apparent quality and influence

Apparent Quality	*Influence*					
	1	2	3	4	5	6
6	1 : 0					
5	2 : 1					
4	5 : 1	5 : 2	3 : 1			
3	5 : 0	5 : 1	4 : 1	4 : 2		
2	5 : 0	5 : 0	5 : 0	5 : 2	4 : 3	
1	5 : 0	5 : 0	5 : 1	5 : 0	4 : 1	3 : 1

To amplify, Table 5.1 shows that only one move has an apparent quality at level 6. This move has an influence of level 1, and does not qualify as a good move. There are 2 moves with an apparent quality of level 5, both of these also with an influence of level 1, and one of these moves is good. Moves at higher influence levels emerge as the apparent quality drops, and for each level of apparent quality, the ratio of available moves to good moves becomes more attractive as the influence level grows. Thus, the ratios improve (roughly) by moving to the right in a given row and by moving upward in a given column, excluding consideration of empty cells. (The table has a few exceptions to this general pattern.)

Consideration of Table 5.1 suggests that the isovalue contours of Figures 5.1 and 5.2 are not precise but fuzzy, and take the form of "probability contours". The ability to develop information of the type depicted in these illustrations gives a useful basis for choice rules in complex decision environments. (See Exercises 1 to 5 at the end of the chapter.)

There are two pitfalls to be avoided. A common mistake made in diversification strategies is to overlook the need for diversifying steps that are *mutually compatible* and thus which do not propel a solution into an unproductive region. This is typically reflected in the fact that once a large distance move is made, the tradeoffs embodied in selecting influential moves change, so that a higher degree of quality must be demanded of a move at a given level of influence (or within a given distance class) in order to qualify the move as attractive. This consideration is illustrated by the change that leads from Figure 5.1 to Figure 5.2. Another common mistake is to overlook the phenomenon where some forms of diversifying moves require a series of simpler supporting moves before their effects can be reasonably determined. Without executing the supporting moves, an evaluation of a given diversifying move cannot be made intelligently. Often logical restructuring and look-ahead analysis is important to exploit this phenomenon, as illustrated in Chapter 3. The choice of a diversifying move should be deferred in such cases until several candidates have been subjected to such a look-ahead analysis.

Empirical studies are called for to identify the depth to which a look-ahead analysis should proceed, and to determine the number of candidates that should be subjected to such analysis in various settings. A strategy that allows previous solutions to be revisited if a threshold of quality is not soon achieved can serve as an approximate form of look-ahead.

Empirical studies are also called for to identify tradeoffs between quality and influence for particular problem classes and at particular stages of diversification (whether or not a look-ahead strategy is used). Such studies essentially would be designed to identify the structure of graphs and tables of the type illustrated in Figures 5.1, 5.2 and Table 5.1. (Deviations from the simple illustrative structures should be expected, and may usefully add to our knowledge of the relationship between these important factors.) Recency-based and frequency-based memory can be used to uncover and characterize situations in which evaluations for large distance moves should be preferable to those of smaller distance moves. The learning approach of target analysis, discussed in Chapter 9, has particular bearing on this issue.

5.2 The Principle of Persistent Attractiveness

A useful strategic idea is embodied in the Principle of Persistent Attractiveness, which may be expressed as follows. Attributes that persist in being attractive to bring into a solution, as a result of belonging to moves with high evaluations — but which rarely or never become incorporated into a solution, because the moves that would accomplish this do not have evaluations quite high enough evaluation to be chosen — should be given an inducement that upgrades the chance they will be selected. (A corresponding version of the principle applies to creating an inducement to remove attributes from solution that persist in being attractive to be dropped. The terminology "persistent attractiveness" carries with it the connotation also of being "persistently unselected.")

As in the case of diversification approaches, the application of this principle requires attention to the fact that attributes which have persisted in being attractive may not necessarily be attractive in combination. Hence, when such an attribute is brought into a solution (as a result of an inducement), care must be taken to avoid bringing in other such attributes that may cause a contrasting change in structure or a notable deterioration in solution quality. One way to handle this is by establishing a separate aspiration level that a given move must satisfy to be chosen, regardless of the amount that its evaluation is elevated by inducements. Similarly, each time an inducement causes an attractive attribute to be brought into solution, the inducements for other attractive (but unselected) attributes may appropriately be scaled back. The degree to which this occurs depends upon the level of diversification that is sought — where as customary, greater levels of diversification may be triggered by evidence that the search is stalled in a particular region or has failed to find improvement for a significant duration.

Issues that arise in identifying persistent attractiveness are illustrated in Table 5.2. Five different attributes with varying degrees of persistent attractiveness are shown, by tracking the number of times these attributes have been contained in the top 5 and the top 10 moves, throughout a succession of 200 iterations. (This iteration value is chosen purely for illustrative purposes.) We say an attribute is "contained" in (or occurs in) a move if executing the move would cause the attribute to become part of the resulting solution.

Table 5.2 Attributes contained in top moves, chosen and unchosen, over a span of 200 iterations.

	Times attribute appeared in		Times	Iterations
Attributes	Top 5 moves	Top 10 moves	selected	in solution
1	3	4	2	180
2	14	17	8	130
3	24	29	2	12
4	21	27	0	0
5	8	50	0	0

As shown by the table, each of the attributes 1, 2 and 3 have been selected at various times to become part of the solution, and have been maintained in the solution for various numbers of iterations over the 200 iteration sequence. For simplicity in our discussion, we suppose that an attribute can occur in only a single available move on any given iteration.

The attributes are ordered so that their persistent attractiveness increases as the index of the attribute increases. For example, during the 200 iteration history, attribute 1 occurred in 3 of the top 5 moves and 4 of the top 10 moves, and was also selected twice, which indicates that its level of (persistent) attractiveness is not very high. Moreover, this attribute was part of the current solution for 180 out of 200 iterations, so that the attribute can scarcely be considered to suffer from lack of inclusion. Attribute 2 occurred in 14 of the top 5 moves and in 17 of the top 10 moves, and was selected 8 times, which indicates a lower rate of success than attribute 1, as measured by the number of times the attribute was selected relative to the number of times it was contained in one of the higher evaluation moves. Also, this attribute was not contained in the current solutions for as many iterations as attribute 1. Nevertheless, attribute 2 also does not warrant a high measure of persistent attractiveness. Attribute 3, on the other hand, clearly deserves a higher measure than attributes 1 and 2.

Attributes 4 and 5 of Table 5.2, which were not chosen or maintained in the solution, and yet which occurred among the top 5 and top 10 moves with a frequency that rivals those of attributes which were brought into the solution, evidently qualify to receive a higher measure of persistent attractiveness than attributes 1, 2 and 3. The question of the relative (persistent) attractiveness of attributes 4 and 5 depends on whether appearing in the top 10 moves should be given more or less weight than appearing in the top 5 moves. This in turn depends on the effectiveness of the move evaluations, as reflected by their ability to differentiate promising from unpromising moves. (This ability also determines whether it may be preferable to consider instead, for example, the top 3 or top 7 moves.) Measures of persistent attractiveness using information such as that of Table 5.2 are considered in Exercises 6 to 8 at the end of the chapter.

5.3 The Principle of Persistent Voting

While persistent attractiveness refers to evaluations over a span of iterations, persistent voting refers to evaluations within a single iteration, but performed by multiple decision rules. As noted in Chapter 1, the idea of integrating the criteria of multiple rules, as suggested by the voting metaphor, provides a useful basis for creating better choices. One way to achieve this integration is by establishing a parametric weighting of functions that underlie different evaluations and searching over different parameter values to identify effective choices (as discussed in Section 1.8 of Chapter 1.). The voting perspective suggests a different type of scheme which is more closely related to using different rules to provide rankings for alternative choices. A simple way to analyze the possibilities provided by the combined use of such rules is shown in Table 5.3.

For illustrative purposes, we suppose that 4 different decision rules are used to provide evaluations, and that each rule is allowed to cast exactly one vote for its first choice, one vote for its second choice, and so on through its fifth choice. These five top-ranked choices are indicated under the heading "Voting Rank" in Table 5.3. The only moves included in the table are those that received at least 2 votes to be included in one of these top ranks, and we have arbitrarily labeled these moves by the indexes 1 through 6.

Table 5.3 Votes to place moves in top 5 ranks by 4 different decision rules.

	Voting Rank					*Total Votes —*
Move	1	2	3	4	5	*Top 5 Ranks*
1	2	0	0	0	0	2
2	1	1	0	0	2	4
3	1	0	0	1	0	2
4	0	3	0	1	0	4
5	0	0	2	0	0	2
6	0	0	1	1	1	3

For example, the column for the first choice by the rules (i.e., for the Voting Rank of 1) shows that 2 of the 4 rules voted for Move 1, while one rule voted for Move 2 and one voted for Move 3. The column for Voting Rank 3 only shows votes by three of the four rules, because one of the rules voted for a move that did not receive at least 2 votes in total among the top 5 ranks (and hence this move does not appear in the table).

The moves are ordered lexicographically in the table based on the votes received. Thus Move 1, which receives more Rank 1 votes than any other, appears first. Moves 2 and 3 receive the same number of Rank 1 votes, but Move 2 receives more Rank 2 votes, and so appears next. The lexicographic ordering gives one way of deciding which move ought to be selected by the votes. Alternatively, Move 2 might be preferred to Move 1 if a voting rank of 2 is not considered to be very different from a voting rank of 1, since Move 2 also receives 2 additional votes not received by Move 1. Similarly, on such a basis Move 4 might be preferred to both Move 1 and Move 2 (in which case it would clearly be preferred to all moves shown).

In common with the Principle of Persistent Attractiveness, the Principle of Persistent Voting is one that deserves fuller examination. Simple experimentation should easily identify options that offer initial promise, and provide a foundation for later refinement. Once again, the learning approach of target analysis discussed in Chapter 9 affords a more sophisticated way to take advantage of these principles.

5.4 Compound Move, Variable Depth and Ejection Chains

The issues of influence, and their relevance for combining the goals of intensification and diversification, do not simply concern isolated choices of moves with particular features, but rather involve coordinated choices of moves with interlinking properties. The theme of making such coordinated moves leads to consideration of *compound moves*, fabricated from a series of simpler components.

As mentioned in Section 1.8 of Chapter 1, procedures that incorporate compound moves are often called *variable depth methods*, based on the fact that the number of components of a compound move generally varies from step to step. One of the simpler approaches, for example, is to generate a string of component moves whose elements (such as edges in a graph or jobs in a schedule) are allowed to be used or "repositioned" only once. Then, when the string cannot be grown any larger, or deteriorates in quality below a certain limit, the best portion of the string (from the start to a selected end point) provides the compound move chosen to be executed. This simple design constitutes the usual conception of a variable depth strategy, but the TS perspective suggests the merit of a somewhat broader view, permitting the string to be generated by a more flexible process. For example, by using TS memory it is possible to avoid the narrowly constrained progression that disallows a particular type of element from being re-used.

Within the class of variable depth procedures, broadly defined, a special subclass called *ejection chain procedures* has recently proved useful. Early forms of ejection chain procedures are illustrated by alternating path methods for matching and degree-constrained problems in graph theory (Berge, 1962). A compound move in this setting, which consists of adding and dropping successive edges in an alternating path, not only has a variable depth but also exhibits another fundamental feature. Some components of the compound move create conditions of imbalance and infeasibility that must be "resolved" by other components. Accordingly, the move is generated by complementary stages that introduce certain elements and eject others. One step of the move creates a disturbance (such as violating a node degree by adding an edge) which must be removed by a complementary step (restoring the node balance by dropping an edge). Add /drop moves, under conditions where each add move creates a form of disruption or infeasibility that can only be restored by a limited number of drop move alternatives, provide a foundation for creating an ejection chain which can be conveniently illustrated by the Min k-Tree problem introduced in Chapter 2. An example is depicted in Figure 5.3.

Fig. 5.3 Simple ejection chains for a Min k-Tree problem instance (k = 3).

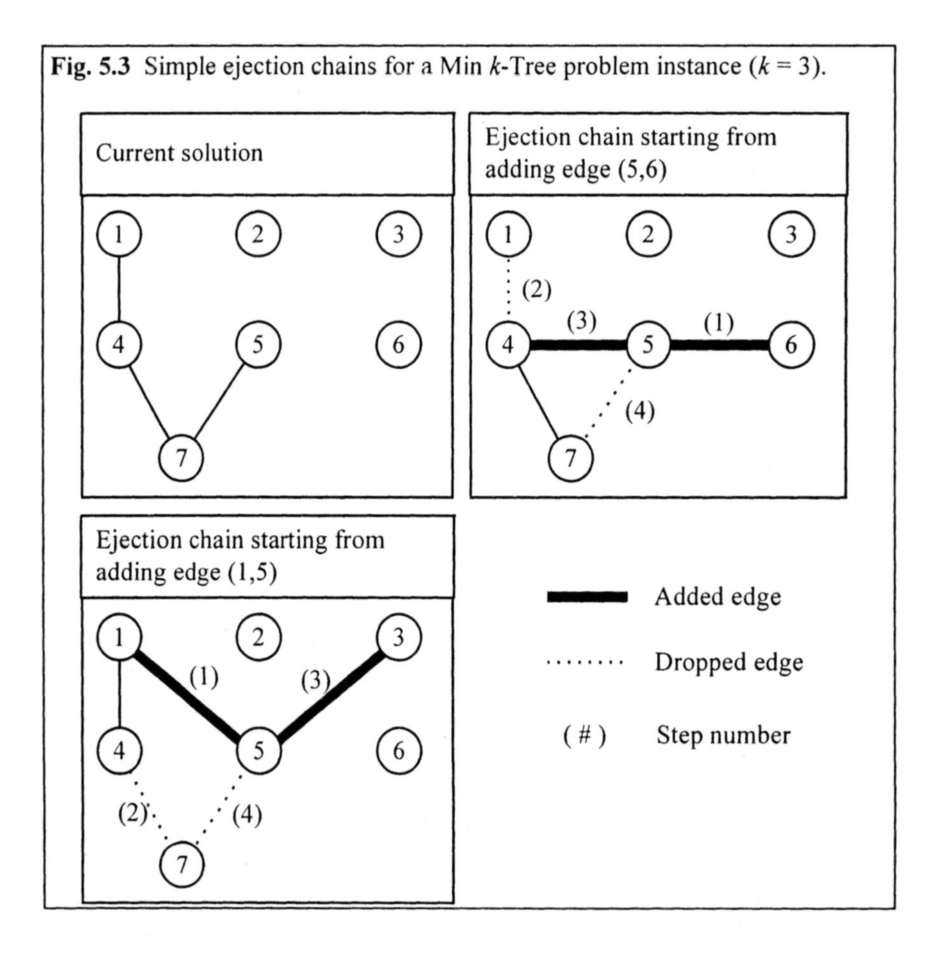

Figure 5.3 shows that in a hypothetical iteration, the current solution (for $k = 3$) consists of the edges (1,4), (4,7) and (5,7). The exploration for a compound move consisting of an ejection chain is launched first from a step that adds edge (5,6). This creates an imbalance (since the tree now has too many edges) and it becomes necessary to eject (here, "drop") an edge to restore the balance. Only a single edge can be chosen for this step (if the restoration seeks to produce a tree structure), and hence the ejection proceeds by dropping edge (1,4) in step (2), as shown. The next add step consists of introducing edge (4,5) at step (3), which permits only two responses for the ejection step in order to re-establish a tree, and in this case the process is concluded at step (4) by dropping edge (5,7). Similarly, a compound move of the same depth (i.e., 4 steps) can be constructed starting from the step that adds edge (1,5). The rule used to determine the stopping criterion during the construction of the compound moves evidently may yield moves of variable depth.

While Figure 5.3 illustrates some of the features of ejection chains, its constructions represent somewhat trivial instances of such chains. The reason is that these constructions afford nothing structurally that is not already provided by executing a normal sequence of add / drop moves. As an alternative, a sequential fan candidate list strategy, as discussed in Chapter 4, would achieve a variable depth analysis of such moves at a more sophisticated level than provided by simply generating a few alternatives and picking the best. (The standard variable depth paradigm, in spite of claims in the literature regarding its generality, only generates a single sequence of linked moves and only varies the choice of the point that terminates the sequence, within the boundaries currently explored). Ejection chains, by contrast, offer the ability to generate solutions that cannot be obtained by juxtaposing collections of standard moves, and set the stage for an analysis that isolates an effective sequence from which a subsequence can be selected.

A more advanced ejection chain construction for the Min k-Tree problem, for example, would apply add / drop moves to a *reference structure* which is more complex than the k-tree structure, but which is designed to allow the k-tree structure to be readily recovered at any step. A straightforward way to do this is to operate on trees of p edges, for $p > k$, with well-defined rules for isolating k-tree trial solutions. The outcome accordingly yields k-trees that could not be obtained (in the same order) by any normal succession of add / drop moves applied to the k-tree structure. The use of reference structures more generally can add considerable power to ejection chain approaches (Chapter 7).

Ejection chain approaches apply naturally to a variety of settings more elaborate than that of adding and dropping edges in graphs. A key principle is that a strategic collection of partial moves generates a critical (or fertile) condition to be exploited by an answering collection of other partial moves. Typically, as in the previous illustration, this occurs in stages that trigger the ejection of elements (or allocations, assignments, etc.) and hence reinforces the ejection chain terminology. In such cases, intermediate stages of construction fail to satisfy usual conditions of feasibility, such as fulfilling structural requirements in a graph or resource requirements in a schedule.

A number of processes that can be interpreted to involve an alteration between a critical condition and a triggered response are evidently somewhat different than ejection chain approaches. An example comes from network flows, in the classical out-of-kilter algorithm (Ford and Fulkerson, 1962). A linked sequence of probing and adjustment steps is executed until achieving a "breakthrough," which triggers a chain of flow changes, and this alternation is repeated until optimality is attained. In contrast to ejection chain approaches, however, such processes involve macro strategies rather than embedded strategies. More importantly, they do not encompass the freedom of choices for intermediate steps allowed in heuristic procedures. Finally, they do not involve special memory or probabilistic links between successive phases to overcome local optimality conditions when a compound move no longer generates an improvement. (The original characterization of variable depth methods also gives no provision for a means to proceed when a compound move fails to improve the current solution.)

Within the heuristic setting, ejection chain approaches have recently come to be applied with considerable success in several problem areas, such as generalized assignment, clustering, planar graph problems, traveling salesman problems and vehicle routing. (See, for example, Dorndorf and Pesch, 1994; Laguna et al. 1995; Pesch and Glover, 1995; Rego and Roucairol, 1996; Rego, 1996a, 1996b.) Such strategies for generating compound moves, coupled with TS processes both to control the construction of the moves and to guide the master procedure that incorporates them, offer a basis for many additional heuristics.

5.5 The Proximate Optimality Principle

In certain settings, search processes can acquire additional power by an implementation linked to a concept called the *proximate optimality principle* (POP). We have already discussed some of the features of the POP idea in association with strategic oscillation in Chapter 4, and we elaborate on these further here. The proximate optimality principle stipulates that good solutions at one level are likely to be found close to good solutions at an adjacent level, and it applies to both simple and compound moves. The term "level" can refer to a stage of a constructive or destructive process, or can refer to a given measure of distance from a specified boundary. The challenge is to define levels and moves that make this formulation usefully exploitable. An important part of the idea is the following intuition. In a constructive or destructive process — as in generating new starting solutions, or as in applying strategic oscillation — it can be highly worthwhile to seek improvements at a given level before going to the next level.

The basis for this intuition is as follows. Moves that involve (or can be interpreted as) passing from one level to another are based chiefly on knowledge about the solution and the level from which the move is initiated, but rely on an inadequate picture of interactions at the new level. Consequently, features can become incorporated into the solution being generated that introduce distortions or undesirable sub-assemblies. Moreover, if these are not rectified they can build on themselves — since each level sets the stage for the next, i.e., a wrong move at one level changes the identity of moves that look attractive at the next level. Consequently, there will be a tendency to make additional wrong moves, each one reinforcing those made earlier. Eventually, after several levels of such a process, there may be no way to alter earlier improper choices without greatly disrupting the entire construction. As a result, even the application of an improvement method to the resulting solution may find it very hard to correct for the previous bad decisions.

The POP notion is therefore exploited in tabu search by remaining at each successive level for a chosen number of iterations, and then restoring the best solution found to initiate a move to the next level. For example, a level may be characterized by compelling values of a function to fall in a given range, or by requiring a given number of elements to be included in a partial construction. Then the process can be controlled to remain at a specified level by employing a neighborhood whose moves maintain the functional values (or number of elements) within the specified limits.

In the context of the Min k-Tree Problem introduced in Chapter 2, the POP principle can be exploited by controlling the number of edges in the solution. The best solution at a level l (i.e., with $l \neq k$ edges) is found before starting the search at a different level. The search moves to a different level by performing a single add or delete move, and it is maintained at the same level by performing swap moves. The POP directly applies, for instance, to the Min k-Tree problem depicted in Figure 2.3 of Chapter 2. The optimal solution of this problem instance when $k = 3$ has a total weight of 25, given by the sum of the weights on edges (8,9), (8,10) and (9,12). Starting from the optimal solution for $k = 3$, the corresponding optimal solution for $k = 4$ can immediately be found by executing the move that adds the edge (10,11) with weight 9 for a total optimal weight of 34. This situation is shown in Figure 5.4.

Of course, it is fortuitous in this example that the optimal solution for $k = 4$ turns out to be adjacent to the optimal solution for $k = 3$, but the example serves to illustrate that "proximate connections" between good solutions can often occur.

The POP notion additionally supports the idea of applying restarting strategies and strategic oscillation approaches by pausing at periodic intervening levels of construction and destruction, in order to "clean up" the solution at those levels. Such an approach is not limited in application to constructive and destructive processes, of course, but can also be applied to other forms of strategic oscillation. Further, the process of "pausing" at a particular level can consist of performing a tight series of strategic oscillations at this level.

Fig. 5.4 An illustration of POP.

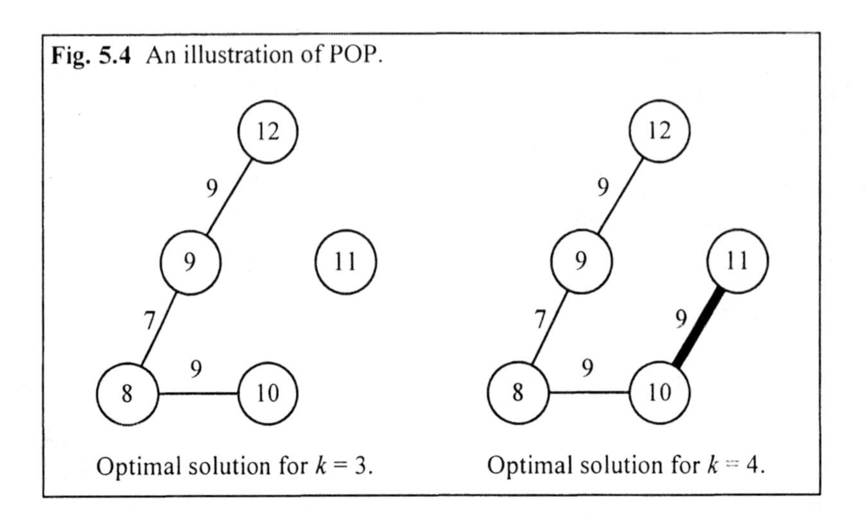

To date, there do not seem to be any studies that have examined this type of approach conscientiously, to answer questions such as: (a) how often (at what levels) should clean up efforts be applied? (b) how much work should be devoted at different levels? (Presumably, if a clean up phase is applied at every level, then less total work may be needed because the result at the start of these levels will already be close to what is desired. On the other hand, the resulting solutions may become "too tightly improved," contrary to the notion of congenial structures discussed in Section 5.6.) (c) how can "attractiveness" be appropriately measured at a given level, since the solution is not yet complete? (d) what memory is useful when repeated re-starts or repeated oscillation waves are used, to help guide the process? (e) what role should probabilities have in these decisions? (f) is it valuable to carry not just one but a collection of several good solutions forward at each step, as in a sequential fan candidate list strategy? (An interesting question arises in a parallel application, related to the sequential fan candidate list strategy: what kinds of diversity among solutions at a given level are desirable as a base for going to the next level?)

Answers to the foregoing questions are relevant for providing improved procedures for problems in scheduling, graph partitioning, maximum weighted cliques, p-median applications and many others.

Interesting similarities and contrasts exist between the POP concept of "level" and the simulated annealing notion of "temperature." Each refers to a succession of adjacent states. However, temperature is a measure of energy, as reflected in an objective function change, and the SA mechanism for incorporating this measure is to bias acceptance criteria to favor moves that limit "negative change," according to the temperature value. The SA mechanism also does not seek to maintain a construction at a given level within a design for isolating and carrying forward best solutions to an adjacent level. The types of levels encompassed by the POP notion cover a wide range of strategic alternatives (by comparison to temperature, for example), as derived in reference to numbers of variables or constraints, parametric representations of costs or requirements, hierarchies of aggregation or disaggregation, and so forth.

The POP concept may be viewed as a heuristic counterpart of the so called Principle of Optimality in dynamic programming. However, it does not entail the associated curse of dimensionality manifested in the explosion of state variables that normally occurs when dynamic programming is applied to combinatorial problems. If the premise underlying this concept is valid, it suggests that systematic exploration of effective definitions of "levels" and "closeness", and the design of move neighborhoods for exploring them, may disclose classes of strategies that usefully enlarge the options currently employed.

We note a final caveat in considering ways to exploit the POP notion. Proximity should be applied, as in the vocabulary building framework, to good components of solutions as well as to good solutions. Moreover, measures of closeness should not be restricted to settings determined by commonly used neighborhoods for moving from a solution to another. The Min k-Tree problem provides a useful example. In the case of the "difficult" problem of Exercise 9 in Chapter 3, the distance between good solutions and the optimal solution is substantial when measured by number of swap moves (particularly if these moves are prioritized by apparent attractiveness). Nevertheless, upon isolating candidates for good solution components, an optimal solution is found by generating a shortest path between these components. Within the domain of NP-hard problems, it must be expected that counterexamples can be found to every exploitation strategy that can be devised, but it may also be anticipated that an appropriate characterization of proximity for a given class of problems can lead to strategies that are useful across a wide range of instances.

5.6 The Principle of Congenial Structures

An important supplement to the POP notion is provided by the Principle of Congenial Structures. The key idea is that there often exist particular types of solution structures that provide *greater accessibility to solutions of highest quality* — and, as an accompaniment, there also frequently exist special evaluation functions (or "auxiliary objective functions") that can guide a search process to produce solutions with these structures.

This principle is illustrated by an application of tabu search in work force scheduling (Glover and McMillan, 1986), where improved solutions were found by modifying a standard objective function evaluation to include a "smoothing" evaluation. The smoothing evaluation in this case was allowed to dominate during early-to-middle phases of generating starting solutions, and then was gradually phased out. However, the objective function itself was also modified by replacing an original linear formulation with a quadratic formulation (in particular, replacing absolute deviations from targets by squared deviations). The use of quadratic evaluations reinforced the "smoothness" structure in this setting and, contrary to conventional expectation, produced solutions generally better for the linear objective than those obtained when this objective was used as an evaluation function.

A more recent application disclosing the importance of congenial structures occurred in multiprocessor scheduling (Hubscher and Glover, 1994). The notion of a congenial structure in this instance was used to guide phases of *influential diversification*, which made it possible to effectively "unlock" structures that hindered the ability to find better solutions, with the result of ultimately providing improved outcomes.

In the context of the Min k-Tree problem, a modified evaluation for "dynamic" moves (i.e., those that results in replacing a node i of the tree by a new node j) may include the net degrees of the nodes participating in the exchange. We refer to the net degree $\delta(i)$ of a node i as the number of edges connected to i that are not connected to any other node in the current partial tree that excludes i and j. Then, a modified evaluation for a move that drops edge (i,i') and adds edge (j,j'), where i and j' are nodes in the current tree, may be formulated as:

$$MoveValue = w(j,j') - w(i,i') + \alpha\,(\delta(i) - \delta(j))$$

where $w(i,j)$ is the weight associated with edge (i,j) and α is an inducement factor. The value of α may be set to zero when $w(j,j') - w(i,i') < 0$ and to a value strictly greater than zero otherwise. This guarantees that the most improving move is always selected, if at least one improving move is available. This evaluation gives an incentive to nodes with high degree, even if the connecting edge to the current tree has a relatively large weight. To illustrate this, consider the solution that consists of the edges (1,2), (1,4), (3,4) and (3,6) for the Min k-Tree problem in Figure 2.3 of Chapter 2. This solution has a total weight of 63. Suppose that two dynamic moves are being examined: Move A that drops edge (1,2) and adds edge (6,7) and Move B that also drops edge (1,2) but adds edge (6,9). Both of these moves are nonimproving, and Move A has a better evaluation than Move B when α is set to zero. However, if $\alpha = 3$ the move values are:

$$MoveValue(\mathrm{A}) = 6 - 1 + (0 - 1) = 4$$
$$MoveValue(\mathrm{B}) = 16 - 1 + 3\,(0 - 4) = 3$$

and Move B is the preferred move. Note that $\delta(2) = 0$ because the edges (1,2) and (2,3) that connect to node 2 are also connected to nodes 1 and 3, which are part of the current partial tree. Similarly, $\delta(7) = 1$ because the edge (7,8) cannot be reached from any node in the current partial tree. The inclusion of node 9 as a result of executing Move B gives access to the edges (5,9), (8,9), (9,11) and (9,12) and therefore the net degree $\delta(9)$ for this node equals 4. The "connectivity factor" $\delta(\bullet)$ can of course be defined in more subtle ways, but the use of such factor illustrates how congenial structures can be isolated to provide better access to improved solutions.

This issue of appropriately characterizing the nature of congenial structures for different problem settings, and of identifying evaluation functions (and associated procedures) to realize these structures, deserves fuller attention. Specific aspects of this issue are examined next.

5.6.1 Congenial Structures Based on Influence.

The influence concept can play an important role in identifying (and creating) congenial structures. This concept is manifested in a number of settings where solution components can be viewed as falling roughly into two categories, consisting of *foundation components* and *crack fillers* (see Section 4.1 of Chapter 4). The crack fillers are those that can relatively easily be handled (such as jobs that are easily assigned good positions or variables that are easily assigned good values) once an appropriate way of treating foundation components has been determined.

Typically, crack fillers represent components of relatively small "sizes," such as elements with small weights in bin packing problems, edges with small lengths in routing problems, jobs with small processing times in scheduling problems, variables with small constraint coefficients in knapsack problems, etc. (There are clear exceptions, as in the example problem of Exercise 9 in Chapter 3, where the two smallest weight edges are "strongly determined" components of good solutions.) Hypothetically, an approach that first focuses on creating good (or balanced) assignments of foundation elements, as by biasing moves in favor of those that introduce larger elements into the solution, affords an improved likelihood of generating a congenial structure. For example, among competing exchange moves within a given interval of objective function change, those that involve larger elements (or that bring such elements into the solution), may be considered preferable during phases that seek productive forms of diversity. Such moves tend to establish structures that allow more effective "endgames," which are played by assigning or redistributing the crack fillers. (The periodic endgames arise figuratively in extended search with transition neighborhoods, and arise literally in multistart methods and strategic oscillation.)

Approaches of this type, which provide a simple approximation to methods that seek to characterize congenial structures in more advanced ways, have some appeal due to their relatively straightforward design. Such approaches, for example, can provide a starting point for exploiting relationships such as those discussed in Section 5.1. As a particularly simple illustration, if an improving move exists, choices may be restricted to selecting such a move with a greatest level of influence. More generally, a set of thresholds can be introduced, each representing an interval of evaluations. Then a move of greatest influence can be selected from those that lie in the highest nonempty evaluation interval. Such approaches motivate a quest for appropriate thresholds of objective function change versus influence change, particularly in different regions or phases of search. Studies that establish such thresholds can make a valuable contribution.

5.6.2 Congenial Structures Based on Improving Signatures.

Another way to generate congenial structures arises by making use of an *improving signature of a solution*. This approach has particular application to searches that are organized as a series of improving phases that terminate in local optimality, coupled with intervening phases that drive the search to new vantage points from which to initiate such phases. (The improving phases can be as simple as local search procedures, or can consist of tabu search methods that use aspiration criteria to permit each sequence of improving moves to reach a local optimum.)

As a first approximation, we may conceive the improving signature **IS**(x) of a solution x to be the number of solutions $x' \in \mathbf{N}(x)$ that are better than x, i.e., that yield $f(x') > f(x)$ for a maximization objective. We conjecture that, in the process of tracing an improving path from x, the probability of reaching a solution significantly better than x is a function of **IS**(x). More precisely, the probability of finding a (near) global optimum on an improving path from x is a function of **IS**(x) and the objective function value $f(x)$. (Our comments are intended to apply to "typical" search spaces, since it is clearly possible to identify spaces where such a relationship does not hold.)

An evident refinement occurs by stipulating that the probability of finding a global optimum depends on the distribution of $f(x')$ as x' ranges over the improving solutions in **N**(x). Additional refinements result by incorporating deeper look-ahead information, as from a sequential fan candidate list strategy. From a practical standpoint, we stipulate that the definition of the improving signature **IS**(x) should be based on the level of refinement that is convenient in a given context.

With this practical orientation, the first observation is that **N**(x) may be too large to allow all its improving solutions to be routinely identified. Consequently, we immediately replace **N**(x) by a subset **C**(x) determined by a suitable candidate list strategy, and define **IS**(x) relative to **C**(x). If we restrict **C**(x) to contain only improving solutions, this requires identifying values (or bounds on) $f(x')$ for $x' \in \mathbf{C}(x)$. Consequently, in such candidate list approaches, knowledge of these values (as well as the size of **C**(x)) is automatically available as a basis for characterizing **IS**(x).

We follow the convention that larger values of **IS**(x) are those we associate with higher probabilities of reaching a global optimum from x. (Hence for example, in a maximization setting, **IS**(x) may be expressed as an increasing function of $f(x)$ and the size of **C**(x), or as a weighted sum of the values $f(x') - f(x)$ for $x' \in \mathbf{C}(x)$.) By this design, explicit TS memory can be used to keep a record of solutions with largest **IS**(x) values, permitting these solutions to be used as a basis for launching additional improving searches. Attributive TS memory (or probabilistic TS rules) can then be applied, as an accompaniment, to induce appropriate variation in the paths examined. Approaches based on these notions are directly relevant to the Pyramid Principle, discussed next.

5.7 The Pyramid Principle

A natural goal of search is to maximize the percentage of time devoted to exploring the most profitable terrain. Our lack of prior knowledge about the features of such terrain leads to *diffused* strategies such as those of simulated annealing, which by design spend large periods of time in unattractive regions, and such as those of random restarting which (also by design) "aimlessly" jump to a new point after each improving phase is completed.

A somewhat different type of strategy is motivated by the Pyramid Principle of improving search paths, which rests on the following observation. Consider an arbitrary improving path from a given starting solution to a local optimum. As the search gets closer to the local optimum, the tributaries that provide improving paths to other local optima become fewer, since all such paths at any given level are contained among those at each level farther from the local optimum.

To formulate this observation and its consequences more precisely, let **LO**(x) denote the set of local optima that can be reached on the union of all improving paths starting from x. Also let **IN**(x) the *improving neighborhood* of x, i.e., $\mathbf{IN}(x) = \{x' \in \mathbf{N}(x): f(x') > f(x)\}$ (assuming a maximization objective). Finally, let **IP**(x) denote the collection of improving paths from x to a local optimum. Then **LO**(x) is the union of the sets **LO**(x') for $x' \in$ **IN**(x), and **IP**(x) is the union of the sets **IP**(x'), each augmented by the link from x to x', for $x' \in$ **IN**(x'). Further,

$$|\mathbf{LO}(x)| \leq \sum_{x' \in \mathbf{IN}(x)} |\mathbf{LO}(x')|,$$

$$|\mathbf{IP}(x)| \geq \sum_{x' \in \mathbf{IN}(x)} |\mathbf{LO}(x')|.$$

The second of these inequalities is strict whenever the first is strict (which usually may be expected), and in general, the number of elements of **IP**(x) can be greatly larger than that of **LO**(x), a discrepancy that grows the farther x is from any given local optimum.

The relevant relationships are completed by defining the *length* of an improving path to be the number of its links, the *distance* $D(x, x'')$ from x to a local optimum x'' to be the length of the longest improving path from x to x'' (under conditions where at least one such path exists), and the *level* $LEV(x)$, to be the greatest distance $D(x, x'')$ as x'' ranges over all local optima accessible by an improving path from x. Then

$LEV(x) = 1 + \underset{x \in \mathbf{IN}(x)}{Max} \{LEV(x')\}$, and for any given improving path, the value $f(x)$ is strictly decreasing function of $LEV(x)$. In addition, $|\mathbf{IP}(x)|$ is nondecreasing function of $LEV(x)$ and a nonincreasing function of $f(x)$.

The Pyramid Principle then can be expressed by saying that the total number of improving paths decreases as $f(x)$ moves closer to a global optimum. If we view the number of such paths as the width of a band that corresponds to different intervals of $f(x)$ values, the band becomes progressively narrower as $f(x)$ approaches its global maximum, hence roughly resembling the shape of a pyramid. Adopting the convention that $\mathbf{IP}(x) = \{x^*\}$ for each globally optimal solution x, where x^* is a dummy "root" solution for these global optima, the apex of the pyramid consists of the point x^*. For many search spaces (such as those with moderate connectivity, and where $f(x)$ takes on multiple values), the rate at which the pyramid narrows as $f(x)$ grows can be dramatic.

Mild assumptions about the structure of improving paths causes this pyramid to generate an analogous pyramidal shape, but inverted, for the probability of finding improving paths to a global optimum as $f(x)$ increases. (The base of the inverted pyramid corresponds to the point where $f(x)$ achieves its maximum value, and the width of this base corresponds to the maximum probability of 1.) Thus, for example, if the size of $\mathbf{IN}(x)$ is approximately randomly distributed, or falls randomly within a particular range for each x at any given $f(x)$ value, then the inverted pyramid structure may be expected to emerge. Under such circumstances, the search can be significantly more productive by a strategy that undertakes to "keep close" to the globally optimum value of $f(x)$. (Such a strategy, of course, stands in marked contrast to the strategies of simulated annealing and random restarting.)

The foregoing analysis is somewhat pessimistic, and potentially myopic, for it implicitly supposes no information exists to gauge the merit of any improving move relative to any other (starting from given level of $f(x)$). Hence, according to its assumptions, all improving moves from a "current" solution x should be given the same evaluation and the same probability of selection. However, it is reasonable to expect that the search space is not so devoid of information, and better strategies can be designed if a sensible means can be identified to extract such information. In particular, the Pyramid Principle is likely to benefit significantly when applied together with the Principle of Congenial Structures. In combination these two principles clearly have implications for the design of parallel processing strategies. Additional considerations relevant to such strategies derive from probabilistic TS implementations, examined in Chapter 7.

5.8 The Space/Time Principle

The Space/Time Principle is based on the observation that the manner in which space is searched should affect the measure of time. This principle depends on the connectivity of neighborhood space, and more precisely on the connectivity of regions that successively become the focus of the search effort.

The idea can be illustrated by considering the hypothetical use of a simple TS approach for the traveling salesman problem, which is restricted to relying on a short term recency-based memory while applying a candidate list strategy that successively looks at different tour segments. (A "segment" may include more that one subpath of the tour and its composition may vary systematically or probabilistically.) In this approach, it may well be that the search activity will stay away from a particular portion of a tour for an extended duration — that is, once a move has been made, the search can become focused for a period in regions that lie beyond the *sphere of influence* of that move (i.e., regions that have no effective interaction with the move). Then, when the search comes back to a region within the move's sphere of influence, the tabu tenure associated with the move may have expired! Accordingly, since no changes have occurred in this region, the moves that were blocked by this tabu tenure become available (as if they had never been forbidden). Under such circumstances, the recency-based tabu memory evidently becomes ineffective. It is not hard to see that more complex scenarios can likewise exert an erratic influence on memory, creating effects that similarly distort its function and decrease its effectiveness.

In problem settings like that of the TSP, where a form of *spatial decomposition* or *loose coupling* may accompany certain natural search strategies, the foregoing observations suggest that measures of time and space should be interrelated. This space/time dependency has two aspects: (1) The clock should only "tick" for a particular solution attribute if changes occur that affect the attribute. (2) On a larger scale, longer term forms of memory are required to bridge events dispersed in time and space. This includes explicit memory that does not simply record best solutions, but also records best "partial" (or regional) solutions.

For example, in the TSP, after obtaining a good local optimum for a tour segment that spans a particular region, continuing the procedure may produce a current solution (tour segment) that is in that region. Then, when improvement is subsequently obtained in another region, the current solution over the full tour — which includes the current partial solution for the first region — is not as good as otherwise would be possible. (This graphically shows the defect of considering only short term memory and of ignoring compound attributes.)

This same sort of "loose coupling" has been observed in forestry problems by Lokketangen and Hasle (1997) who propose similar policies for handling it. Quite likely this structure is characteristic of many large problems, and gains may be expected by recognizing and taking advantage of it.

5.9 Discussion Questions and Exercises

Quality/Influence Tradeoffs (Exercises 1 - 5).

1. Consider the isovalue contours of effective quality shown in Figures 5.1 and 5.2. Suppose that an optimal solution to a given problem is known beforehand.

   (a) What types of experiments could be conducted to identify an approximate form for such isovalue contours for that problem, using particular measures of (apparent) quality and influence?
   (b) How could you use frequency memory to provide a "historical measure" of influence? Discuss the relevance of the fact that this type of influence can change, depending on the choices that may be made to respond to it.

2. By extension of Exercise 1, how would you take account of the fact that the shape of the isovalue contours is likely to change over time, not only as a result of historical influence measures (as in part (b) of Exercise 1) but also as a result of different choice rules that respond to current estimates of tradeoffs between quality and influence (as illustrated by the difference between Figures 5.1 and 5.2). In particular, consider differences in choice rules that are triggered according to whether the current search appears productive or unproductive.

3. In anticipation of applying different search phases, which use different rules for exploiting quality/influence tradeoffs, discuss why it may not be necessary to identify clear or "meaningful" isovalue contours at many points of the search, but only at certain key points.

   (a) Indicate how the experiments developed in Exercises 1 and 2 might be designed to help identify such key points.
   (b) Describe how you might seek to determine a "probabilistic contour" based on expected values, or on expected values and variances, for different phases of search.
   (c) How might experiments be designed to test alternative measures of quality and influence, as a basis for finding measures that yield more usefully exploitable information? (Assume a small set of problems is available which is highly representative of the class of problems to be solved, and an optimal solution is known for each.)

4. Discuss how the experiments considered in the preceding exercises might instead be adapted to generate information of the type shown in Table 5.1.

   (a) Would information such as that of Table 5.1 be more highly exploitable than the information from an effort to generate isovalue contours? Would such information be more convenient to generate?
   (b) Discuss the relevance of the fact that the information in Table 5.1 already is anticipated to deviate to some extent from an "idealized" form.

(c) How would you modify the table, or modify the analysis of the table, in the case where an attribute can appear in more than one move?

5. The structure of the information recorded in a series of tables such as Table 5.1 may harbor a different pattern of tradeoffs between quality and influence than the pattern illustrated by Table 5.1.

   (a) Do you expect that different patterns (and patterns of greater or lesser consistency) might be generated by different measures of quality and influence?
   (b) Since a given pattern does not need to take the form illustrated in Table 5.1 to be exploitable, does this further suggest that generating such tables would be preferable to trying to map isovalue contours? (Consider that isovalue contours need not conform to the pattern illustrated in Figures 5.1 and 5.2 to be exploitable. Since a good deal of economic analysis is based on the consideration of isovalue contours, does this affect your conclusions?)

Persistent Attractiveness (Exercises 6 - 8).

6. Consider an expanded version of Table 5.2 which appears as Table 5.4 below, in which three additional attributes (6, 7 and 8) are included.

Table 5.4 Attributes contained in top moves, chosen and unchosen (over a span of 200 iterations).

Attributes	Times attribute appeared in Top 5 moves	Times attribute appeared in Top 10 moves	Times selected	Iterations in solution
1	3	4	2	180
2	14	17	8	130
3	24	29	2	12
4	21	27	0	0
5	8	50	0	0
6	58	77	3	36
7	43	68	0	0
8	6	92	0	0

Discuss how you might evaluate the persistent attractiveness of the attributes 6, 7 and 8 relative to the other attributes.

7. Describe tests that might be performed to determine how to take advantage of the Principle of Persistent Attractiveness. In particular, consider the potential usefulness of creating a table that indicates the number of times that attributes occur in the top k moves for each value of k from 1 to k_{max} (where, for example, $k_{max} = 20$)? (Discuss the relevance of using such a table as a basis for

identifying a smaller table to be used as part of a routine problem solving procedure.)

8. How might the tests of Exercise 7 be amended to account for the effects of different phases of search, as where some phases more strongly emphasize intensification and others more strongly emphasize diversification? Should different sets of iterations spanned by different phases be used to construct different tables such as Table 5.2, which are "phase-specific"? Discuss the potential relevance of maintaining more than one such table, spanning iterations not only of a current phase but also of other intervals (e.g., including an interval that covers all iterations of the search).

Persistent Voting (Exercises 9 - 10).

9. By analogy with Exercise 7, what experiments might be performed to test how to take advantage of the Persistent Voting Principle? Formulate your response to include consideration of different search phases, as discussed in Exercise 8.

10. How would you modify the experiments of Exercise 9 to examine alternatives for allowing different choice rules to have different numbers of votes (including fractional ones)? Consider how differing numbers of votes (or equivalently, different weightings of the votes) might be changed in different search phases.

Parallel Processing and Experimental Design (Exercises 11 - 13).

11. Discuss how each of the key concepts of the preceding exercises (quality/influence tradeoffs, persistent attractiveness and persistent voting) can be exploited by parallel processing.

    (a) How can time-dependent considerations such as changing tradeoffs be partly translated into space-dependent considerations by applying different response strategies on different processors?
    (b) What advantages may parallel processing offer through an ability to simultaneously apply different measures of quality and influence and different weights for voting on preferred outcomes? (Consider how the voting metaphor can be shifted from applying to the search from a given processor to apply to integrating the searches of multiple processors.)

12. In each of the exercises that ask for designing experiments, what value do you think might attach to knowing optimal solutions in advance? What differences would be entailed in the types of experiments that would be performed? (These considerations are basic issues of the target analysis learning approach.)

13. Identify one or more experiments you might perform to test each of the following principles, and to determine how you might take advantage of them.

Identify how the tests may overlap, so that the same test may provide information about more than one principle. Also, in each case, discuss the effect that knowing an optimal solution in advance may have on the nature of the tests that are designed (as in Exercise 12).

(a) the Proximate Optimality Principle
(b) the Principle of Congenial Structures (based on influence, using thresholds, and on improving signatures, keeping track of **IS**(x) values).
(c) the Pyramid Principle
(d) the Space/Time Principle

6 TABU SEARCH IN INTEGER PROGRAMMING

We have already observed that tabu search can be used to modify decision rules for basis exchange (pivoting) methods, and therefore can be applied to solving mixed zero-one integer programming problems and various nonlinear programming problems by this means. Such applications exploit the fact that optimal solutions to these special zero-one discrete and nonlinear problems can be found at extreme points, and that pivoting methods provide a mechanism to move from one extreme point to another. In addition, however, tabu search can also be applied to solve integer and nonlinear programming problems in other ways. In this section we focus on integer programming (IP) problems, including *mixed* integer programming (MIP) problems where some variables can take on continuous values, as well as *pure* IP problems, where all variables must receive integer values.

The most commonly used methods for integer programming problems, which are incorporated into commercial software, are *branch and cut* procedures based on branch and bound approaches augmented by the use of cutting planes. The fundamental branching moves, which assign different values or bounds to integer variables — to generate alternatives that arise in the tree search strategies of branch and bound — can also readily be embedded in tabu search. The outcome of this embedding yields trajectories that are not possible within the tree search framework. When TS is used to guide branching strategies, it becomes natural to include variations of branching moves that are not considered in usual branch and bound methods.

Similarly, in common with branch and cut procedures, tabu search can be used to take advantage of cutting planes and to guide their application. TS can also be used to incorporate "false" cutting planes within the search, treating them as tabu restrictions over intervals devoted to searching specific regions. The possibilities for integrating tabu search with branching and cutting plane processes are made additionally cogent because TS provides a direct way to exploit theorems that interrelate search methods with procedures that generate valid and conditionally valid inequalities.

At present, these multiple ways of applying tabu search to solve IP problems are largely unexplored. This may be due to the fact that practical special cases of integer programs can be handled expediently by tailored TS implementations. As a result, there is not a widespread need to create general purpose implementations. (It is often true that a tailored implementation can be made more effective than a general one.) Nevertheless, general purpose methods have substantial value and consequently we are motivated to examine some of the basic opportunities for using TS to design them (Glover and Laguna, 1997b, 1997c).

We first consider an approach for guiding branching decisions in a "TS Branching" procedure, which can be used either as an alternative to branch and bound or as a supplement to it. (In a supplementary role, for example, such an approach can be used to generate trial solutions with the goal of enabling a branch and bound approach to find an optimal solution at an earlier stage.)

6.1 A Tabu Branching Method

The branching moves typically used in general purpose methods for pure and mixed integer programming arise by solving a linear programming relaxation where all variables are allowed to be continuous. An integer variable x_j that receives a fractional (non-integer) value in a solution to such a relaxed problem then gives rise to the two branching possibilities expressed by the logical disjunction:

$$x_j \leq v \text{ or } x_j \geq v + 1,$$

where v and $v + 1$ are the integers that bracket the current fractional value of x_j. The branch and bound approach makes use of these two possibilities to generate corresponding branches of a logical decision tree. At each node of the tree, an analysis is performed to identify which variable x_j and which branching alternative for the variable will be selected for immediate exploration. The imposition of the branching inequality then generates a more constrained linear program which is solved to continue the process, while the alternative branch is saved to be explored later. Various strategies and auxiliary procedures for generating and exploring such a decision tree are described in Nemhauser and Wolsey (1988), Parker and Rardin (1988), Savelsbergh (1994), Magee and Glover (1996), Beasley (1996). (Also, see Section 6.5 for an alternative perspective on creating the branches that such procedures include in their repertoire.)

Tabu search can make use of the same techniques for choosing preferable branching alternatives but is not compelled to rely on the strict progression of decisions that branch and bound must follow to exhaustively explore, or implicitly rule out, all nodes of the search tree. In particular, a tabu search based method does not have to revert to "backtracking" or "jump tracking" to resume the search from earlier stages of the tree, but can control the search to establish a favorable tradeoff between factors such as the quality of information generated and the effort required to exploit this information.

The following types of moves are relevant to a tabu search based approach:

Restriction: impose a branch of the form $x_j \leq v$ or $x_j \geq v + 1$
Relaxation: relax (undo) a branch previously imposed
Reversal: impose a branch that complements a branch previously imposed (replacing $x_j \leq v$ by $x_j \geq v + 1$, or vice versa).

This set of moves provides options that differ slightly from those customarily available to branch and bound, and the alternatives are managed by different rules. When these moves are guided by tabu search, there is no need to "go back" to an earlier stage (of an implicit tree) to determine available alternatives at various junctures of the search. However, a strategic oscillation design may well go back by "undoing" moves in a pattern that can not be duplicated by branch and bound (since there is no requirement to recover a previously generated node of a tree).

To clarify the relationship of possibilities that exists between tree search methods and TS, it may be noted that branch and bound (or branch and cut) can only implement a reversal move in the very limited circumstance where the reversed branch meets one of the leaves (current end nodes) of the decision tree. (All other reversals require branch choices following the complemented branch to be dropped, to create a sequence that currently terminates with the complemented branch. See Exercises 1 and 2 at the end of this Chapter.) Tabu search, by contrast, can reverse a branch regardless of the stage it was imposed. In this sense, TS is able to maintain the search process at (or close to) a "full resolution level" of the tree — i.e., at a level where the branches yield integer values for all integer variables (or drive the solution infeasible, identifying the need for a relaxation or reversal). This may be viewed as a process whereby the tree is allowed to reconstitute itself, by reshuffling the order of decisions that lead to a current leaf node.

Figure 6.1 shows the three types of tabu branching moves from a tree perspective. The branches depicted in this figure are represented sequentially, as might be done in a last-in-first-out (LIFO or backtracking) representation, where only the current branch sequence is maintained (along with memory of branches in this sequence whose complements have already been examined). For simplicity, the variables are indexed to correspond to the sequence of the branches.

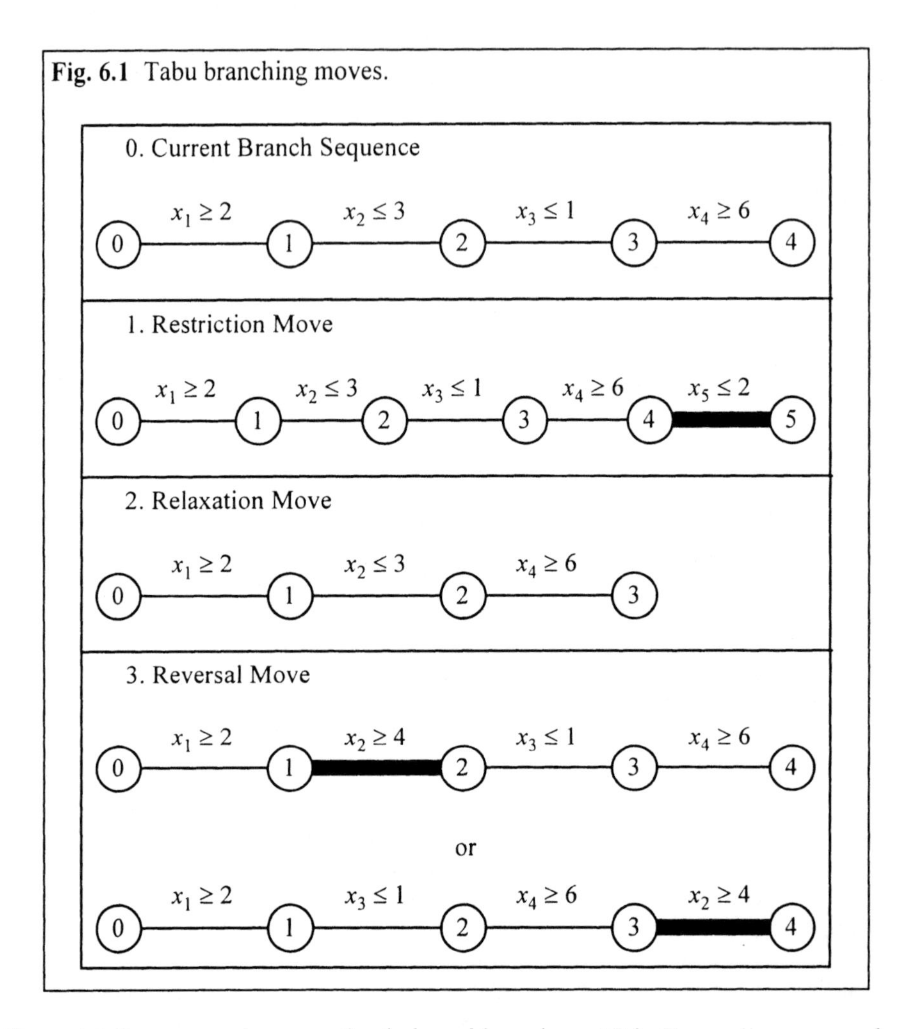

Fig. 6.1 Tabu branching moves.

The restriction move, shown as the darkened branch $x_5 \leq 2$ in Frame 1, corresponds exactly to a type of move that would be employed in branch and bound. The relaxation move, shown in Frame 2, collapses the current set of branches by removing one, in this case $x_3 \leq 1$, and has no precise counterpart in branch and bound. The reversal move, shown in Frame 3, likewise has no precise branch and bound counterpart. This move is represented in two ways, first by leaving the reversed branch in place, and second by shifting the branch to the end of the sequence. Relative to the use of recency based memory, the latter representation is more nearly accurate. However, since tabu tenures and aspiration criteria may not entail that the "latest move" is the last to be free of tabu restrictions, clearly there is no precise meaning to a sequential representation of a tabu branching procedure, in the contrast to the case for branch and bound. The ability to make decisions that are not constrained by sequence, in the more limited sense that sequence is interpreted in branch and bound, is an important characteristic of TS in this setting. Figure 6.1 also makes clear that the tabu branching moves can be employed in a manner to keep the branch depth (i.e., the number of branches currently imposed) close to a full resolution level, since none of the moves is compelled to revert to a somewhat shallower level as would be required by a tree search framework.

Viewing the branching process from a tree perspective, however, discloses another important contrast with branch and bound. The ability to conduct a large part of the search close to a full resolution level has implications for the quality of information used to make choices of new moves. The quality of bounds and other information that compose choices made at earlier levels of a tree is not as good (by the criteria of accuracy and reliability) as the quality available at deeper levels of the tree. Consequently a TS trajectory that chooses moves to stay close to a full resolution level generally provides better information at each step than is available to a tree search approach. Moreover, branch and bound must "live with" decisions made at earlier levels, as it explores descendants of the corresponding nodes.

There is a trade-off, however, because evaluations of choices at a deep level (to take advantage of options not available to branch and bound) can require greater effort than evaluations at earlier levels. Another "cost" of a more flexible framework, which can be quite significant for problems that are easy to solve by implicit enumeration, is that TS relinquishes the ability to categorically eliminate conditionally imposed branches once and for all. For this reason, approaches that marry a TS strategy with a branch and bound strategy provide useful possibilities for exploration.

The concept of the Proximate Optimality Principle can be applied in a TS method that uses branching moves to move away from a full resolution level, by a design that can make such movement appropriate. Both initially and during waves of strategic oscillation, the TS approach can interrupt the use of restriction moves (that build to a particular depth) or relaxation moves (that go back to a particular depth) to execute an extended series of reversal moves. Thus the method can stay at a given depth, which may be strategically varied, while seeking to improve the composition of alternatives selected. This modified composition at various depths can be used as a foundation for moving to other depths, following the types of possibilities for strategic oscillation already discussed.

The three types of tabu branching moves naturally invite the use of correspondingly different types of tabu tenure. For example, one type of tabu tenure will affect the period following a relaxation move before an associated restriction move can be made (i.e., a restriction move that duplicates the condition previously relaxed). Other move possibilities similarly generate other forms of tenure. Such tenures may not apply to all intervening iterations, however. The tenure to prevent a reversal move from being "re-reversed" (to impose the previous complementary bound condition) may be based, for example, only on iterations involving other reversal moves or restriction moves, but without counting iterations consisting of relaxation moves.

Tabu branching is also well suited to be used with path relinking. Starting from any base solution, partial or complete, the values of integer variables in a matched collection of one or more target solutions immediately imply a set of candidate branching moves to move the base solution "toward" the collection. These can include reversal moves relative to branches that determine the current base solution, and relaxation moves can also be appropriate at intermediate steps to aid in evaluating which branch from a set of candidates is preferable to impose next. It is generally desirable to keep the evolving base solution close to a full resolution level, and if necessary to impose selected additional branches to create trial solutions at that level, in order to generate a sequence that contains a high proportion of complete solutions along the path. As usual, aspiration criteria give a basis for divergences from the path, and for early termination to reinstate a guidance structure that is independent of the target solutions.

The areas described next provide additional possibilities for investigation, which can supplement the options made available by tabu branching.

6.2 Tabu Search and Cut Search

Several frameworks exist for generating valid cutting planes. We focus on one that is useful for establishing ties with tabu search, especially in conjunction with a tailored form of the scatter search approach described in Chapter 9. Once more, variants that incorporate elements of exact methods afford an opportunity to assure an optimal solution will be obtained in a finite number of steps. Our development is based on the *convexity cut* (or *intersection cut*) ideas that were first introduced in the context of concave programming by Tuy (1964), and in the context of integer programming by Balas (1971) and Young (1971). We follow the characterization of Glover (1972) which extends the original integer programming derivations to encompass general convex sets.

The convexity cut framework adopts a geometric (vector space) perspective. As in implementations of branch and bound, the natural setting for applying this approach is in conjunction with solution methods for linear programming — and extreme point methods in particular. Cutting planes may be visualized in this setting to arise by extending edges (extreme rays) of the polyhedral cone associated with a linear programming basis. (For convenience, we use the term cone to refer to what is sometimes called a half-cone.) The vertex of this cone, which we denote by $x(0)$, corresponds to an LP basic solution, and the operation of extending an edge from $x(0)$ corresponds to assigning positive values to a selected nonbasic variable, holding each of the other nonbasic variables at zero. More generally, in applying a bounded variable version of the simplex method, an edge is extended by assigning values to a parameter that either increases the nonbasic variable from its lower bound or decreases the variable from its upper bound.

An example of a two-dimensional LP polyhedral cone (that will serve as the basis for a later illustration) is shown in Figure 6.2.

Fig. 6.2 Example LP cone.

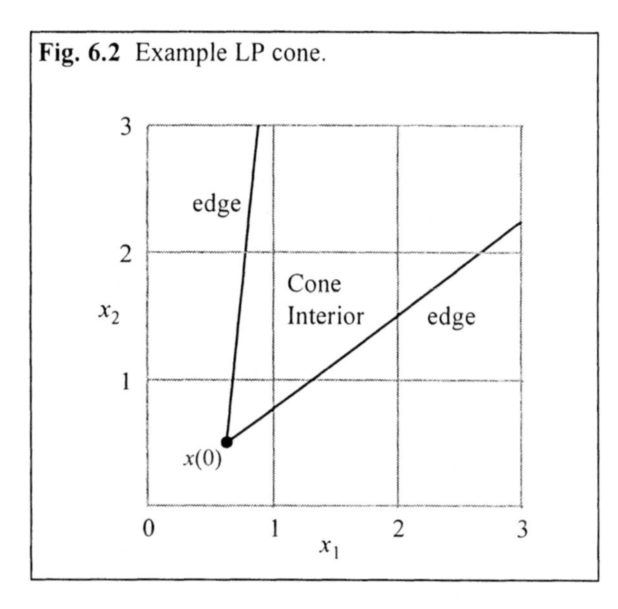

The variables x_1 and x_2 represent variables that are required to receive integer values, and hence we have shown the grid that identifies the location of the integer points on the plane. In this example x_1 and x_2 correspond to basic variables. (The nonbasic variables that determine the edges of the cone are not identified, but might for instance be denoted by x_3 and x_4.) Feasible LP solutions lie within the region spanned by the edges of the cone, that is, the region that includes these edges together with the cone interior, which may be further limited by the inclusion of constraining half spaces not shown. (Such half spaces correspond to inequalities $x_j \geq 0$ and $x_j \leq U_j$ for additional basic variables other than x_1 and x_2. In the example shown, the constraint inequalities $x_1 \geq 0$ and $x_2 \geq 0$ do not limit the cone, but if upper bounds for x_1 and x_2 were given by $x_1 \leq 3$ and $x_2 \leq 3$, then the feasible region would be restricted to consist just of the portion of the cone that appears in Figure 6.2.)

We may identify the relevant components for describing cutting planes and search processes in this setting as follows. Associated with a given current basic extreme point solution $x(0)$, as obtained by the bounded variable simplex method, let $\{x_j : j \in \mathbf{NB}\}$ denote the current set of nonbasic variables and let $\{x_j : j \in \mathbf{B}\}$ denote the current set of basic variables. The set of all variables will be indexed by $\mathbf{N}$

(hence $\mathbf{N} = \mathbf{NB} \cup \mathbf{B}$), and the set of integer variables will be indexed by **I**. For explicitness, we stipulate that each variable x_j, $j \in \mathbf{N}$ must satisfy the bounds $U_j \geq x_j \geq 0$, where U_j may be infinity. The polyhedral LP cone, of which $x(0)$ is the vertex, is the region spanned by the edges

$$x(h) = x(0) - D_h u_h, \text{ for } u_h \geq 0, h \in \mathbf{NB}$$

where D_h is the "current tableau" vector associated with the nonbasic variable x_h, and u_h is the parameter identifying the change in the value of x_h (from its lower or upper bound value which it receives at $x(0)$). We assume throughout the following that $x(0)$ is feasible for the LP problem, though this assumption can be relaxed. (In general, we require that $x(0)$ is a feasible extreme point for a region that results by discarding some of the constraints that define the original LP feasible region.)

The customary LP basis representation explicitly identifies only the subset of the entries d_{hj} of D_h that are associated with the current basic variables x_j. We use the full vector representation here, understanding that the entries of D_h are zero for all nonbasic variables except x_h, which has a coefficient of 1 or -1. We choose the sign convention for entries of D_h that yields a coefficient for x_h of $d_{hh} = 1$ if x_h is currently at its lower bound at the vertex $x(0)$, and of $d_{hh} = -1$ if x_h is currently at its upper bound at $x(0)$. (Hence, u_h respectively corresponds to x_h or $U_h - x_h$.) By this convention, if $x(0)$ is a feasible extreme point of the LP problem, then the feasible extreme points adjacent to $x(0)$ are points $x(h)$ that occur for (specific) nonnegative values for u_h, and for strictly positive values except under degeneracy.

The basic convexity cut result that makes use of this representation is founded on a simple construction. In common with the application of branching moves, we assume that the LP vertex $x(0)$ does not yield integer values for all the integer variables $x_j, j \in \mathbf{I}$. Then the steps of this construction are as follows.

6.2.1 Convexity Cut Construction

Step 1. Identify a closed convex region whose interior includes $x(0)$ but no feasible integer solutions. (A simple example is to rely on the convex region given by solutions that satisfy the inequality $v + 1 \geq x_j \geq v$, where $x_j, j \in \mathbf{I}$, is a variable whose value $x_j(0)$ is fractional at the LP vertex, and v and $v + 1$ are the integers immediately below and above this fractional value.)

Step 2. Extend each edge of the LP polyhedral cone until it meets the boundary of the convex set. (In the case of the example indicated, this results by identifying a positive value for each u_h, and hence for each nonbasic variable x_h, that causes x_j to receive the value v or $v + 1$. The value is found by solving for u_h in $x_j(h) = x_j(0) - d_{hj} u_h$, where $x_j(h) = v$ or $v + 1$ according to whether d_{hj} is positive or negative. In case d_{hj} is 0, u_h is infinity.)

Step 3. Pass a hyperplane through the endpoints of the edges of the LP basis cone where they intersect the boundary of the convex set. Letting u_h^* identify the value of u_h that corresponds to the point of intersection for edge h (as determined in Step 2), the hyperplane can be expressed as the set of points

$$\sum_{h \in \mathbf{NB}} \left(\frac{1}{u_h^*} \right) u_h = 1$$

where $1/u_h^* = 0$ if u_h^* is infinity. (This can be expressed in terms of the nonbasic variables x_h by substitution using the identity $u_h = x_h$ or $u_h = U_h - x_h$, according to whether x_h is nonbasic at its lower or upper bound in the current LP solution $x(0)$.)

The result of the preceding construction creates two half spaces associated with the hyperplane. One replaces "=" by "≤" in the defining hyperplane equation, and includes all points that lie on the side of the hyperplane that contains the LP vertex (which lies in the interior of this half space). The other replaces "=" by "≥" in the hyperplane equation, and includes all points that lie on the other side of the hyperplane (thereby excluding the LP vertex and "cutting off" the basic solution that assigns x_j a fractional value).

Convexity Cut Theorem.
The half space

$$\sum_{h \in \mathbf{NB}} \left(\frac{1}{u_h^*} \right) u_h \geq 1$$

that excludes the LP vertex contains all feasible integer solutions, and the associated hyperplane therefore qualifies as a valid cutting plane.

An example of two different convexity cuts is shown in Figure 6.3. In this figure we use the two simple convex regions $1 \leq x_1 \leq 2$ and $1 \leq x_2 \leq 2$, both of which contain $x(0)$ in their interior. (They clearly do not contain any feasible integer solutions in their interior since there are no integer values that lie strictly between the adjacent integers 1 and 2.) The intersections of the edges of the LP cone with the boundaries of the region $1 \leq x_1 \leq 2$ occur at the two points labeled A, and the intersection with boundaries of the region $1 \leq x_2 \leq 2$ occur at the two points labeled B. Thus, two cutting planes are generated, one identified by the dotted line through the "A points" and the other by the dotted line through the "B points." Cutting planes such as these generated from convex regions of the form $v \leq x_j \leq v + 1$ can be shown to correspond to those introduced for mixed integer programming by Gomory (1960). The two cuts of this particular example are not especially strong. In this instance, a convex region that yields a better cut is $2 \le x_1 + x_2 \le 3$. (Such a region can be identified by a transformation of the coordinate system, applying ideas subsequently described.)

Fig. 6.3 Convexity cuts.

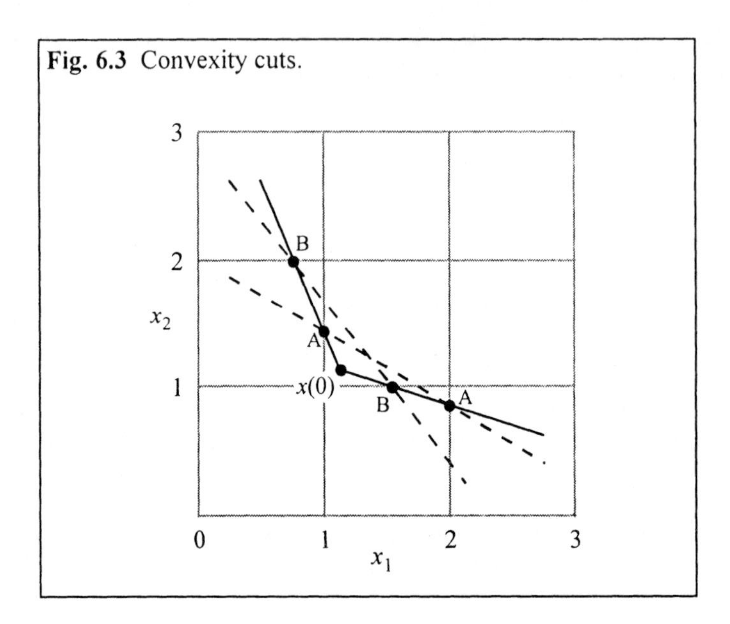

By drawing on the perspective of the convexity cut framework, we will now show how to go beyond these constructions to provide special options for generating and exploiting trial solutions, while simultaneously generating stronger cuts.

6.3 Cut Search

The convexity cut result can be refined in several ways, including a formulation that allows for negative edge extensions (Glover, 1975a; Sherali and Shetty, 1980). However, we are interested in a particular development embodied in a *cut search* theorem that gives a direct bridge between cutting planes and search methods, and affords additional possibilities for integrating tabu search with branch and bound or branch and cut procedures.

To express this theorem, we assume that an assignment of values to the integer variables $x_j, j \in \mathbf{I}$, automatically yields values for all variables $x_j, j \in \mathbf{N}$. This is true (by definition) in the case of pure integer programs, and can be made true for mixed integer programs by a step of solving an LP problem, for any given assignment of values to the integer variables, to obtain conditionally optimal values for the continuous variables. In fact, as we will see, the cut search development makes it possible to solve a simple restricted mixed integer program to obtain a best solution from a given collection of implicit candidate solutions, and to simultaneously generate a cutting plane by reference to this collection.

Define an *admissible* value for a variable x_j, $j \in \mathbf{I}$, to be an integer that satisfies the lower and upper bounds of 0 and U_j. Correspondingly, define an *admissible (coordinate) hyperplane* to be a hyperplane of the form $x_j = v$, for some $j \in \mathbf{I}$, where v is an admissible value for x_j. Let C denote a specific collection of admissible hyperplanes, and for each variable x_j, $j \in \mathbf{I}$, identify the set V_j of values v for which admissible hyperplanes are contained in C; i.e.,

$$V_j = \{v\text{: the hyperplane } x_j = v \text{ belongs to } C\}.$$

By means of these sets V_j, we identify the set of solutions $X(C)$ generated by the collection C, which we define to consist of those solutions in which every integer variable x_j receives one of the values in V_j:

$$X(C) = \{x\text{: } x_j \in V_j, \text{ for all } j \in \mathbf{I}\}.$$

Finally, let X^* denote a set of integer trial solution vectors. We are particularly interested in the situation where the set X^* contains all solutions generated by C, hence where $X(C) \subset X^*$.

Our development, which slightly re-states results of Glover (1972), is based on the notion of extending edges of the LP cone, or rays consisting of convex combinations of these edges, and to identify C as the collection of admissible hyperplanes that are encountered as a result of such extensions. In particular, if we let D denote a convex combination of the vectors D_h, $h \in \mathbf{NB}$, and let u denote a nonnegative scalar, then we can represent such a ray by

$$x = x(0) - Du.$$

The admissible hyperplanes that are successively intersected may then be identified to consist of those that occur at positive values of u for which

$$x_j = x_j(0) - d_j u \text{ is an admissible value for } x_j$$

for some $j \in \mathbf{I}$, where d_j is the jth component of D. (Note that an integer variable x_j which is currently nonbasic, and which therefore receives an admissible value at $x(0)$, will continue to receive this value for positive values of u on any ray that gives D_j a zero weight in the convex combination to yield D. Hence the associated hyperplane $x_j = v$ is automatically included in C.) The values of u that yield successive intersections with additional admissible hyperplanes are determined directly from the preceding equation by sorting (relative to $j \in \mathbf{I}$) to determine each "next" value of u. As this process is implemented, each new admissible hyperplane encountered is added to C, while simultaneously keeping track of the set $X(C)$ of solutions generated by C.

The simplest approach for determining X^* is therefore to take $X^* = X(C)$; i.e., to use the generation process itself to create the trial solutions to be considered. However, we may take advantage of having generated trial solutions previously which we can add to X^* and which can also be used to assure the special requirement that X^* includes $X(C)$.

The option of extending more than a single ray simultaneously provides useful latitude in determining the collection C, since the successive intersections of different rays may encounter different admissible hyperplanes (or encounter such hyperplanes in different sequences). A straightforward but particularly important feature of the sets V_j that results from these extensions is embodied in the following constructive result.

Lemma A. Each set V_j can be expressed in the form $V_j = \{x_j\text{: } L'_j \leq x_j \leq U'_j \text{ and } x_j \text{ integer}\}$, where L'_j and U'_j are integers that satisfy $L'_j \geq 0$ and $U'_j \leq U_j$. By construction, the values L'_j and U'_j may initially be given by setting:

$L'_j = U'_j = x_j(0)$	if $x_j(0)$ is an integer and $d_j = 0$; and otherwise
$L'_j = v+1$ and $U'_j = v$,	where v and $v+1$ are the integers that bracket $x_j(0)$.

(This yields $L'_j = U'_j + 1$ for noninteger $x_j(0)$, to render V_j empty.) Thereafter, whenever an extended ray intersects an admissible hyperplane not previously encountered, then for some $j \in \mathbf{I}$ this hyperplane has the form:

$$x_j = \begin{cases} L'_j - 1 & \text{if } d_j > 0 \\ U'_j + 1 & \text{if } d_j < 0 \end{cases}$$

Consequently, L'_j is decreased or U'_j is increased to include the value identified by this hyperplane.

The convenient form imparted to the set of solutions $X(C)$ by the preceding observation has useful implications for exploiting this set by generating cutting planes. In particular, the construction of Lemma A, and the nature of the process that gives rise to this construction, leads to a basic theorem which says that it is legitimate to form a cutting plane by extending every edge of the cone until it intersects an admissible hyperplane not contained in C. If the edges of the cone provide the rays that are used to generate C, this means that each such edge can be extended strictly farther than it was extended in the process of generating C. (Not only is it permitted to go beyond admissible hyperplanes encountered by its own extension, but also beyond those encountered by the extensions of all other edges or rays.)

We can summarize this by defining the *potentially admissible* set $P_j, j \in \mathbf{I}$, to consist of admissible values for x_j that are not contained in V_j. Specifically, by the nature of our construction, we can restrict P_j to consist of the integers $L_j' - 1$ and $U_j' + 1$, excluding integer values that lie outside the bounds for x_j. Then we define a *new potential value* of the parameter u_h to be the value that extends the edge h to its first intersection with an admissible hyperplane not in C, given by

$$\text{Min}(u > 0: x_j = x_j(0) - d_{hj}\, u \in P_j, \text{ for some } j \in \mathbf{I}).$$

If P_j is empty, or if no values in P_j can thus be attained for positive values of u, then we define the new potential value of u_h to be infinity. We assume that the set X^* of solutions examined includes $X(C)$, and say that a cut is valid, *conditional on* $X(C)$, if it excludes $x(0)$ but no feasible integer solutions with the possible exception of some elements of $X(C)$.

Cut Search Theorem.

A valid convexity cut conditional on $X(C)$ is given by the half space

$$\sum_{h \in \mathbf{NB}} \left(\frac{1}{u_h^*} \right) u_h \geq 1$$

where each u_h^* is assigned the new potential value for u_h.

The preceding theorem implicitly determines the conditional cut by reference to the convex region $\{x\text{: } L_j'' \leq x_j \leq U_j'', j \in \mathbf{I}\}$, where $L_j'' = L_j' - 1$ and $U_j'' = U_j' + 1$ with the provision that $L_j'' = -\infty$ if $L_j' = 0$ and $U_j'' = \infty$ if $U_j' = U_j$. It is easy to see that this region contains all points of $X(C)$, but no other feasible integer solutions, in its interior. The theorem more broadly is valid for generating cuts and trial solutions for any problem where feasible solutions are determined by intersections of specified hyperplanes or half spaces, as in minimizing a concave function over a polyhedron, where an optimal solution is assured to lie at some extreme point (determined by the intersection of particular sets of boundary hyperplanes). The form of C then differs in evident ways from the structure it acquires for IP problems, and auxiliary processing may be needed to generate members of $X(C)$ (as by pivoting to identify extreme points implied by the collection C).

In the integer programming context, the theorem allows the following refinement:

Corollary: Let $x_j^*(h) = x_j(0) - d_{hj}\ u_h^*$. If there exist integers a_0 and a_j, $j \in \mathbf{I}$, such that

$$a_0 \leq \sum_{j \in \mathbf{I}} a_j x_j^*(h) \leq a_0 + 1 \qquad \text{for all } h \in \mathbf{NB},$$

then the value of u_h^* for each $h \in \mathbf{NB}$ can be redefined (further increased or left unchanged) by determining the value that yields

$$\sum_{j \in \mathbf{I}} a_j x_j(h) = a_0 \text{ or } a_0 + 1.$$

The simplest form of the refinement given in the preceding Corollary occurs by identifying a single variable x_j, $j \in \mathbf{I}$ and integers a_0 and $a_0 + 1$ such that

$$a_0 \leq x_j^*(h) \leq a_0 + 1 \qquad \text{for all } h \in \mathbf{NB}.$$

Then u_h^* can be redetermined to yield $x_j^*(h) = a_0$ or $a_0 + 1$ for the specified $j \in \mathbf{I}$. The refinement may also be advantageously applied by reducing a few of the u_h^* values to permit the conditions of the Corollary to hold, thus allowing other values to be increased. (An example of this is provided later.) We remark that the Corollary implicitly determines the cut from a union of two convex regions. While such a union is not convex, it covers the cone, and hence contains a convex region of the appropriate form.

An illustration of the Cut Search Theorem is provided in Figure 6.4 for the two dimensional example previously shown in Figure 6.2. Only the LP cone is shown, without considering other constraints that may affect the feasible region. The edges of the cone emanating from the vertex $x(0)$ are shown as heavy lines, to the point where each encounters its first intersection with an admissible coordinate hyperplane. (We have elected to extend edges of the cone rather than other rays, since there are only two edges to consider in two dimensions. Likewise we have arbitrarily chosen to stop extending each edge at its first intersection with an admissible hyperplane.) In this case, the admissible hyperplanes that are intersected are for different variables, and hence the initial empty $X(C)$ enlarges to contain the circled point (1,1).

Fig. 6.4 Cut search theorem illustration.

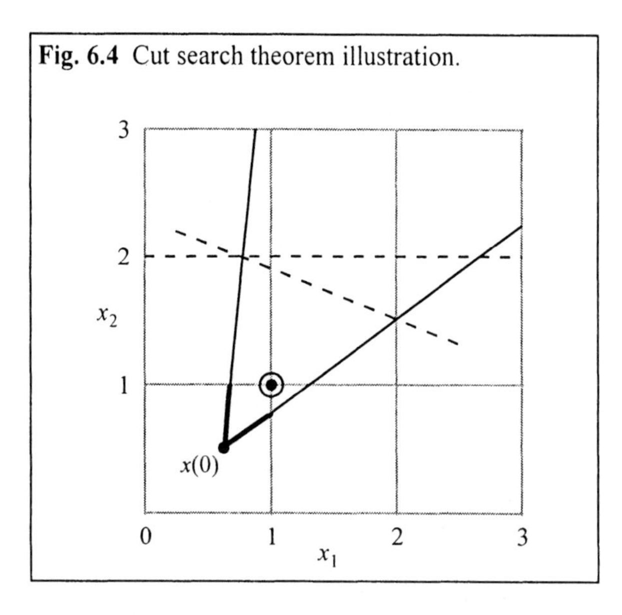

After examining the solution given by the circled point, the further extension of the edges that yields the cutting plane of the Cut Search Theorem is shown by the light (solid) lines. The cut itself is shown by the dotted lines determined by the intersection of each edge with a new admissible hyperplane that has not previously been encountered. In fact, there are two such cuts. The second arises by the condition specified in the Corollary to the theorem, for the simple case $a_0 \le x_j^*(h) \le a_0 + 1$, $h \in$ **NB**, for $j = 2$. That is, the endpoints of both edges that determine the first cut lie within the same convex region, determined by $1 \le x_2 \le 2$. This makes it possible to extend one of the edges more deeply than in the first cut, until both edges meet the boundaries of the region $1 \le x_2 \le 2$ (in this case, meeting the hyperplane $x_2 = 2$).

Several observations can be made about the Cut Search Theorem, as follows. The last two (Observations 6 and 7) are especially relevant for exploiting the cut search ideas within a branch and cut framework.

Observation 1. The cut can be introduced at intermediate stages of generating C and $X(C)$, before deciding to terminate the generation process. New trial solutions to be added to X^*, as a result of belonging to $X(C)$ but not X^*, may be then tested immediately to see if they satisfy the current cutting plane inequality. (When the inequality is violated, this test can avoid more lengthy efforts to check such trial solutions for feasibility. If the edges of the example in Figure 6.4 were to be extended beyond their first intersection to "probe for" new solutions to add to $X(C)$, this test could make use of the cut generated from the first intersections to immediately eliminate all points that satisfy $x_2 < 3$.)

Observation 2. In the case of pure integer programs, the process of checking new solutions of $X(C)$ to be added to X^* can also be accelerated by taking advantage of the structure of $X(C)$. Let e_j denote the unit vector with a 1 in position j and 0's elsewhere. Then for any given $x' \in X(C)$ a subset of solutions of $X(C)$ can be represented in the form

$$x = x' + qe_j,$$

as q ranges over $|V_j|$ integer values (including $q = 0$). If x' is infeasible, then the solutions $x' + qe_j$ are infeasible either for all $q > 0$ or for all $q < 0$. Similarly, if x' is better than (has a better objective value than) $x' + qe_j$ for any $q > 0$ ($q < 0$) then x' is better than $x' + qe_j$ for all $q > 0$ ($q < 0$). (The objective function coefficient of x_j in an initial LP basis representation, where all integer variables are nonbasic, immediately determines whether $q > 0$ or $q < 0$ yields a worse solution. See Observation 7 below for a more comprehensive way to exploit such relationships, by solving a restricted integer programming subproblem.)

Observation 3. The theorem offers a basis to determine when a method should rely more heavily on cutting than on searching, and vice versa. Specifically, a cut can always be generated by the theorem for the case where $X(C)$ is empty (or is entirely contained in a set X^* previously generated). In this situation, no new trial solutions need to be examined. The transition point at which $X(C)$ will leave the state of being empty occurs where V_j is empty for exactly one $j \in \mathbf{I}$, and $X(C)$ will become nonempty as soon as this last V_j acquires an element as a result of intersecting an admissible hyperplane $x_j = v$. Several "last sets" V_j may become nonempty simultaneously, when several admissible hyperplanes are intersected at once. The numbers of elements in the sets V_k, $k \in \mathbf{I} - \{j\}$ disclose the size that $X(C)$ will achieve when V_j becomes nonempty. (The calculation must be adjusted if a collection of sets simultaneously qualify to become the set V_j.) Since $X(C)$ will include solutions for every combination of values $x_k \in V_k$ (joined with the value assigned to x_j when V_j first becomes nonempty), it is possible that the first nonempty $X(C)$ will be large. The size of this first nonempty $X(C)$ may be used as a measure of the relative strength of cuts based on different strategies for selecting and extending a collection of rays.

When $X(C)$ contains more elements (outside of a pre-existing X^*) than can be examined conveniently, this signals that it is preferable to generate the convexity cut from the convex region $v \le x_j \le v + 1$, where x_j lies strictly between v and $v + 1$, and where either $x_j = v$ or $x_j = v + 1$ will be a "transition intersection" that causes $X(C)$ to become nonempty.

Observation 4. As noted in reference to the example of Figure 6.3, convexity cuts from regions of the form $v \le x_j \le v + 1$ are equivalent to the mixed integer cutting planes of Gomory (1960). Computational experience has suggested that these cutting planes can be productively applied for a limited number of iterations, after which their strength tends to deteriorate. Identifying a member of these cutting planes that corresponds to a "last empty" V_j, as in Observation 3, gives a basis for choosing among them. More particularly, monitoring the size of the first nonempty $X(C)$ provides an indication of the turning point where it becomes preferable to generate stronger cuts by building a set X^* and examining its members.

Observation 5. Deeper edge extensions and stronger cuts may result by transforming current basic integer variables x_j into new integer variables y_j, which have the form

$$y_j = x_j + \sum_{h \in \mathbf{NB} \cap \mathbf{I}} a_h x_h$$

where the coefficients a_h are chosen to be integers. This transformation replaces the LP basis equation

$$x_j = x_j(0) - \sum_{h \in \mathbf{NB}} d_{hj} x_h$$

by

$$y_j = x_j(0) - \sum_{h \in \mathbf{NB} \cap \mathbf{I}} \left(d_{hj} - a_h\right) x_h - \sum_{h \in \mathbf{NB} \setminus \mathbf{I}} d_{hj} x_h \ .$$

Consequently, integer values a_h for $h \in \mathbf{NB} \cap \mathbf{I}$ can be chosen to make each resulting coefficient d_{hj} - a_h a positive or negative fraction (with absolute value less than 1), thus potentially causing the intersections for integer values of y_j to occur at deeper levels than those for integer values of the corresponding x_j. (The idea of using such integer transformations as a way to yield stronger cuts is implicit in early approaches for generating cutting planes for pure and mixed integer programming. In Section 6.5 we show how a unified perspective of cutting and branching allows this idea to create stronger branching alternatives as well.)

For the purpose of identifying the set $X(C)$, and hence trial solutions to be added to X^*, each x_j value, $j \in \mathbf{B}$, can be recovered from the corresponding y_j value by the equation

$$x_j = y_j - \sum_{h \in \mathbf{NB} \cap \mathbf{I}} a_h x_h \ .$$

Bounds on admissible y_j values may not be known, and so all integer values of y_j variables intersected may need to be transformed into x_j values and checked in this manner.

This approach for generating deeper extensions of some edges increases the relevance of Observations 3 and 4. In particular, some of the nonbasic integer variables x_h, whose coefficients have been successfully reduced may encounter a series of integer values, in order for their associated edges to reach the intersection point with a region $v \le y_j \le v + 1$. This causes the associated sets V_h to enlarge. In addition, the specific set V_j (where x_j receives a value indirectly at the same time as y_j reaches v or $v + 1$) is one of the later sets, if not "the last," to become nonempty. This is the type of situation where a first nonempty $X(C)$ may be large. Hence, as long as transformations exist that indeed can make the first non-empty $X(C)$ contain many elements, simple convexity cuts for $X(C)$ empty are likely to be reasonably strong and worth implementing.

Observation 6. Additional unimodular transformations of integer variables beyond those of Observation 5, that generate integer combinations which include more than a single basic variable, provide a basis for creating coordinate systems that allow the space to be more effectively probed by extending rays. Instances of such transformations that restrict attention to coordinate hyperplanes for a single integer variable

$$y = \sum_{j \in \mathbf{I}} a_j x_j$$

for selected integers a_j can be used to exploit the Corollary to the Convexity Cut Theorem.

The relevance of a transformed coordinate system is illustrated in Figures 6.5 and 6.6. These figures are based on the same LP problem (hence both figures show the same LP cone), and employ the same rules for extending rays to probe the space (selecting the edges of the cone in each case). Figure 6.5 shows the result of extending the two edges for a standard 2-dimensional coordinate system. As indicated in the diagram, starting from the LP vertex, Edge 1 leads downward to the right and Edge 2 leads upward to the left. The rule for extending these two edges is:

- Extend Edge 1 to its first intersection with an admissible coordinate hyperplane.

- Alternate the choice of edges (hence choosing Edge 2 on the second step, Edge 1 on the third, etc.), to extend the chosen edge to its first intersection with an admissible hyperplane not yet encountered. At each intersection, identify the new solutions generated by the intersected hyperplanes.

- Stop after the third intersection for Edge 1 and generate the cut from the next intersection of each edge (intersection #4 for Edge 1 and intersection #3 for Edge 2).

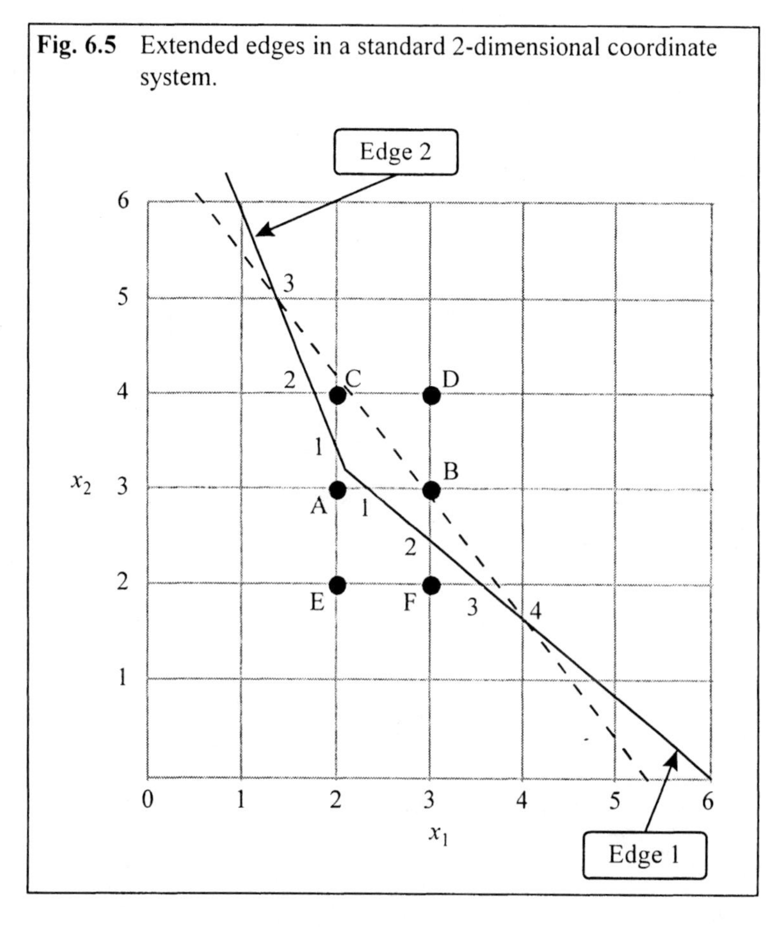

Fig. 6.5 Extended edges in a standard 2-dimensional coordinate system.

The four intersections for Edge 1 and the three intersections for Edge 2 are identified in Figure 6.5 by the numbers placed beside these intersections. The circled points indicate those that are generated by the intersections, as indicated in Table 6.1.

Table 6.1 Identification of points in Figure 6.5.

Points	*Added at the time of intersection*
A	#1 for Edge 2 (given #1 for Edge 1)
B	#2 for Edge 1
C and D	#2 for Edge 2
E and F	#3 for Edge 1

It may be noted that three of the points generated (A, E and F) lie outside the cone and hence are infeasible. (Some points inside the cone may also be infeasible, but we do not show other constraints that may affect feasibility.) The cutting plane shown significantly dominates those that may be generated by the options for keeping $X(C)$ empty (in this case, the cutting plane that joins intersection #1 of Edge 1 with intersection #2 of Edge 2, and the cutting plane that joins intersection #1 of Edge 2 with intersection #2 of Edge 1). Nevertheless, the cut is rather shallow. A possibility for improvement (though not strict improvement) exists by retracting the extension of Edge 2 sufficiently to allow both edge extensions to lie between hyperplanes $x_1 + x_2 = a_0$ and $x_1 + x_2 = a_0 + 1$, for an appropriate a_0. Then in this case the cut $x_1 + x_2 = a_0 + 1$, which goes diagonally through the points C and D and the other diagonal points below, results by applying the Corollary as previously indicated.) Still, we seek a way to do better.

Fig. 6.6 Hyperplanes approximately parallel to faces of the cone.

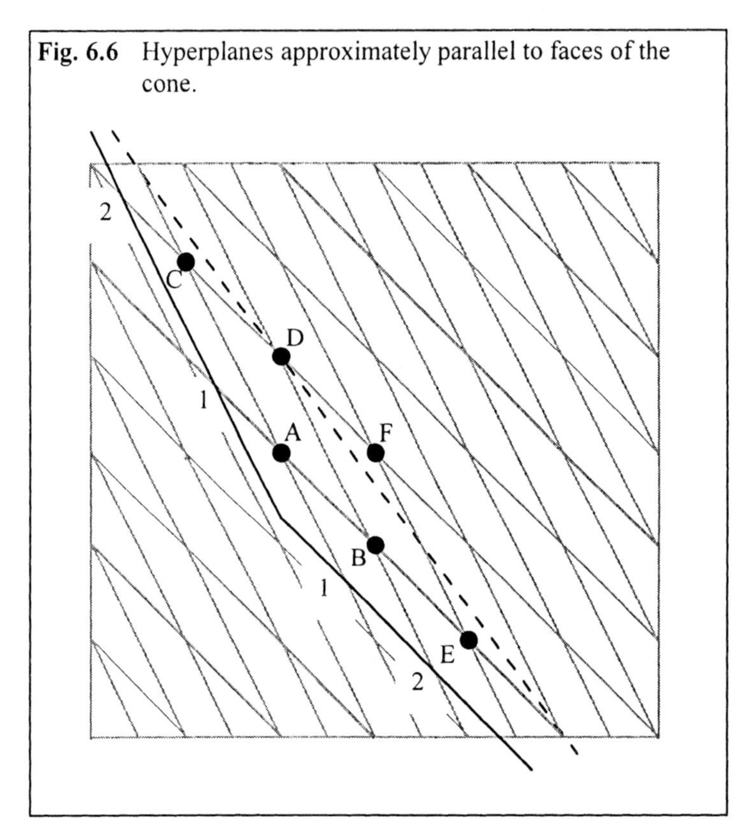

Figure 6.6 shows the result of creating a coordinate system whose defining hyperplanes are approximately parallel to the corresponding hyperplanes that determine faces of the cone. (In two dimensions, of course, edges and hyperplanes coincide.) The rule for extending the edges, and the intersections that generate the points A through F are exactly the same as in the example of Figure 6.5. However, the identities of the points A through F are much different in this transformed coordinate system. In this case, all of the points lie within the cone, and identify a generally preferable collection of candidate solutions. Moreover, the cutting plane that results is considerably stronger than the one of Figure 6.5. (Fuller details of this

example can be verified by generating a larger graph, which is not to be confined to the limited space available for Figure 6.6.)

Another benefit of the transformed coordinate system is worth noting. If a single ray is extended approximately through the middle of the cone in Figure 6.5, and stopped shortly after passing point D of this figure, then point D is the only point generated. However, if this same ray is extended the same distance in Figure 6.6 (roughly in the direction of point F, which is the same as the point D of Figure 6.5), then it will generate all six of the points A through F of this figure. The transformed system gives a richer set of possibilities from the standpoint of searching as well as cutting.

A framework for determining such a unimodular transformation of coordinates (and the inverse transformation for recovering the associated values of the original variables) is given in Glover (1964), together with rules to assure that the variables that define the new system are nonnegative. The "corner polyhedron" (group theory) development of Gomory (1965) and Gomory and Johnson (1972) also introduces considerations related to determining such a transformation. (Illustrations of how such a transformed system can be created, together with observations for supplementing the process of probing the hyperplanes, are given in Exercises 5 to 16 at the end of this Chapter.)

Observation 7. The structure of the set $X(C)$, and more particularly the sets V_j as identified in Lemma A, suggests an alternative approach that defers examination of new members of $X(C)$. A best element of $X(C)$ can be determined in a single step by solving an MIP problem which imposes the restrictions

$$x_j \in V_j \text{ for } j \in \mathbf{I}.$$

These restrictions can be controlled by limiting the set of rays selected to be extended, and by limiting their extensions (just as normally would be done in a process that examines elements of $X(C)$ as they are generated). However, the resulting MIP problem to dispose of $X(C)$ as an aggregate, without examining its members individually, may be able to handle a set $X(C)$ that is much too large to be examined by a process of "itemization." For example, in a 0-1 MIP problem, if the restrictions $x_j \in V_j$ for $j \in \mathbf{I}$ permit 20 integer variables to receive either the value 0 or 1, while constraining other variables to a single value, then the set $X(C)$ will contain approximately a million members, which is likely more than can be conveniently examined individually. Yet an MIP problem with only 20 zero-one variables can typically be solved very quickly. The resulting solution provides a candidate for an optimal solution (if the problem contains a feasible solution), while permitting the cutting plane of the Cut Search Theorem to be imposed.

To facilitate such a process, the restricted MIP problem can be solved by post-optimizing from the current LP basic solution. Basic variables x_j that receive fractional values and whose sets V_j contain a single integer value effectively impose a collection of "fixed branch" steps. The approach can be applied recursively (since the restricted problem is an MIP problem that may be further restricted). It may also be coupled with tabu branching, as described earlier, either to dispose of restricted problems heuristically, or to supplement (and accelerate) an exact method for solving these problems. The potential speed of solving the restricted problems suggests that several different problems may be generated and solved at each LP vertex produced by a heuristic or exact branching process.

The preceding observations disclose that the goals of finding good solutions and generating good cutting planes are interrelated in a variety of subtle ways. We now turn our attention more fully to the first of these goals, to take a closer look at strategies for extending rays to probe the solution space, and thereby to generate useful candidates to include in X^*. We consider a special approach for generating solutions based on identifying a subset of intersections with admissible hyperplanes relative to a directional rounding concept. This leads to a natural variant of the scatter search strategy, which creates solutions for inclusion in X^* by tracing paths that implicitly link multiple rays, and which also gives rise to additional theorems that interrelate search processes and cutting planes.

6.4 Star Paths for Integer Programs

We are motivated to place increased emphasis on finding good candidate solutions due to the following observation: it is entirely possible, by selecting appropriate rays for probing the space, to generate a small set of candidates $X(C)$ which includes an optimal solution, yet which may not yield a particularly strong cutting plane. In fact, as will be seen, there is a convex collection of rays each of which finds an optimal solution, providing a trajectory along which each integer variable moves monotonically closer to its optimal value. For problems in zero-one variables, such rays find an optimal solution as the first solution generated, at the point where $X(C)$ transitions from being empty to nonempty. Clearly, a set $X(C)$ that has a single element is unlikely to generate a very strong cutting plane, and the fact that an optimal solution can be contained in such a set suggests that the quest for good cutting planes and the quest for good solutions may embody somewhat different considerations. The challenge we address, therefore, is to create a strategy that increases the chance of selecting one of the "right" rays to generate an optimal or near-optimal solution.

This shift of focus leads to a corresponding shift of strategy. Instead of generating solutions from the union of hyperplanes intersected by a collection of rays, we generate solutions from hyperplanes intersected by each ray individually. More precisely, our approach is organized to examine a special subset of the union of the hyperplanes intersected by different rays. The link that provides this organization is a special variant of scatter search.

Our development is based on results of Glover (1995a) for zero-one problems, which we reformulate to allow several of the basic ideas also to be extended to more general IP problems. We first establish connections to the cut search framework. Consider once again the generic representation of a (half) ray of the LP cone:

$$x = x(0) - D\,u,\ \ u \geq 0$$

where D is a convex combination of the vectors D_h, $h \in \mathbf{NB}$.

In the present development we change our frame of reference to consider rays that are generated by identifying specific *focal points* through which these rays are extended. If x' is any point that lies in the LP cone, distinct from $x(0)$, then we can construct an alternative representation of rays of the cone by defining

$$Ray(x') = \{x = x(0) + (x' - x(0))\,u : u \geq 0\}$$

The vector $x' - x(0)$ equals a negative multiple of some D that is a convex combination of the D_h vectors. We are especially interested in focal points that lie on cutting planes, and more precisely that lie within the restricted region of a cutting plane where it intersects the LP cone and the LP feasible region.

For specific values u_h^*, $h \in \mathbf{NB}$, which define a valid cutting plane by the Convexity Cut Theorem (or a conditionally valid cutting plane by the Cut Search Theorem), we identify as before the endpoints of the associated extended edges, given by

$$x^*(h) = x(0) - D_h\,u_h^*,\ \ h \in \mathbf{NB}$$

We allow reference to cutting planes with negative edge extensions, by the simple expedient of replacing both negative and infinite u_h^* values by large positive finite values, which also yields a valid cutting plane. (More refined ways of dealing with cutting planes that have negative edge extensions result by solving a linear program that includes such cuts, thereby giving a second LP vertex for guiding the search, as will be discussed later.)

Associated with these positive u_h^* values, we identify the face F^* created by the intersection of the hyperplane through the points $x^*(h)$ with the LP cone, which is the convex combination of the points $x^*(h)$:

$$F^* = \left\{ x = \sum_{h \in \mathbf{NB}} \lambda_h x^*(h) : \sum_{h \in \mathbf{NB}} \lambda_h = 1, \lambda_h \geq 0 \right\}$$

Let X_{LP} and X_{IP} respectively denote the sets of LP feasible and IP (or MIP) feasible points. (Again we note the assumption that values for continuous variables are directly determined from values for the integer variables.) We are interested in choosing focal points x' that lie on F^* and on $F^* \cap X_{\mathrm{LP}}$. This provides a more convenient and effective way of identifying rays for probing the solution space than by simply generating convex combinations of the vectors D_h, $h \in \mathbf{NB}$.

Our first result concerns the set of solutions generated by $Ray(x')$ as it successively intersects admissible hyperplanes. Since $Ray(x')$ is uniquely determined by the focal point x', we denote this set of solutions by $X(x')$.

It is entirely possible that $X(x')$ will be empty, since $Ray(x')$ may fail to intersect any admissible hyperplanes $x_j = v$ for some $j \in \mathbf{I}$. This occurs in the situation where $x_j(0)$ is fractional, and $x'_j = x_j(0)$. Then the value of x_j on $Ray(x')$, given by

$$x_j = x_j(0) + (x'_j - x_j(0))\,u,\ u \geq 0$$

is equal to $x_j(0)$ for all u. For this special situation we will treat $X(x')$ as nonempty by stipulating that $Ray(x')$ intersects the indeterminate hyperplane $x_j = \#$, where the symbol # refers to an unspecified value. We assume a decision rule will be applied (as implicit in later developments) to cause this value to become either the integer immediately below or immediately above $x_j(0)$. (Once the value for # is determined, the alternative value is excluded, hence only one hyperplane will be intersected by x_j on $Ray(x')$.) For the moment, for uniformity and to simplify the statement of certain results, we may suppose that # is always the integer obtained by rounding $x_j(0)$ to its nearest integer neighbor, handling the assignment $x_j(0) = 0.5$ by rounding down.

Define an admissible x vector to be one such that each x_j, $j \in \mathbf{I}$ receives an admissible value (an integer that satisfies the bounds for x_j). Then we state the following observation.

Lemma B. The composition of $X(x')$, for $x' \neq x(0)$, is given by the set of admissible x vectors such that

$x_j(0) < x_j \leq U_j$	if $x'_j > x_j(0)$
$0 \leq x_j < x_j(0)$	if $x'_j < x_j(0)$
$x_j = x_j(0)$	if $x'_j = x_j(0)$ and $x'_j(0)$ is an integer
$x_j = \#$	otherwise.

Moreover, $X(x) = X(x')$ for all $x \in Ray(x')$, such that $x \neq x(0)$. Finally, for any two focal points x' and x'' distinct from $x(0)$, $X(x') \cap X(x'') \neq \varnothing$ implies $X(x') = X(x'')$.

For purposes of visualization, Lemma B implies that the regions where different focal points x' all generate the same set $X(x')$ may be identified by the orthants that result when the coordinate system for x_j, $j \in \mathbf{I}$ is translated to make $x(0)$ the origin. That is, the hyperplanes that determine the boundaries of the orthants are given by $x_j = x_j(0)$ for each $j \in \mathbf{I}$. (Unlike a customary coordinate system, such a translated orthant may not include points of all the boundary hyperplanes that delimit it. Whether points of particular hyperplanes $x_j = x_j(0)$ are included or excluded from particular orthants depends on whether $x_j(0)$ is an integer and on the rule for determining #.) This partitioning of the space by a translated coordinate system is useful for illustrating additional properties associated with $X(x')$, as shown subsequently.

Lemma B also has an important procedural implication: by its characterization, there is no need to extend $Ray(x')$ to sequentially identify the admissible hyperplanes it intersects in order to know the composition of $X(x')$. Such a "bit by bit" extension of the ray is only relevant for the goal of identifying the set $X(C)$, where C is the union of the admissible hyperplanes intersected by multiple rays. Cut search procedures, as we have seen, typically do not extend each ray to its limit when generating $X(C)$, and so only produce a subset of the admissible values for x_j that are contained in any given set $X(x')$. The sequential determination of $X(x')$ is relevant where we may wish to stop before the entire set is generated. Then it can be useful to determine sequences by which different partial extensions of rays may intersect the same hyperplanes, thus allowing the set $X(C)$ to remain small while probing more deeply into the space. In the present setting, however, our concerns are different.

6.4.1 Main Results and Implications for Zero-One Problems

We first focus on zero-one IP problems, where $U_j = 1$ for all $j \in \mathbf{I}$, and later show how to extend our results to general IP problems. Within the zero-one setting, Lemma B motivates the definition of a *directional rounding* operator $\delta(x'_j)$, $j \in \mathbf{I}$, relative to the components of a specific focal point x', as follows.

$$
\begin{array}{ll}
\delta(x'_j) = 1 & \text{if } x'_j > x_j(0) \\
\delta(x'_j) = 0 & \text{if } x'_j < x_j(0) \\
\delta(x'_j) = x_j(0) & \text{if } x'_j = x_j(0) \text{and } x'_j(0) \text{ is an integer} \\
\delta(x'_j) = \# & (\text{for } \# = 0 \text{ or } 1) \text{ otherwise.}
\end{array}
$$

The directional rounding terminology derives from the fact that, to be LP feasible, $x(0)$ must yield values such that $x_j(0)$ lies between 0 and 1 for each $j \in \mathbf{I}$, and hence $\delta(x'_j)$ effectively constitutes a rounding of $x_j(0)$ "in the direction of" x'_j. There is no requirement in this definition that x'_j itself must be between 0 and 1.

By extension, we define the directional rounding $\delta(x')$ of the vector x' to be the point x^o determined by the assignment

$$x^o_j = \delta(x'_j), j \in \mathbf{I}.$$

By Lemma B, it is clear that $X(x')$ contains a single admissible solution in the case of a zero-one problem, and that this solution is precisely $\delta(x')$.

An illustration of directional rounding for different focal points is shown in Figure 6.7. An arrow connects the vertex $x(0)$ to each focal point x' that is directionally rounded, and another arrow identifies the integer point $\delta(x')$ that is the result of this rounding.

The points that take the role of a focal point x' are labeled A, B, C, and D in Figure 6.7. We have included the translated coordinate system described in connection with Lemma B, shown as the dotted lines through the point $x(0)$. This frame of reference discloses how focal points within the different translated orthants are mapped into associated zero-one points by directional rounding. Points that lie on the dotted lines separating the orthants directionally round to points determined by the rule for defining the value #.

Fig. 6.7 Examples of directional rounding.

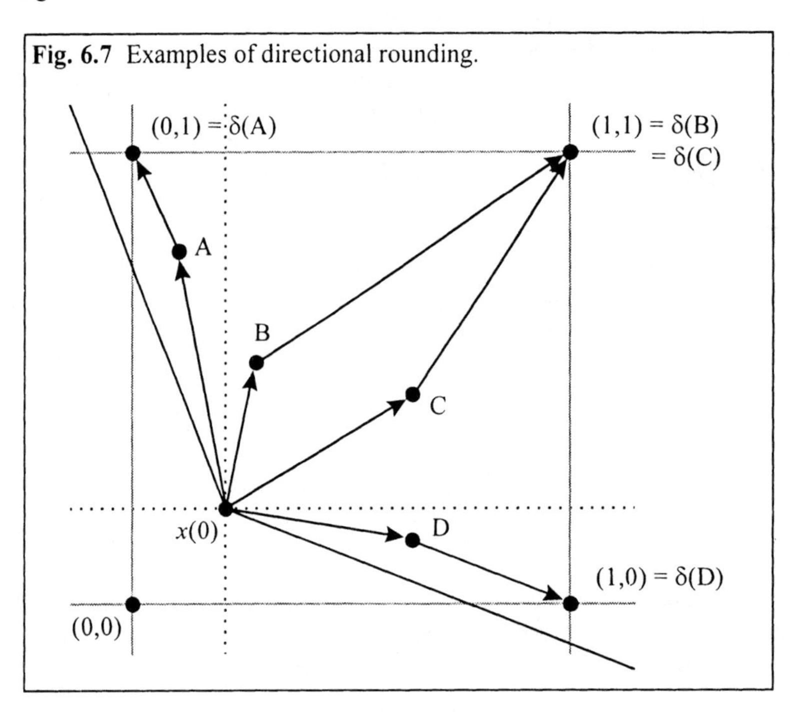

We now go beyond the consequences of Lemma B and examine results that lead to methods for searching the space of zero-one solutions. As a foundation for our first main result, which is not restricted to the zero-one setting, it is useful to think of $X(x')$ as a mapping of the focal point x' into a set of candidate integer solutions. Associated with this mapping we may identify an inverse mapping, applied to an integer candidate solution x^{o}, whose range is the set of all focal points x' that generate x^{o}. As we henceforth restrict x' to lie on F^{*} or $F^{*} \cap X_{LP}$, we refer to the elements generated by this inverse mapping as the *image* of x^o on F^* or $F^* \cap X_{LP}$. Then we define

$$(Y)\text{-}Image(x^o) = \{x' \in Y: x^o \in X(x')\}$$

for $Y = F^*$ or $Y = F^* \cap X_{LP}$.

Feasible Convex Image Theorem.

Assume $u_h^* > 0$ for each $h \in \mathbf{NB}$. Then $(F^*)\text{-}Image(x^o)$ is a convex set for each admissible x^o, and is a non-empty convex set for each admissible x^o in the LP cone.

In addition, if the values of u_h^* determine a valid cutting plane, then $(F^* \cap X_{LP})\text{-}Image(x^o)$ is a non-empty convex set for each $x^o \in X_{IP}$.

Finally, for any two admissible points x^1 and x^2,

$(Y) - Image(x^1) \cap (Y) - Image(x^2) \neq \varnothing$ implies

$(Y) - Image(x^1) = (Y) - Image(x^2)$, for $Y = F^*$ and $Y = F^* \cap X_{LP}$.

We observe that, although the non-empty convex sets identified in the theorem may not be closed, it is clear by Lemma B that they contain non-empty closed polyhedral regions that map into the indicated admissible points.

This result has particular significance in the case of zero-one problems, due to the fact that each focal point x' then maps into a single point $\delta(x')$ by directional rounding. In this setting, the theorem implies that a search for feasible and optimal zero-one solutions may be restricted to examining focal points that lie on F^*, and more particularly that lie on $F^* \cap X_{LP}$ if F^* is determined by a valid cutting plane.

Figure 6.8 provides a 2-dimensional example that identifies the images of the vertices of the unit hypercube on F^* and on $F^* \cap X_{LP}$. F^* constitutes the line segment that joins the two points A^* and D^*, which are selected to be endpoints of edges of the LP cone that generate a valid cutting plane. $F^* \cap X_{LP}$ is the interior portion of this line segment that joins the points A and D, which lie on the boundary of the unit hypercube. The convex regions that compose the images of these zero-one points are the interior line segments AB, BC, etc., as identified in Table 6.2.

Fig. 6.8 F^* and $F^* \cap X_{LP}$ images of zero-one points.

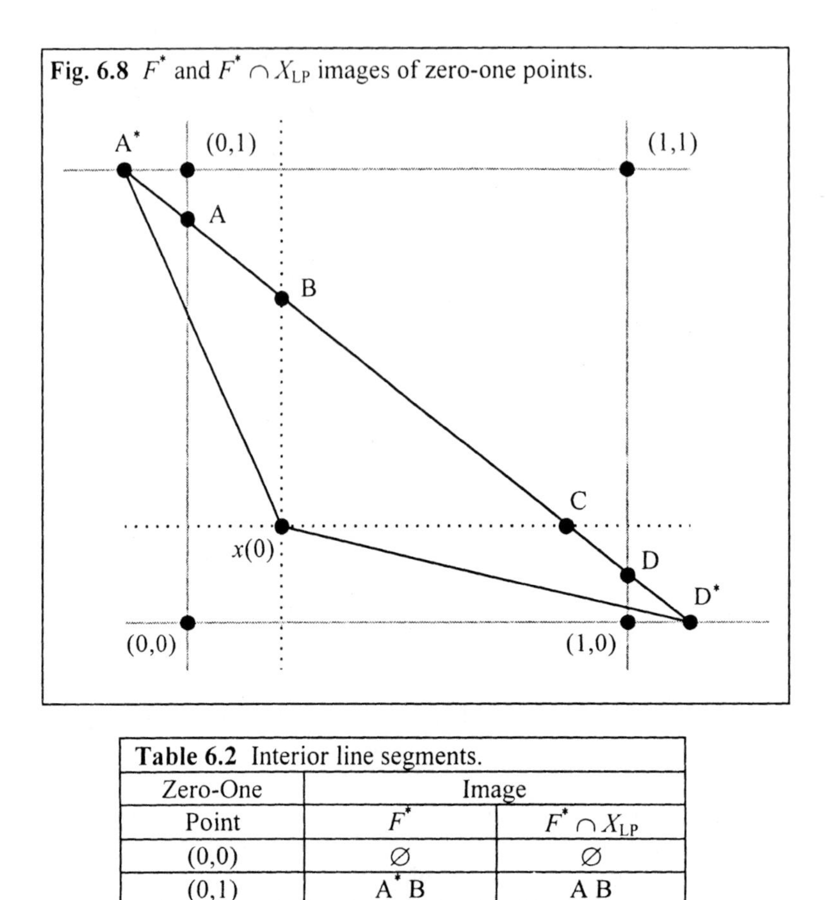

Table 6.2 Interior line segments.

Zero-One	Image	
Point	F^*	$F^* \cap X_{LP}$
(0,0)	∅	∅
(0,1)	A* B	A B
(1,1)	B C	B C
(1,0)	C D*	C D

Once again we have superimposed a translated coordinate system on the diagram, depicted by dotted lines that intersect at the translated origin $x(0)$. The labeled points on the line segment A*D* correspond to intersections of F^* with the hyperplanes that delimit the translated coordinate system, and with hyperplanes of the unit hypercube. (The latter hyperplanes in this case identify boundaries of X_{LP}, assuming for this example that there are no LP constraints other than those given by the LP cone and the bounds on the zero-one variables.) It may be noted that the infeasible point (0,0) has no image on F^*, i.e., it's convex image is empty. Both of the feasible points (0,1) and (1,1) have non-empty convex images, as the theorem specifies. In addition, the infeasible point (1,0) also has a non-empty convex image. It may be observed that the addition of certain hyperplanes (lines in this 2-dimensional example) corresponding to constraints that exclude (0,1) and (1,0) from the feasible region may eliminate part or all of the convex image of these points in the region $F^* \cap X_{LP}$. However, there is no constraint hyperplane (more precisely, half space) that can both render (1,1) infeasible and eliminate its image from $F^* \cap X_{LP}$.

The existence of a nonempty convex region associated with each feasible zero-one IP solution x^o, so that every focal point in this region directionally rounds to x^o, suggests an approach of searching for integer solutions by examining focal points that are contained in different subregions of F^* and $F^* \cap X_{LP}$. In particular, since F^* and $F^* \cap X_{LP}$ are themselves convex, the Feasible Convex Image theorem motivates a strategy of generating focal points on these regions as linear convex combinations of other points. Thus the theorem directly provides a basis for applying the scatter search approach, where the directional rounding operator conveniently accomplishes the generalized rounding process stipulated by scatter search to map fractional values of integer variables into integer values. We summarize the general form of such a version of scatter search as follows.

6.4.2 Scatter Search and Directional Rounding for Zero-One Problems

Step 1. Start with an LP feasible extreme point $x(0)$.

Step 2. Identify positive values u_h^*, $h \in$ **NB**, and identify the points $x^*(h)$ as initial reference points for scatter search.

Step 3. Generate convex combinations of the reference points to create focal points x'.

Step 4. Apply directional rounding to the focal points to create points $\delta(x')$ as candidates for seeking an optimal zero-one solution.

(Selected focal points may then guide the determination of new reference points to repeat the process.)

A specific way to implement such a scatter search approach is the concern of our following development. We remark that when the values u_h^* (and associated points $x^*(h)$) generate a valid cutting plane, as in Figure 6.8, it is easy to restrict consideration to focal points that lie in $F^* \cap X_{LP}$. This can be accomplished by generating convex combinations of pairs of reference points x' and x'' on the boundary of F^* (e.g., where x' and x'' are selected from the points $x^*(h)$ themselves or from the faces of F^*, which consist of convex combinations of all but one of the points $x^*(h)$, $h \in$ **NB**). Then an elementary computation identifies a line segment L, contained within the line segment that joins x' and x'', so that L lies entirely within $F^* \cap X_{LP}$ and its endpoints lie on the boundary of X_{LP}. (The computation identifies lower and upper bounds on the parameter that defines the convex combinations of x' and x'', which are directly implied by the LP constraints. Then the maximum lower bound and minimum upper bound for this parameter provide the endpoints of L specified. When the bounds are inconsistent with the 0 and 1 bounds of the parameter, no feasible convex combination of x' and x'' exists.) This observation, coupled with others that follow, lead us to consider an approach that generates convex combinations of such pairs of points.

6.4.3 Special Implications of LP Optimality

A special outcome holds for zero-one problems, by means of the directional rounding operator, which occurs in the situation where $x(0)$ is an optimal LP extreme point. Let X^* denote the subset of $F^* \cap X_{LP}$ whose points map into feasible integer solutions by directional rounding, i.e.,

$$X^* = \{x \in F^* \cap X_{LP}: \delta(x) \in X_{IP}\}.$$

(X^* thus is the union of the images $(F^* \cap X_{LP})$-*Image*(x), as x ranges over feasible IP solutions.) Let $Conv(X^*)$ denote the convex combination of all pairs of points in X^*. Clearly $Conv(X^*) \subset F^* \cap X_{LP}$, and by the Feasible Convex Image Theorem, each point of X_{IP} can be obtained by directionally rounding associated points of $Conv(X^*)$. We now give a result which essentially says that an optimal focal point (i.e., a focal point that maps into an optimal zero-one point) can be found among the extreme points of $Conv(X^*)$.

> *Optimal Focal Point Theorem.*
> Assume $x(0)$ is an optimal extreme point for the LP relaxation of the zero-one problem, and F^* is determined by values $u^*(h)$, $h \in \mathbf{NB}$, that generate a valid cutting plane. Then if $X_{IP} \neq \varnothing$, there exists an extreme point x' of $Conv(X^*)$ such that $\delta(x')$ is an optimal element of X_{IP}.

To take advantage of the preceding theorem, and the Feasible Convex Image Theorem, we seek to avoid unnecessary examination of focal points that map into the same zero-one solution, or that map into zero-one solutions which are dominated by solutions derived from other focal points already examined. In addition, we focus the generation of focal points so that they lie in a heuristically determined subregion of F^* (or $F^* \cap X_{LP}$).

6.4.4 The Creation of Star-Paths

We consider a strategy for generating zero-one solutions based on constructing paths within F^* and directionally rounding points on these paths. More precisely, in overview, we first construct specially designed paths between selected points of $x^*(h)$, $h \in \mathbf{NB}$ and other *matched boundary points* of F^* (continuing to suppose the face F* is determined from a valid cutting plane). Then these paths are mapped by directional rounding to create associated paths of zero-one solutions called star-paths. The elements of the star-paths provide candidate solutions to check for IP feasibility, and to be used for the phase of scatter search that seeds other heuristic processes.

We let the directional rounding operator refer to sets as well as vectors by defining $\delta(X) = \{\delta(x): x \in X\}$, where X is any collection of points in the LP cone. We thereby differentiate between a given path $P \subset F^*$ and the set of points obtained by directionally rounding from $x(0)$ to P, that is, the set $\delta(P)$. While P represents a continuous trajectory linking two elements of F^*, the trajectory represented by $\delta(P)$ is not continuous but broken, and consists of points that are mapped through the surface defined by F^* to produce a projection of P onto the zero-one vertices of the unit hypercube. We call the collection of points $\delta(P)$, a *star-path.* This terminology is motivated by envisioning the surface F^* containing P as lying above $x(0)$, as in a number of our diagrams, and hence the feasible zero-one points of $\delta(P)$ also lie above $x(0)$, displaced at least as far as F^* and characteristically beyond, as if composing a collection of "stars" scattered across the space of zero-one solutions. (By this image infeasible points, which may lie below F^*, represent "fallen stars.")

We focus attention on the case where the path P consists of a line segment joining two focal points x' and x'' on F^*, given by

$$P(x',x'') = \{x = x' + (x'' - x')\lambda : 1 \geq \lambda \geq 0\}$$

Then the associated star-path $\delta(P)$ may be identified as the set of points

$$\delta(P) = \{\delta(x): x \in P(x',x'')\}.$$

We are particularly interested in the situation where x' and x'' are boundary points of F^*. Our motivation for this stems from the following result, which is implied by the theorems of the preceding section.

> *Corollary:* Assume x^{opt} is an optimal IP solution and x' is any boundary point of F^*. Then there is another boundary point x'' of F^*, such that x^{opt} belongs to the star-path derived from x' and x''; i.e., $x^{opt} \in \delta(P)$, where $P = P(x',x'')$. Moreover, there exists such an x'' that $x^{opt} \in \delta(P \cap X_{\text{LP}})$.

As noted in Section 6.4.2, the lower and upper bound values for λ that precisely determine $P \cap X_{\text{LP}}$ as a line segment within $P = P(x',x'')$ are determined by a straightforward calculation. We next consider the choice of points to generate star-paths in accordance with the observation of the preceding Corollary.

6.4.5 Choosing Boundary Points for Creating Star-Paths

Let $X^*(\mathbf{NB}) = \{x^*(h): h \in \mathbf{NB}\}$. The points of $X^*(\mathbf{NB})$ are natural candidates to be included among the boundary points for generating star-paths, since $X^*(\mathbf{NB})$ provides the foundation for generating F^*. Further, we may initially pair each point $x^*(h)$ from this set with the point $y^*(h)$ that is the center of gravity of the remaining points of $X^*(\mathbf{NB})$, that is, where $y^*(h)$ is the midpoint of the lower dimensional face spanned by the points of $X^*(\mathbf{NB}\text{-}h)$ (the set $\{x^*(k)\text{: } k \in \mathbf{NB},\ k \neq h\}$). By such pairing, the path $P(x^*(h),y^*(h))$ traverses the interior of F^* to reach a boundary point "equidistant" from the points of $X^*(\mathbf{NB}\text{-}h)$. Thus the star-path $\delta(P)$, for $P = P\left(x^*(h), y^*(h)\right)$, is biased toward containing solutions anticipated to satisfy the IP feasibility conditions.

We may improve this bias, and simultaneously incorporate objective function considerations as well as feasibility considerations, as follows. Let $z(h)$ denote a modified objective function value for the point $x^*(h)$ that includes a penalty for infeasibility; that is, $z(h)$ is the objective function value for the IP problem when $x = x^*(h)$, increased by an amount that measures the relative infeasibility of $x^*(h)$ if $x^*(h)$ is not feasible. To create the point $y^*(h)$ as a weighted combination of the points of $X^*(\mathbf{NB}\text{-}h)$, we account for the $z(h)$ values by introducing a positive valued function f that preserves their relative ordering. Assuming a minimization objective, we stipulate that

$$f(z(h)) > 0 \qquad \text{for all } h \in \mathbf{NB}, \text{ and}$$
$$f(z(h)) \geq f(z(k)) \text{ if } z(k) \geq z(h) \qquad \text{for all } h, k \in \mathbf{NB}.$$

(The inequality $z(k) \geq z(h)$ is reversed for a maximization objective.)

Weights $w_k(h)$, $k \in \mathbf{NB}\text{-}h$, to generate the $y^*(h)$ points are then given by identifying a normalizing constant $C(h) = \sum_{k \in \mathbf{NB}-h} f(z(h))$. Then we define

$$w_k(h) = \frac{f(z(h))}{C(h)}, \quad k \in \mathbf{NB} - h.$$

As a result, this gives

$$y^*(h) = \sum_{k \in \mathbf{NB}-h} w_k(h) x^*(k), \quad h \in \mathbf{NB}.$$

The properties stipulated for the function f include the case where $f(z(h)) = 1$ for all $h \in \mathbf{NB}$, which generates each $y^*(h)$ as the center of gravity of the points of $X^*(\mathbf{NB}\text{-}h)$.

Computationally, it is burdensome to have to generate the set of $|\mathbf{NB}| - 1$ weights $w_k(h)$ for each $h \in \mathbf{NB}$, which involves on the order of $|\mathbf{NB}|^2$ weights overall. It is possible to do much better as a result of the following observation.

Remark. Define $f_0 = \sum_{k \in \mathbf{NB}} f(z(h))$ and create associated weights $w_h = \dfrac{f(z(h))}{f_0}$, $h \in \mathbf{NB}$. Then identify the weighted center of gravity y for $X^*(\mathbf{NB})$ given by

$$y = \sum_{h \in \mathbf{NB}} w_h x^*(h).$$

The ray from $x^*(h)$ through y, given by

$$\{x = x^*(h) + (y - x^*(h))\lambda: \lambda \geq 0\}$$

then contains the points of the path $P(x^*(h),y^*(h))$. Moreover, the truncated ray that results by restricting λ to satisfy $\lambda \leq \dfrac{1}{1-w_h}$ is identical to this path, with

$$y^*(h) = \frac{y - w_h x^*(h)}{1 - w_h}.$$

The preceding Remark shows that determining the weighted center y and generating points on the truncated ray from $x^*(h)$ through y makes it possible to avoid identifying each of the different sets of weights $w_k(h)$, resulting in computational effort of $O(|\mathbf{NB}|)$ instead of $O(|\mathbf{NB}|^2)$. Alternately, $y^*(h)$ can be computed as indicated in the Remark and the path $P(x^*(h),y^*(h))$ can be identified directly.

We note one further amendment for determining focal points that may serve as endpoints of paths P to generate star-paths. The Optimal Focal Point Theorem suggests the merit of selecting a focal point from a region of $F^* \cap X_{\mathrm{LP}}$ that is attractive in terms of the objective function evaluation. We have already noted how this evaluation may be used to influence weights used to select path endpoints. We can also solve the LP problem produced by adjoining the cutting plane which gives F^*, and the resulting solution will be a point on $F^* \cap X_{\mathrm{LP}}$ that is attractive to become a path endpoint. When cutting planes can be generated which are able to make significant progress toward obtaining an integer solution they can be introduced and treated as part of the LP problem itself. An appropriate point to transition from using such cutting planes to augment the LP representation, and to use them instead as a basis for generating star-paths (and for identifying a strategic focal point to become and endpoint of paths in F^* and $F^* \cap X_{\mathrm{LP}}$), is an issue for empirical investigation.

6.4.6 An Efficient Procedure for Generating the Star-Paths

Given means for determining the paths $P = P(x^*(h),y^*(h))$ for creating associated star-paths $\delta(P)$, as addressed in the preceding section, it remains to provide an approach for identifying solutions generated by these star-paths.

An infinite number of λ values exist between 0 and 1 to generate P by the definition

$$P(x^*(h),y^*(h)) = \{x = x^*(h) + (y^*(h) - x^*(h))\lambda: 0 \le \lambda \le 1\}.$$

Each x on P gives a directionally rounded solution $\delta(x)$ of the collection $\delta(P)$. However, $\delta(P)$ contains only a limited number of distinct points, consisting of vertices of the unit hypercube, and hence (infinitely) many of the λ values map into the same point of $\delta(P)$.

We will show that it is possible to identify only a small finite number of λ values that generate the star-path elements. In addition, these elements can be generated highly efficiently and without duplication. As a basis for this we demonstrate that the star-path can be represented as a mapping of P onto a collection of distinct, successively adjacent, vertices of the zero-one hybercube.

Let $\delta°(\lambda)$ identify the points of $\delta(P)$ as a function of the parameter λ. That is, for a given value of λ, which yields the point x of $P(x', x'')$ given by $x = x' + (x'' - x')\lambda$, we define $\delta°(\lambda) = \delta(x)$. Let $\delta_j^{\circ}(\lambda)$ denote the jth component of $\delta°(\lambda)$; i.e., $\delta_j^{\circ}(\lambda) = \delta(x_j)$, for $x_j = x + \Delta_j \lambda$, where $\Delta_j = x_j'' - x_j'$. Define the subsets **I**(0), **I**(+) and **I**(-) of **I** to consist respectively of those $j \in$ **I** such that $\Delta_j = 0$, $\Delta_j > 0$ and $\Delta_j < 0$. Finally, for $j \in \mathbf{I}(+)$ or $j \in \mathbf{I}(-)$, identify the special λ value given by

$$\lambda(j) = \frac{x_j(0) - x_j'}{\Delta_j}.$$

These definitions allow a precise characterization of $\delta°(\lambda)$ as follows.

Lemma C. The elements of $\delta°(\lambda)$ are given by

(a) $\delta_j^{\circ}(\lambda) = \delta(x_j')$, $\quad j \in \mathbf{I}(0)$

(b) $\delta_j^{\circ}(\lambda) = \begin{cases} 0 & \text{if } \lambda < \lambda(j) \\ 1 & \text{if } \lambda \ge \lambda(j) \end{cases} \quad j \in \mathbf{I}(+)$

(c) $\delta_j^{\circ}(\lambda) = \begin{cases} 0 & \text{if } \lambda \ge \lambda(j) \\ 1 & \text{if } \lambda < \lambda(j) \end{cases} \quad j \in \mathbf{I}(-)$

Lemma C does not depend on the assumption that λ is restricted to satisfy $0 \le \lambda \le 1$, but applies to the case where $P(x',x'')$ is the infinite line joining x' and x'', and not just the segment between these points. In addition, the lemma introduces a specific "tie breaking" rule to handle the case $\lambda = \lambda(j)$, where the original definition of directional rounding requires such a rule to choose between a value of 0 or 1.

To take advantage of Lemma C, let $\theta(1),\ldots, \theta(r)$ be a permutation of the indexes of $\mathbf{I} - \mathbf{I}(0)$ so that $\lambda(\theta(1)) \leq \lambda(\theta(2)) \leq \ldots \leq \lambda(\theta(r))$, where $r = |\mathbf{I} - \mathbf{I}(0)|$. Also let $\lambda(0)$ be any value of λ such that $\lambda(0) < \lambda(\theta(1))$. (It is acceptable to take $\lambda(0) = -\infty$.) By convention, we will suppose that the values $\lambda(\theta(q))$ $q = 1,\ldots, r$, are all distinct so that $\lambda(\theta(q)) < \lambda(\theta(q+1))$ for all $q < r$. This convention allows a maximum number of elements of the star-path $\delta(P)$ to be created, and also leads to characterizing these elements as adjacent vertices of the unit hypercube defined relative to the integer components of x. We will show that this convention is trivially easy to impose; that is, no explicit perturbation needs to be introduced to allow the $\lambda(\theta(q))$ values to be treated as distinct in case there are tied values.

Star-path Theorem.

The star-path $\delta(P)$, for $P = P(x',x'')$, contains precisely $r + 1$ distinct points, which can be generated by the rule of Lemma C when λ takes the values $\lambda(0)$, $\lambda(\theta(1))$, ..., $\lambda(\theta(r))$. The points $\delta^\circ(\lambda)$ constitute successively adjacent vertices of the unit hypercube, linked to each other by the following relationship.

For any arbitrary value of $\lambda < \lambda(\theta(r))$, let $\lambda_{\text{next}} = \lambda(p)$, where $p = \theta(q)$ for $q = \text{Min}(k : \lambda(\theta(k)) > \lambda)$. Then $\delta^\circ(\lambda)$ and $\delta^\circ(\lambda_{\text{next}})$ are associated by the rule

$$\delta_j^\circ(\lambda_{\text{next}}) = \delta_j^\circ(\lambda) \qquad \text{for } j \neq p, j \in \mathbf{I}$$

$$\delta_p^\circ(\lambda_{\text{next}}) = 1 - \delta_p^\circ(\lambda).$$

The fact that the points $\delta^\circ(\lambda)$ are successively adjacent vertices of the unit hypercube, as specified in the preceding theorem, implies that it is unnecessary to create a numerical shift of tied values of $\lambda(\theta(q))$ by an explicit perturbation in order to allow the specified points of the star-path to be generated. The following simple method based on this result produces exactly the desired points $\delta^\circ(\lambda)$, as λ ranges over any interval $\lambda_{\text{start}} \leq \lambda \leq \lambda_{\text{end}}$. Consequently, this applies to the special case where $P(x',x'')$ is a line segment generated by $0 \leq \lambda \leq 1$, and also applies to the representation of the path given in the earlier Remark, where λ can exceed the value 1.

Star-Path Generation Method

Step 0. Let $\lambda = \lambda_{\text{start}} - \varepsilon$, for a small positive value of ε, and generate the solution vector $x^\circ = \delta^\circ(\lambda)$ by Lemma C. If $\lambda \geq \lambda(\theta(r))$, x° is the only vector to be generated and the procedure stops. Otherwise, identify $p = \theta(q)$, where $q = \text{Min}(k: \lambda(\theta(k)) > \lambda)$.

Step 1. Generate the next x^o vector by setting $x_p^o = 1 - x_p^o$, without changing any other elements $x_j^o, j \in \mathbf{I}, j \neq p$.

Step 2. Set $q = q + 1$. If $q > r$ or if $\lambda(\theta(q)) > \lambda_{end}$, stop. Otherwise, set $p = \theta(q)$ and return to Step 1.

Setting $\lambda = \lambda_{start} - \varepsilon$ in Step 0 generates the maximum number of points for the star-path, in case λ_{start} coincides with one of the $\lambda(\theta(q))$ values. (Otherwise, the reference to ε is unnecessary.) Note that $\delta^o(\lambda)$ is only computed once, in Step 0. Thereafter, each new vector x^o results simply by changing the value of the single element x_p^o in Step 1. This corresponds to applying the formula of the Star-path Theorem for $x^o = \delta^o(\lambda_{next})$.

Also, there is no requirement in Step 2 that the $\lambda(\theta(q))$ values be strictly increasing. The procedure simply increments q, and all tie breaking is entirely implicit. Finally, the vector $x^o = \delta^o(\lambda)$ can be checked very efficiently to determine if it is a feasible zero-one solution, due to the fact that exactly one element of x^o changes at each execution of Step 1. Thus, the feasibility check can be based on a marginal calculation, rather than evaluating a complete new vector at each step. (All of the $x^*(h)$, $y^*(h)$ pairs will generate the point $\delta(y)$, where y is the weighted center of gravity identified previously. This duplication can easily be avoided, if desired, by accounting for the λ value to be skipped.)

The earlier results have demonstrated that selecting λ values within the range from 0 to 1 suffice if $x' = x^*(h)$ and $x'' = y^*(h)$, for $x^*(h)$ and $y^*(h)$ appropriately matched. Further, it is readily possible, as previously remarked, to identify an LP feasible line segment within $P(x^*(h),y^*(h))$, if such a segment exists, as a basis for determining a smaller range of λ values. However, we note it may be useful to allow consideration of points generated over a wider range of λ values, using the preceding method.

In the case where $x(0)$ is an optimal LP extreme point, the Optimal Focal Point Theorem implies there exist star-paths for which it is possible to restrict attention to the first and last feasible zero-one solutions encountered on the path. Thus, an option is stop the execution of the Star-Path Generation Method as soon as a first feasible zero-one solution is found, and then apply the method in reverse, starting from $\lambda = \lambda_{end} + \varepsilon$, and again stopping at the first feasible zero-one solution found (or upon reaching the point where the forward pass stopped). However, this option should be considered with caution, because in spite of the extreme point property of the Optimal Focal Point Theorem, there also can exist star-paths in which an optimal solution is encountered at a point other than the first or last feasible solution of the path. (See Exercise 14.)

6.4.7 *Adaptive Re-Determination of Star-Paths Using Tabu Search Intensification Strategies.*

There are many possibilities for generating star-paths, which derive from using different cutting planes for generating the set of points $X^*(\mathbf{NB})$ and from different choices of a function f for generating weights to produce the $x^*(h)$, $y^*(h)$ pairs. We show that the outcomes of implementing such approaches can be embedded in a tabu search intensification process that determines new star-paths adaptively. In this way the scatter search method effectively learns how to modify itself to take advantage of information generated from previous efforts.

The basic notion is the following. Each star-path trajectory is based on focal points in F^* that are directionally rounded to produce the star-path points. The preceding section shows these focal points do not have to be explicitly identified (except for the first). Nevertheless, the λ value that produces each star-path point is always known, and can be used to identify the associated focal point of F^*. (We may take liberties by allowing some focal points to lie outside of F^*.) Moreover, in general, a range of λ values is known that produces each star-path point. A midpoint of this range can be selected to give a representative focal point.

As a result, we may consider a collection E of elite zero-one candidate points that are generated from various star-paths. This gives rise to an associated collection $F^*(E)$ of focal points on F^* that create the elements of E (where each $x \in F^*(E)$ yields a point $\delta(x)$ of E). The criteria for identifying E can be based on objective function values penalized for infeasibility or can more broadly include reference to diversification.

The elements of $F^*(E)$ identify a set of preferred focal points, and their convex combinations may be viewed as defining a *preferred focal region* as the basis for an intensification process. Moreover, the points of $F^*(E)$ can be treated exactly as the points $X^*(\mathbf{NB})$ to generate a weighted center of gravity y. Then the point y can be used to generate paths $P(x^*(h),y)$, which are extended beyond y to create a match between $x^*(h)$ and an implicit point $y^*(h)$. Intensification occurs as a result of tracing these paths through the region spanned by $F^*(E)$, but diversification also occurs because the paths "extrapolate beyond" this region.

By the Star-Path Theorem and the Star-Path Generation Method accompanying it, there is no need to precisely identify a λ value that will generate a point $y^*(h)$, since it suffices to take $x' = x^*(h)$ and $x'' = y$, and to generate a sequence $\lambda(\theta(q))$ that will encompass all relevant possibilities. The extreme ends of the sequence are likely to be irrelevant, and there is no need to duplicate the use of $x^*(h)$ as a focal point. Hence the star-path generation may be applied by starting at a $\lambda(\theta(q))$ value larger than 0 (since $\lambda = 0$ treats $x' = x^*(h)$ as a focal point), and to continue until persistent deterioration in the star-path elements occurs. The points $x^*(h)$ themselves can be replaced on subsequent rounds by shifting them toward the most recent y, in which case negative starting λ values may be relevant.

As the adaptive procedure is repeated, a standard tabu search process can be used to avoid duplicating the composition of the collection E, and hence of the focal set $F(E)$. Moreover, it is possible to apply the approach in a compound manner. We identify a way to do this as follows, effectively permitting a given collection E to generate alternative collections by a nested form of adaptation.

6.4.8 Strategy for Augmenting a Collection of Preferred Solutions

A particularly simple instance of scatter search can serve as a basis for compounding the adaptive approach for generating star-paths, with the goal of producing new zero-one solutions as candidates to be included in a preferred collection E. The approach is a strongly focused form of intensification that treats members of E as reference points, and gives rise to additional candidates for membership in E by a process of creating weighted combinations. (A fuller description of scatter search is given in Chapter 9.)

Let S denote an index set for a chosen subset of the solutions in E, and identify the members of this subset by $X(S) = \{x(s), s \in S\}$. For simplicity, consider the result of rounding in a nearest neighbor sense. If all points are weighted equally, thereby producing a center of gravity of $X(S)$, the result of nearest neighbor rounding yields a point x whose component x_j, $j \in \mathbf{I}$ receives the value taken by the majority of the components $x_j(s)$, $s \in S$. Thus if S contains an odd number of elements, the point x is uniquely determined.

This observation suggests the following strategy for generating trial solutions x as candidates to augment the collection E, or to replace its inferior elements.

Simple Scatter Search Approach to Augment E

1. Let $S(k)$ be the index set for the "k best" solutions from the collection E.
2. Choose one or more even values of k, and consider each of the k subsets of $S(k)$ consisting of k-1 of its elements.
3. Represent each of the chosen subsets of solutions by $\{x(s), s \in S\}$, where S contains an odd number of elements. From each subset, generate a trial point x where, for each $j \in \mathbf{I}$: $x_j = 1$ if $x_j(s) = 1$ for the majority of $s \in S$, and $x_j = 0$ otherwise.
4. For each trial point x, test whether it passes a threshold of attractiveness to be admitted to the collection E (e.g. whether it has an evaluation better than the worst or average member of E).

By way of illustration, the foregoing procedure can select $k = 4$ and $k = 6$ in Step 2, thus generating each of the 4 subsets of 3 solutions associated with $S(4)$, and each of the 6 subsets of 5 solutions associated with $S(6)$. Then, in Step 3, $S(4)$ and $S(6)$ together generate 10 trial points. If k is chosen to be relatively small as in this example, subsets of k-3 elements can also be chosen for even values of k and subsets of k-2 elements can be chosen for odd values of k. For instance, 20 additional trial points can be generated in Step 3 by letting S range over all 3 element subsets of the 6 best solutions, and 21 additional trial points can be generated by letting S range over all 5 element subsets of the 7 best solutions. Some of the trial points produced by the method may duplicate others, but such duplications may be avoided (without bothering to identify and weed out the duplicates) by requiring each selected point to be better than its predecessor in the threshold test of Step 4.

Trial solutions can be obtained other than by majority vote by weighting solutions $x(s)$ in relation to their objective function values $z(s)$. If all of the elements of $X(S)$ are feasible, let $z(0)$ denote the optimum objective function value for the original LP solution. Then $\Delta z(s) = z(s) - z(0)$ is positive for all $s \in S$, or else an optimal solution is known. (If some elements of $X(S)$ are infeasible, simply select $z(0)$ to be smaller than the minimum $z(s)$ value.) Create a convex combination of the points of $X(S)$ using the weights $\frac{\Delta z(s)}{D}$, where $D = \sum_{s \in S} \Delta z(s)$. Then x can be generated by rounding this outcome.

6.4.9 *Star-Paths for General Integer Programs*

The star-path construction can be extended to IP problems in general integer variables, with upper bounds U_j that may differ from 1, by defining the directional rounding operator relative to a *unit hypercube envelope* of a focal point. The mapping $\delta(x')$ remains the same as previously given in the special case where x' belongs to the zero-one unit hypercube, but differs in a natural way for focal points in other hypercubes.

To define the extended mapping appropriately, we continue to suppose the values of variables x_j, $j \in$ **N-I**, are determined once the values for the integer variables are specified, and determine $\delta(x')$ for a given focal point x' by specifying the values $\delta(x'_j)$ for each $x'_j, j \in$ **I**. Let v'_j and v'_j+1 be integer values that bracket the value of x'_j, and let L'_j=Max$(0, v'_j)$ and U'_j=Min(U_j, v'_j+1). (Hence L'_j and U'_j are just the adjacent integers v'_j and v'_j+1 unless x'_j violates one of its bounds.) Then $\delta(x'_j)$ is defined exactly as for the zero-one case by replacing the values 0 and 1 respectively with L'_j and U'_j.

By this general definition, a single ray can contain more than one relevant focal point. However, all relevant points can be easily identified by noting where the values L'_j and U'_j change as the edge is extended.

In the general case, # can represent either L'_j and U'_j, and a decision rule such as the one implicit in the Star-Path Generation Method is assumed to select precisely one of these values. Figure 6.9 illustrates the integer points $x°$ generated by this definition, indicated by enclosing these points in circles, as x' ranges over focal points on a face F^* determined by a valid cutting plane. The image of each such point $x°$ on F^* (that is, the region (F^*)-*Image*($x°$)) is also identified for each of these points by the divisions of F^* denoted by the letters A through E. Thus, the line segment AB is the image of the point labeled AB, the line segment BC is the image of the point labeled BC, and so forth. (The rule for defining # determines whether the point B itself should map into the integer point labeled AB or BC.)

Fig. 6.9 Star-path for general integer variables.

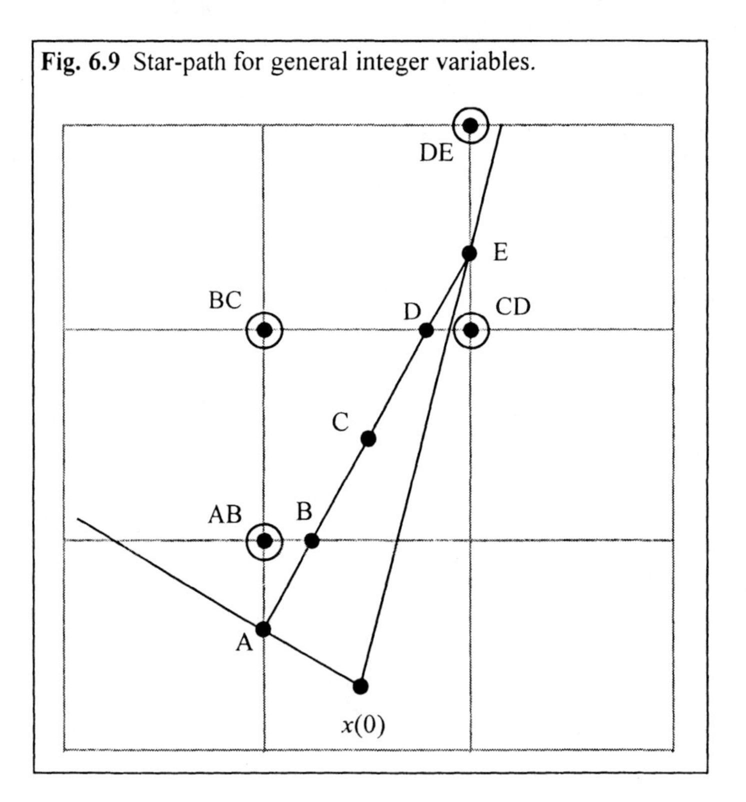

The integer points of this figure no longer all belong to the same hypercube, as they do in the zero-one case. (In fact, only the point AB lies on the unit hypercube that contains $x(0)$.)

The Star-Path Theorem directly generalizes to apply to this extended definition of $\delta(x)$. Each integer point successively generated on a star-path is adjacent to its predecessor, and the Star-Path Generation Method also generalizes to permit the integer points of the star-path to be generated efficiently (Exercise 23).

6.4.10 Summary Considerations

The star-path constructions make it possible to produce solution combinations with special properties for integer programming problems. The basis for this approach results by focusing on two aspects of the scatter search framework, the characterization of reference points that generate solution combinations and the mechanism for transforming fractional elements into integer elements. Parallel processing can take a useful role in exploiting the star-path construction by generating and coordinating the set of points $X^*(\mathbf{NB})$, and in choosing a function f to generate weights for producing additional matched points $y^*(h)$. Parallel processing also can take a role in simultaneously determining parameters associated with different elements of $X^*(\mathbf{NB})$, and in determining the specific star-paths associated with the $x^*(h)$, $y^*(h)$ pairs.

These processes can be usefully applied by selecting $x(0)$ from different extreme points of the LP feasible region. Trial solutions generated from star-paths, whether feasible or not, can be used to create a modified objective function, weighting costs to produce optimal LP extreme points for the modified objective that lie in the vicinity of these trial solutions. Then these extreme points are natural candidates to serve as $x(0)$. Likewise, the trial solutions may be subjected to heuristic modification, following the standard scatter search design.

6.5 Branching on Created Variables

We conclude this chapter by examining straightforward ideas which can have a significant impact on the ability to solve integer programming problems more effectively. The underlying concepts can be used both to enhance the application of tabu branching and to directly improve the design of branch and cut procedures.

We begin by considering the standard approach for generating the branching steps in a branch and bound approach. For convenience, we introduce a slight change of notation and disregard upper bounds on the variables. (The connection with previous notation and the associated handling of upper bounds will be apparent.) Let s denote a current basic variable that is constrained to be integer-valued (i.e., s corresponds to one of the variables x_j, $j \in \mathbf{B} \cap \mathbf{I}$), and identify the current LP basis representation for s by the equation

$$s + \sum_{h \in \mathbf{NB}} a_h x_h = a_0 .$$

Assuming a_0 is not an integer, we may consider branching on s by compelling it alternately to be bounded below and above by the integers on either side of a_0; that is, we seek to create the branches represented by the two inequalities

$$s \le \lfloor a_0 \rfloor \qquad \text{(Down Branch)}$$
$$s \ge \lceil a_0 \rceil \qquad \text{(Up Branch)}$$

The terms $\lfloor u \rfloor$ and $\lceil u \rceil$, for an arbitrary scalar u, respectively denote the largest integer $\le u$ and the smallest integer $\ge u$.

Identify the two associated integer-valued slack variables for these inequalities by $s' = \lfloor a_0 \rfloor - s \ge 0$ and $s'' = s - \lceil a_0 \rceil \ge 0$. Substituting the resulting expressions for s (i.e., $s = \lfloor a_0 \rfloor - s'$ and $s = s'' + \lceil a_0 \rceil$) in the basis equation, and defining the "fractional parts" of a_0 by $f_0 = a_0 - \lfloor a_0 \rfloor$ and $g_0 = \lceil a_0 \rceil - a_0$, gives the following associated branch equations

$$s' - \sum_{h \in \mathbf{NB}} a_h x_h = -f_0 \qquad \text{(Down Branch Equation)}$$
$$s'' + \sum_{h \in \mathbf{NB}} a_h x_h = -g_0 \qquad \text{(Up Branch Equation)}$$

Since both f_0 and g_0 are positive by the assumption that a_0 is not an integer, the branching equations assign negative values to s' and s'' when the nonbasic variables are 0 (thus causing each of the branches to exclude the current LP vertex).

By taking account of the nonnegativity of s' and s'', the preceding equations can be expressed in standard inequality form

$$\sum_{h \in \mathbf{NB}} a_h x_h \ge f_0 \qquad \text{(Down Branch Inequality)}$$
$$-\sum_{h \in \mathbf{NB}} a_h x_h \ge g_0 \qquad \text{(Up Branch Inequality)}$$

We illustrate these inequalities graphically by an example that provides a basis for illustrating a number of additional ideas later. For this, we refer to the source equation

$$s + 1\tfrac{1}{5}x_1 - 2\tfrac{1}{5}x_2 = 7\tfrac{2}{5}.$$

The corresponding Up and Down Branch Inequalities are

$$1\tfrac{1}{5}x_1 - 2\tfrac{1}{5}x_2 \geq \tfrac{2}{5} \qquad \text{(Down Branch Inequality)}$$

$$-1\tfrac{1}{5}x_1 + 2\tfrac{1}{5}x_2 \geq \tfrac{1}{5} \qquad \text{(Up Branch Inequality)}$$

These two resulting branches are depicted in Figure 6.10. Each branch evidently excludes the origin, which corresponds to excluding the basic solution that assigns 0 values to the nonbasic variables x_1 and x_2.

Fig. 6.10 Illustrative Up and Down branches.

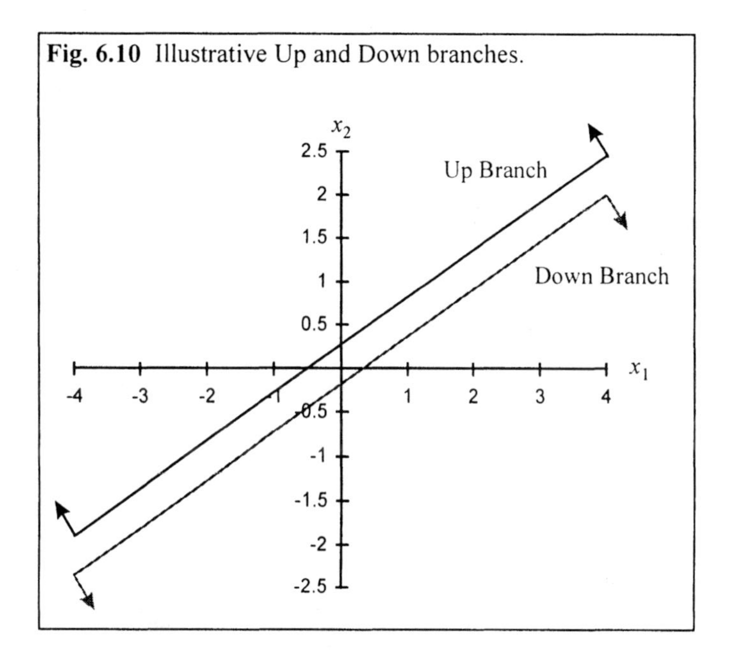

6.5.1 A Perspective to Unite Branching and Cutting

As noted in Section 6.1, the branch and cut philosophy, which seeks to integrate branching procedures and cutting plane procedures in a single approach, provides some of the more effective exact solution methods for pure and mixed IP problems. It is useful to go a step farther, however, by viewing branching processes and cutting plane processes as part of a common framework. From this viewpoint, each process can be characterized as a variant of the other. More particularly, branching can be conceived as *provisional cutting*, and cutting can be conceived as *selectively-infeasible branching* (where one of a pair of branching alternatives is constructed to be infeasible). The relevance of this perspective becomes manifest by considering the notion of branching on created variables (Glover, 1975a, 1975b).

To demonstrate, consider again the source equation

$$s + \sum_{h \in \mathbf{NB}} a_h x_h = a_0 .$$

The variables of this equation are customarily assumed nonnegative, but the branching alternatives ($s \geq \lceil a_0 \rceil$ and $s \leq \lfloor a_0 \rfloor$) depend only on the fact that s is an integer variable. Consequently, two corresponding alternatives arise by replacing s with any other integer variable y, where y is created from s and the nonbasic variables x_h, $h \in \mathbf{I}$. Specifically, we can establish such alternatives by defining $y = s + \sum_{h \in \mathbf{NB}} k_h x_h$, where each k_h is an integer for $h \in \mathbf{I}$ and $k_h = 0$ for $h \notin \mathbf{I}$. Letting $b_h = a_h - k_h$, we therefore obtain the source equation

$$y + \sum_{h \in \mathbf{NB}} b_h x_h = a_0 .$$

By the same construction as before we obtain branches for nonnegative integer variables y' and y'' given by

$$y' - \sum_{h \in \mathbf{NB}} b_h x_h = -f_0 \qquad \text{(Down Branch Equation)}$$

$$y'' + \sum_{h \in \mathbf{NB}} b_h x_h = -g_0 \qquad \text{(Up Branch Equation)}$$

and similarly obtain the associated branching inequalities

$$\sum_{h \in \mathbf{NB}} b_h x_h \geq f_0 \qquad \text{(Down Branch Inequality)}$$

$$-\sum_{h \in \mathbf{NB}} b_h x_h \geq g_0 \qquad \text{(Up Branch Inequality)}$$

The only difference between these inequalities and the earlier branching inequalities is to replace a_h by the quantity b_h, which differs from a_h by an integer amount for $h \in \mathbf{I}$ and is the same as a_h for $h \notin \mathbf{I}$.

From the perspective that conceives a cutting plane to result from a pair of branches in which one is infeasible, we may consider making all b_h nonnegative, which will cause the Up Branch Inequality to become infeasible. By the same token, we may consider making all b_h nonpositve, which will cause the Down Branch Inequality to become infeasible.

Such an outcome is easy to achieve for a pure IP problem, where all variables are integer, since then each k_h can freely be chosen to establish the resulting sign conditions for the b_h coefficients. It is also easy to see which values of k_h are most appropriate. If we undertake to make all $b_h \geq 0$, so that the Down Branch Inequality becomes the cut (the inequality that is not rendered infeasible), then it is desirable to make each b_h as small as possible — since then the Down Branch Inequality will imply any other such cut inequality that has larger coefficients. Similarly, if we undertake to make all $b_h \leq 0$, it is desirable to make the nonnegative values $-b_h$ as small as possible (so that the Up Branch Inequality will imply all associated cut alternatives). The smallest nonnegative b_h values occur by letting $k_h = \lfloor a_h \rfloor$, while the smallest nonnegative $-b_h$ values occur by letting $k_h = \lceil a_h \rceil$. Thus, defining $f_h = a_h - \lfloor a_h \rfloor$ and $g_h = \lceil a_h \rceil - a_h$, and noting that $k_h = \lfloor a_h \rfloor$ yields $b_h = f_h$ and $k_h = \lceil a_h \rceil$ yields $b_h = g_h$, we obtain the following two instances of the branching inequalities:

$$\sum_{h \in \mathbf{NB}} f_h x_h \geq f_0 \qquad \text{(P-Cut: Down Branch Inequality)}$$

$$-\sum_{h \in \mathbf{NB}} f_h x_h \geq g_0 \qquad \text{(Infeasible: Up Branch Inequality)}$$

and

$$-\sum_{h \in \mathbf{NB}} g_h x_h \geq f_0 \qquad \text{(Infeasible: Down Branch Inequality)}$$

$$\sum_{h \in \mathbf{NB}} g_h x_h \geq g_0 \qquad \text{(N-Cut: Up Branch Inequality)}$$

The indicated cuts are labeled the "P-Cut" and the "N-Cut" (P for positive and N for negative) because they are respectively the well known cutting planes of "positive fractional parts" and "negative fractional parts" due to Gomory (1958). The orientation that uncovers these cuts as a direct byproduct of branching construction soon leads to additional consequences of greater significance.

To illustrate the outcomes identified so far, consider again the example source equation that gives rise to the branching inequalities shown in Figure 6.10, i.e.,

$$s + 1\tfrac{1}{5}x_1 - 2\tfrac{1}{5}x_2 = 7\tfrac{2}{3}.$$

Assuming both x_1 and x_2 are integer variables, and choosing $k_h = \lfloor a_h \rfloor$ in one case and $k_h = \lceil a_h \rceil$ in the other, we obtain the two related source equations

$$y + \tfrac{1}{5}x_1 + \tfrac{4}{5}x_2 = 7\tfrac{2}{5}$$

and

$$y - \tfrac{4}{5}x_1 + \tfrac{1}{5}x_2 = 7\tfrac{2}{5}$$

The two branching inequalities for the first of these equations are

$$\tfrac{1}{5}x_1 + \tfrac{4}{5}x_2 \ge \tfrac{2}{5} \qquad \text{(P-Cut: Down Branch)}$$

and

$$-\tfrac{1}{5}x_1 - \tfrac{4}{5}x_2 \ge \tfrac{3}{5} \qquad \text{(Infeasible Up Branch)}$$

While, the two branching inequalities for the second of these equations are

$$-\tfrac{4}{5}x_1 - \tfrac{1}{5}x_2 \ge \tfrac{2}{5} \qquad \text{(Infeasible Down Branch)}$$

and

$$\tfrac{4}{5}x_1 + \tfrac{1}{5}x_2 \ge \tfrac{3}{5} \qquad \text{(N-Cut: Up Branch)}$$

Fig. 6.11 Illustrative P and N Cuts.

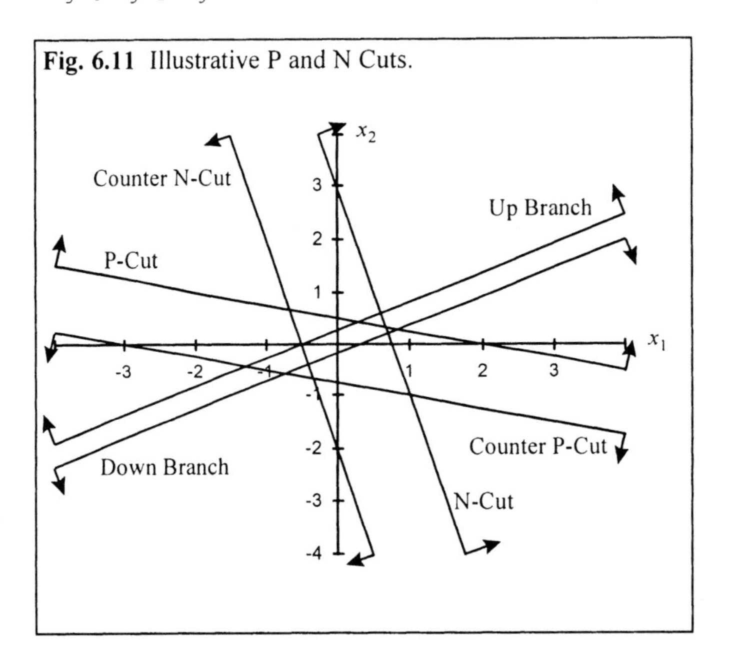

These two sets of branches, and the cutting planes they produce, are shown in Figure 6.11. The infeasible Up Branch that accompanies the P-Cut is called the Counter P-Cut, and the infeasible Down Branch that accompanies the N-Cut is called the Counter N-Cut. For comparison, we also include the original branching inequalities shown in Figure 6.10.

Figure 6.11 shows that the infeasible branches (Counter P-Cut and Counter N-Cut) exclude the nonnegative orthant from their associated regions, and therefore assure that their respective alternative branches are indeed cutting planes. The P-Cut and N-Cut eliminate a portion of the nonnegative orthant, including the origin, and hence render the current basic solution infeasible.

In addition, each of the original branching inequalities clearly eliminates a large part (in fact, an infinite part) of the nonnegative orthant that is not eliminated by either of the cuts. Conversely, the cuts also remove an important segment of the region that is admitted by the original branching inequalities. This illustrates why the strategy of combining branching and cutting can be more powerful than applying either of the two approaches in isolation from the other. We will see, however, that the branching component of such a strategy can be made still more powerful.

6.5.2 Straddle Dominance - Branching on Created Variables

Since cuts can arise by branching on one particular class of created variables, we may suspect that other types of created variables may yield branches with other useful features. Previous observations about cut search (Section 6.3) already provide a basis for confirming this suspicion.

Further examination of Figure 6.11 gives an idea of relevant issues. First of all, we would like to exploit branching inequalities, in the sense of "provisional cuts," to take advantage of the fact that each member of a branching pair generally excludes a larger part of the feasible space than a standard cutting plane. Accompanying this, we would like the region between the branches — the region that is rendered infeasible by *both* of them — to be as large as possible. (The size of this region, determined by its width, gives an indication of the power of a branching pair.)

Ideally, in fact, we would seek a pair of branches that is strong enough to render cutting planes of the type we have illustrated superfluous. The branches derived from the original integer variable s, as shown in Figure 6.11, clearly do not have this property. Each of the two cutting planes eliminates a significant part of the area near the origin that is not eliminated by either branch. A truly appealing branching pair would give each member of the pair the ability to completely eliminate all solutions that are eliminated by at least one of the cutting planes shown in Figure 6.11; that is, the two branches would effectively *straddle* such a cutting plane. We use this observation as the basis for the following definition.

> *Straddle Dominance* — A branching pair *straddle-dominates* a given cutting plane if each member of the pair (conceived as a provisional cut) excludes all points of the nonnegative orthant that are excluded by the cutting plane.

Figure 6.11 provides additional clues about how to pursue the goal of straddle dominance. In order to straddle-dominate the P-Cut or N-Cut shown, both branches of a branching pair must intersect the coordinate axes for the nonbasic variables (here x_1 and x_2) in one of two ways: either an intersection point must be negative, or it must occur at a positive value of the nonbasic variable that is at least as large as the intersection value of the cutting plane. For example, the branching pair of Figure 6.11 does not straddle-dominate either the P-Cut or the N-Cut because the positive intersections of each branch with the x_1 or x_2 axis occur at a smaller value than the cutting plane intersection. (The absence of an intersection, which occurs for a nonbasic variable with a 0 coefficient in the associated branching inequality, is conceived as an infinitely large intersection. We note that the coordinate axes are just the edges of the LP cone, as treated in the cut search and star path frameworks of the preceding sections.)

Stated alternatively, we want the positive intersections of each branching inequality to be at least as "deep" — to occur at least as far from the origin — as the corresponding positive intersections of the cutting plane. The options made available from the strategy of branching on created variables assure that such a goal can be achieved.

A simple algebraic analysis gives the desired result. When branches are produced from a created variable y, we observe that the coefficient b_h in the Down Branch is "paired with" the coefficient $-b_h$ in the Up Branch. Consequently, a negative coefficient on one branch (which automatically dominates the P-Cut and the N-Cut on that dimension) entails with a positive coefficient on the alternative branch. We accordingly seek to make this positive coefficient as small as possible in order to get an intersection at least as deep as the P-Cut or the N-Cut. Creating such an intersection will generally provide a substantially deeper branch than obtained by branching on the original integer variable s.

Suppose we start with $b_h = f_h$ for all h, so that the Down Branch is a P-Cut while the Up Branch is infeasible. Then we apply the following rule: for each h such that $f_h > f_0$, subtract 1 from the current value of b_h (= f_h) to give the new value $b_h = -g_h$. This causes both branches of the branching pair to yield a better intersection for x_h for each b_h changed. The Down Branch will be better because its intersection is negative and the Up Branch will be better because its intersection, given by $\frac{g_0}{g_h}$ (which equals $\frac{1-f_0}{1-f_h}$), must be larger than the previous intersection for the Down Branch, given by $\frac{f_0}{f_h}$. All other positive intersections for the Down Branch will remain the same as in the P-Cut, and the corresponding intersections of the Up Branch will be negative.

The preceding simple rule therefore assures that a branching pair will be created that straddle-dominates the P-Cut. A similar process produces a branching pair that straddle dominates the N-Cut. (Start with $-b_h = g_h$ for all h, so that the Up Branch begins as an N-Cut and then add 1 to each b_h such that $g_h > g_0$ to give $-b_h = -f_h$). But, indeed, these two processes are identical, since as long as we consider only the coefficients where a_h is not an integer (hence f_h and g_h are positive and sum to 1), then $f_h > f_0$ if and only if $g_h < g_0$, and similarly $f_h < f_0$ if and only if $g_h > g_0$.

As a result, the process creates a branching pair that simultaneously straddle-dominates both the P-Cut and the N-Cut.

To summarize, the branching pair that achieves this effect is created by partitioning the index set $\mathbf{I} \cap \mathbf{NB}$ for the nonbasic integer variables into two sets $\mathbf{I}(1)$ and $\mathbf{I}(2)$ so that $f_h \le f_0$ (equivalently $g_h \ge g_0$) for all $h \in \mathbf{I}(1)$ and $f_h \ge f_0$ (equivalently $g_h \le g_0$) for all $h \in \mathbf{I}(2)$. (Indexes for which $f_h = f_0$ and $g_h = g_0$ can be arbitrarily assigned to either set, as can indexes for which $f_h = g_h = 0$.) Then the resulting pair of branches, which we call the Straddle-Dominating or S-D pair, can be represented in inequality form as

$$\sum_{h\in\mathbf{I}(1)} f_h x_h - \sum_{h\in\mathbf{I}(2)} g_h x_h + \sum_{h\in\mathbf{NB}-\mathbf{I}} a_h x_h \ge f_0 \qquad \text{(S-D: Down Branch)}$$

$$-\sum_{h\in\mathbf{I}(1)} f_h x_h + \sum_{h\in\mathbf{I}(2)} g_h x_h - \sum_{h\in\mathbf{NB}-\mathbf{I}} a_h x_h \ge g_0 \qquad \text{(S-D: Up Branch)}$$

We have included the continuous variables x_h for $h \in \mathbf{NB} - \mathbf{I}$ since the branching inequalities are valid for MIP problems as well as pure IP problems. (The P-Cut and N-Cut are derived only for pure problems, but we will immediately see additional implications for MIP problems.)

We return now to our previous example, whose original source equation is

$$s + 1\tfrac{1}{5}x_1 - 2\tfrac{1}{5}x_2 = 7\tfrac{2}{5}.$$

The S-D pair is then given by:

$$\tfrac{1}{5}x_1 - \tfrac{1}{5}x_2 \ge \tfrac{2}{5} \qquad \text{(S-D: Down Branch)}$$

and

$$-\tfrac{1}{5}x_1 + \tfrac{1}{5}x_2 \ge \tfrac{3}{5} \qquad \text{(S-D: Up Branch)}$$

These two branches are graphed in Figure 6.12, which also includes the cutting planes for the original variable s. It is clear that the S-D pair straddle-dominates both the P-Cut and the N-Cut, and excludes a significantly larger region than the two original branches obtained by branching on s (as seen by comparing Figure 6.12 with Figure 6.11).

Fig. 6.12 Illustrative S-D pair.

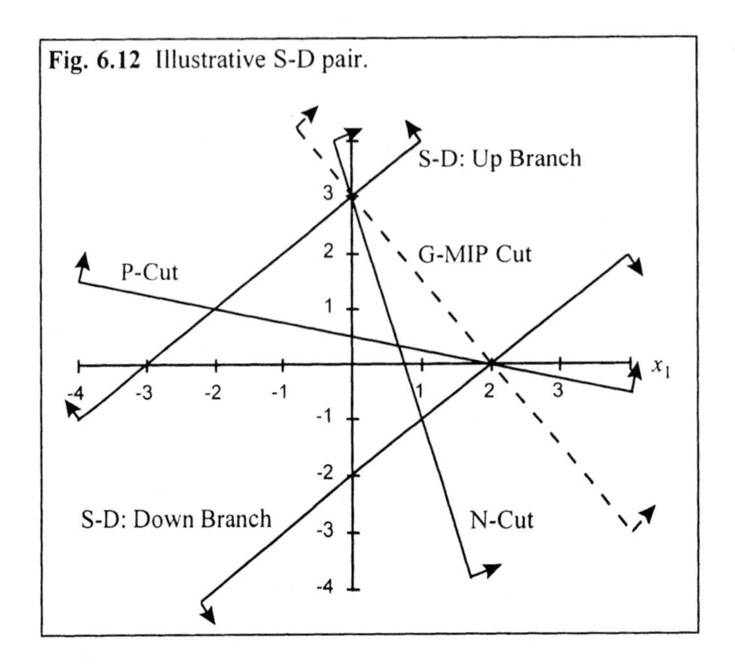

One further feature can be observed from Figure 6.12. The dotted line that joins the intersection points of the S-D pair with the x_1 and x_2 axes is also a cutting plane — and indeed by observations of earlier sections it is the MIP cutting plane proposed in Gomory (1960), which we henceforth call the G-MIP cutting plane (G for Gomory).

As a basis for reference, the inequality form of the G-MIP cut can be written as

$$\sum_{h \in \mathbf{NB}} b_h^* x_h \geq 1$$

where $b_h^* = \dfrac{b_h}{f_0}$ if $b_h \geq 0$ and $b_h^* = -\dfrac{b_h}{g_0}$ if $b_h < 0$, when b_h takes the value that yields an S-D pair.

The relationship between the three cutting planes of Figure 6.12 and the S-D branching pair is always as shown here. That is, the G-MIP cutting plane dominates the P-Cut and the N-Cut (in a pure IP setting) as a result of intersecting each coordinate axis at the deeper of the intersections of these other two cuts.[1] In turn, the

[1] In the case of an all-integer IP problem it is possible to scale the G-MIP cut to yield an integer-valued slack variable by multiplying the inequality by the value D, which equals the absolute value of the determinant of the LP basis matrix. The effect on numerical stability and the strength of future cuts is suspect, since the new basis determinant after a pivot will be a factor of D larger than without the scaling, but no empirical tests of the effect have been

S-D branching pair straddle-dominates all three cuts by likewise sharing the deeper intersections (but additionally including negative intersections). We summarize our main conclusions as follows.

> *Straddle-Dominance Theorem.*
>
> The S-D branching pair straddle-dominates both the P-Cut and the N-Cut for pure IP problems, and also straddle-dominates the G-MIP cutting plane.

The branching result embodied in the preceding theorem is a direct manifestation of observations that show the G-MIP cut is an instance of a convexity cut. For clarity we point out that possibly **I**(1) or **I**(2) may contain all of $\mathbf{I} \cap \mathbf{NB}$, which causes the G-MIP cutting plane to coincide with either the P-Cut or the N-Cut in the pure IP setting. In this case, the straddle-dominance of the S-D pair is not strict, because this pair will also yield one member that coincides with the cut and the other member that is infeasible. This "degenerate" form of straddle dominance implies that the associated pure IP cutting plane is relatively strong. (There of course may be alternatives that generate stronger cutting planes and stronger S-D pairs.)

From the present vantage point, we emphasize again that our key concern is not to identify stronger cutting planes, which we already know how to do by several approaches, but to identify more effective branching alternatives. The straddle-dominance result becomes significant for this latter goal because it provides an indication of "branching power" that is not normally shared by customary approaches that branch on original problem variables.

6.5.3 Added Significance of S-D Branching

Figure 6.12 poses an interesting conundrum. Although, the G-MIP cutting plane dominates the two associated pure integer cutting planes, practical wisdom counsels that it is usually preferable to use the pure IP cuts when possible — i.e., when the problem is a pure IP problem. The reason for this seemingly implausible outcome is the following. The slack variable that transforms the G-MIP cutting plane into an equation (and hence that becomes a new variable in the augmented LP problem) is a continuous variable rather than an integer variable. Since the ability to obtain deeper intersections depends on modifying coefficients of integer variables, the introduction of such a continuous slack variable attenuates the strength of future cuts that may be obtained. On the other hand, the slack variables for the P-Cut and N-Cut (and associated cuts obtained from other integer multiples and combinations of source equations) are integer variables. Consequently, the process of alternating the strength of future inequalities does not occur in a corresponding fashion for these pure IP cuts.

---

conducted. Such a scaling fails to provide an integer-valued slack for the general MIP problem in any event.

The desirability of introducing slack variables that are integer variables rather than continuous variables reinforces the value of using an S-D branching pair. Not only does such a pair straddle-dominate the G-MIP cut, but in addition its slack variables (y' and y'') are integer-valued. Thus, the S-D branching pairs are advantageous for retaining the ability to obtain stronger cuts and branches on future iterations.

The approach of branching on created variables clearly does not need to be limited to branching on S-D pairs. Coefficients b_h that are larger in absolute value than f_h and g_h can be used. The resulting reduction of the depth of one of the two branches suggests that this alternative should be used sparingly (and probably should retain the absolute value of all coefficients to be less than 2). Branching pairs other than an S-D pair in general should be considered only where it appears especially desirable to "branch deeply" relative to a particular variable, by making its coefficient more negative — at the expense of a correspondingly shallower branch relative to that variable when the opposing branch is introduced. However, it is possible to compensate for this weakening of one branch by an extended construction. The next section identifies a way to get further leverage out of the branching options when some variables are given different coefficients than in the S-D pair, provided these coefficients are all changed in the same direction.

6.5.4 Triple Branching

We have noted that the composition of the sets $\mathbf{I}(1)$ and $\mathbf{I}(2)$ is not unique in case there exist one or more variables x_h, $h \in \mathbf{I} \cap \mathbf{NB}$, for which $f_h = f_0$, since the indexes of these variables can be allocated to $\mathbf{I}(1)$ and $\mathbf{I}(2)$ in any way desired. Different choices of such allocations provide different S-D branching pairs. (Variables for which $f_h = g_h = 0$ can also be allocated arbitrarily to $\mathbf{I}(1)$ and $\mathbf{I}(2)$, but such allocations do not affect the S-D pairs since the variables contribute 0 coefficients to both the Down and Up branches.)

We now consider the effect of shifting the allocation between $\mathbf{I}(1)$ and $\mathbf{I}(2)$ in a more general fashion, starting from any given allocation and selecting any subset $\mathbf{I}^*$ of either $\mathbf{I}(1)$ or $\mathbf{I}(2)$ to be transferred to the opposite index set. Thus, if $\mathbf{I}^*$ belongs to $\mathbf{I}(1)$ we replace the value $b_h = f_h$ by $b_h = -g_h$ for each $h \in \mathbf{I}^*$, and if $\mathbf{I}^*$ belongs to $\mathbf{I}(2)$ we replace the value $b_h = -g_h$ to $b_h = f_h$ for each $h \in \mathbf{I}^*$. In fact, we allow a more general change: $\mathbf{I}^*$ may include elements of both $\mathbf{I}(1)$ and I(2), provided we replace b_h by adding a positive integer amount to it for all $h \in \mathbf{I}^*$ or by subtracting a positive integer amount from it for all $h \in \mathbf{I}^*$. (This includes the case of transferring a subset of $\mathbf{I}(1)$ or $\mathbf{I}(2)$ to the other set, which is equivalent to restricting the positive integer to be 1 for all $h \in \mathbf{I}^*$, and requiring that the resulting b_h retains an absolute value less than 1. We have emphasized the "transfer case" because it is probably the preferable type of change in most instances.)

The use of such a modified set of b_h values gives one member of a branching pair the ability to "branch more deeply" than the corresponding member of the S-D pair, and offers additional flexibility in designing branching strategies. The apparent disadvantage of introducing such a modified set of values is that making one branch stronger automatically causes the other to be weakened. However, we show that this disadvantage can be offset by integrating the two new branches with the S-D branching pair.

The most comprehensive way to integrate the new branches with the S-D pair is of course to generate and implement all four possibilities for combining a branch from one pair with a branch from the other pair. Obviously, such a combination retains the full strength of the implications of the S-D pair, but its appeal from this standpoint is countered by doubling the alternatives to be examined. Nevertheless, we will observe that all the information to be derived from such a comprehensive approach can be obtained by restricting attention to just three branches — a "triple branch" approach — where two of the three branches impose a single inequality rather than two inequalities in combination. We summarize this outcome formally as follows.

Triple Branch Theorem.

Let $\mathbf{I}^*$ be a chosen subset of $\mathbf{I} \cap \mathbf{NB}$, and let k'_h denote a positive integer for each $h \in \mathbf{I}^*$. Further, let b'_h denote the b_h value determined by the S-D branching pair for all $h \in \mathbf{NB}$, and let $b''_h = b'_h + k'_h$ for all $h \in \mathbf{I}$ or $b''_h = b'_h - k'_h$ for all $h \in \mathbf{I}^*$. Consider the modified branching pair,

$$\sum_{h \in \mathbf{I}^*} b''_h x_h + \sum_{h \in \mathbf{NB}-\mathbf{I}^*} b'_h x_h \geq f_0 \qquad \text{(New: Down Branch)}$$

$$-\sum_{h \in \mathbf{I}^*} b''_h x_h - \sum_{h \in \mathbf{NB}-\mathbf{I}^*} b'_h x_h \geq g_0 \qquad \text{(New: Up Branch)}$$

Then the following set of three branches encompasses the same feasible alternatives as the set of all combinations of joining one member of the S-D branching pair with one member of the modified branching pair:

Case 1: If $b''_h = b'_h + k'_h$ for all $h \in \mathbf{I}^*$.
- (1) (S-D: Down Branch),
- (2) (New: Up Branch),
- (3) (S-D: Up Branch) and (New: Down Branch).

Case 2: If $b''_h = b'_h - k'_h$ for all $h \in \mathbf{I}^*$.
- (1) (New: Down Branch),
- (2) (S-D: Up Branch),

(3) (New: Up Branch) and (S-D: Down Branch).

The foregoing result is easily established by observing in the first case that the combination of (S-D: Down Branch) and (New: Up Branch) is infeasible (as determined by summing their associated inequalities). Also, (1) and (2) consist of a single branch each because (S-D: Down Branch) dominates (New: Down Branch) and (New: Up Branch) dominates (S-D: Up Branch). The second case is established similarly. (The foregoing theorem is also valid when the S-D pair is allowed to be defined differently, so that the coefficients b_h' have different values than those specified, but such an allowance typically does not provide an interesting alternative.) The outcome of collapsing four branches to three by the preceding theorem is analogous to the outcome of collapsing two branches to one in the derivation of a cutting plane. (In both instances one branch is rendered infeasible, though this occurs in a slightly more subtle way for triple branching.) The operation of summing the inequalities of (3) leads to the following inference.

Corollary: The branching alternative (3) in both Case 1 and 2 implies

$$\sum_{h \in \mathbf{I}^*} x_h \geq 1$$

While the branching alternative (3) can be somewhat more restrictive than the inequality of the Corollary (particularly where all $k_h' = 1$), the latter inequality is interesting because it shows the members of $\mathbf{I}^*$ are themselves subject to a restriction as a result of (3). Specifically, if $\mathbf{I}^*$ is chosen to consist of a single index h, then the associated x_h is compelled to be at least 1 by this alternative. Of course, all three branches independently dominate the G-MIP cut.

To illustrate the triple branching result we expand our earlier example to include an additional nonbasic variable x_3. (Two dimensional examples yield only a "degenerate version" of the result, where one of the branches is infeasible and only two branches survive instead of three.) Specifically, we make use of the source equation

$$s + 1\tfrac{1}{5}x_1 - 2\tfrac{1}{5}x_2 - 2\tfrac{2}{5}x_3 = 7\tfrac{2}{5}.$$

The S-D pair for this equation consists of

$$\tfrac{1}{5}x_1 - \tfrac{1}{5}x_2 - \tfrac{2}{5}x_3 \geq \tfrac{2}{5} \qquad \text{(S-D: Down Branch)}$$

and

$$-\tfrac{1}{5}x_1 + \tfrac{1}{5}x_2 + \tfrac{2}{5}x_3 \geq \tfrac{3}{5} \qquad \text{(S-D: Up Branch)}$$

Choosing $\mathbf{I}^* = \{3\}$, and choosing $b_h'' = b_h' + 1$ for $h \in \mathbf{I}^*$, which retains the absolute value of b_3'' less than 1 (replacing $-g_3 = -\frac{2}{5}$ in the Down Branch by $f_3 = \frac{3}{5}$), we obtain the modified pair of branches given by

$$\tfrac{1}{5}x_1 - \tfrac{1}{5}x_2 + \tfrac{3}{5}x_3 \geq \tfrac{2}{5} \qquad \text{(New: Down Branch)}$$

and

$$-\tfrac{1}{5}x_1 + \tfrac{1}{5}x_2 - \tfrac{3}{5}x_3 \geq \tfrac{3}{5} \qquad \text{(New: Up Branch)}$$

The triple branch outcome consists of

(1) $\tfrac{1}{5}x_1 - \tfrac{1}{5}x_2 - \tfrac{2}{5}x_3 \geq \tfrac{2}{5}$ (S-D: Down Branch)

(2) $-\tfrac{1}{5}x_1 + \tfrac{1}{5}x_2 - \tfrac{3}{5}x_3 \geq \tfrac{3}{5}$ (New: Up Branch)

(3) $-\tfrac{1}{5}x_1 + \tfrac{1}{5}x_2 + \tfrac{2}{5}x_3 \geq \tfrac{3}{5}$ (S-D: Up Branch)

$\quad\;\; \tfrac{1}{5}x_1 - \tfrac{1}{5}x_2 + \tfrac{3}{5}x_3 \geq \tfrac{2}{5}$ (New: Down Branch)

The dominance conditions discussed following the statement of the Triple Branch Theorem are evident by inspecting the example branching pairs. Also, it is clear that alternative (3) implies $x_3 \geq 1$ by summing its component branch inequalities.

Fig. 6.13 Cross section for variables x_1 and x_3.

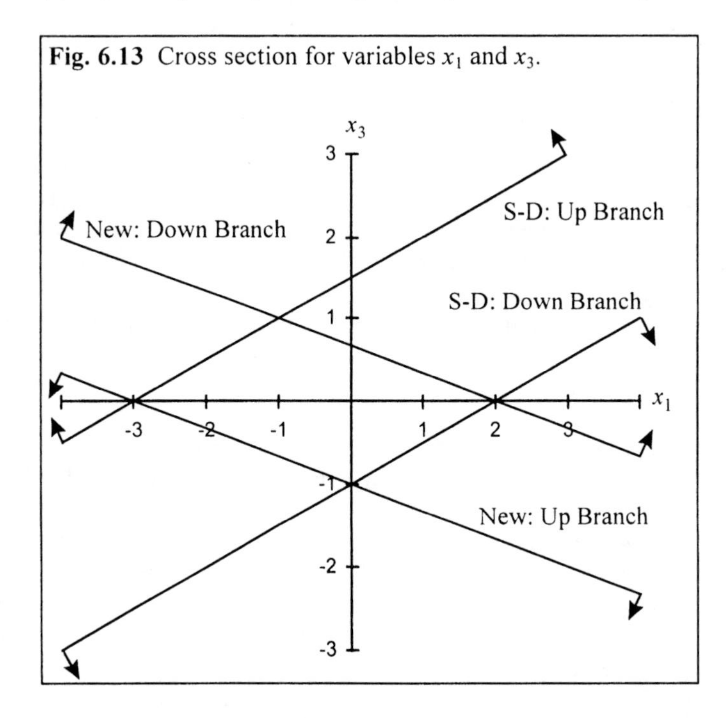

For visualization, we graph the 2-dimensional cross sections of the 3-dimensional space for this example. The cross section for the variables x_1 and x_2 is already shown in Figure 6.12, because these variables have the same coefficients in both the S-D pair and the New pair. Figures 6.13 and 6.14 show the cross sections for the variables x_1 and x_3 and the variables x_2 and x_3.

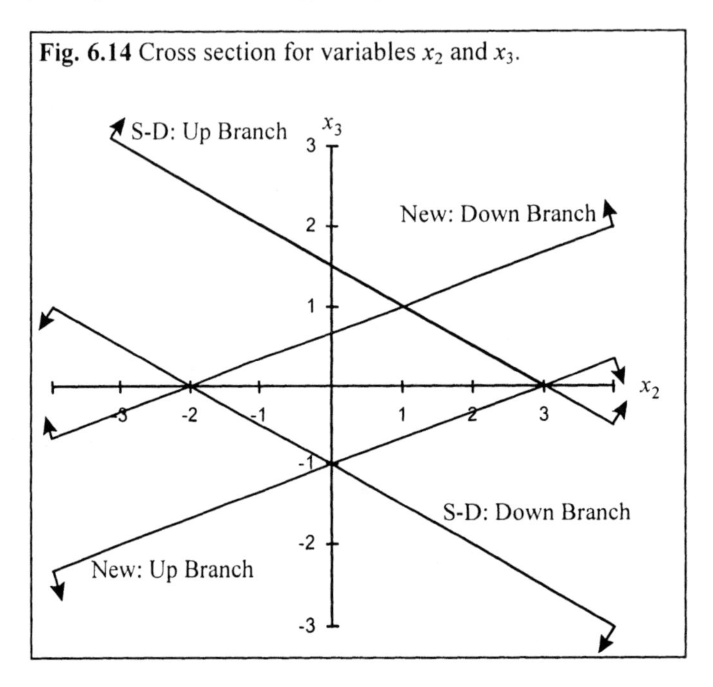

Fig. 6.14 Cross section for variables x_2 and x_3.

For the indicated cross sections, Figures 6.13 and 6.14 confirm graphically that (S-D: Down Branch) dominates (New: Down Branch) and (New: Up Branch) dominates (S-D: Up Branch). Note that the infeasibility of certain cross sections do not imply that these branches are infeasible relative to the nonnegative orthant in 3-dimensional space. These figures also disclose how the combined branches (S-D: Down Branch) and (New: Up Branch) exclude all feasible alternatives on these cross sections.

6.5.5 Conclusions

The strategies proposed for branching on created variables are conceptually straightforward and easy to implement. We emphasize that the straddle-dominance property does not imply that branching relative to an S-D pair is always preferable to cutting. Branching can still potentially create a combinatorial growth of alternatives, whose numbers cutting may help appreciably to reduce. However, since the straddle-dominance property is typically not shared by standard approaches, which branch on original problem variables, an approach that branches on created variables can provide a useful enhancement to branch and cut methods. Similarly, the use of tabu branching, which has the capacity to probe alternative branching sequences more flexibly than branch and bound, provides a further way to benefit from the "deep branching" that gives rise to the straddle-dominance property.

Finally, we note that strategies for branching on created variables are natural accompaniments to the cut search and star-path approaches, which are likewise able to take advantage of transformed variables. The opportunity to generate trial solutions by these approaches provides a useful complementarity to the strategies that seek more effective branches.

6.6 Discussion Questions and Exercises

Tabu Branching (Exercises 1 - 4).

1. Create an illustrative partial branch and bound tree, restricting the leaf nodes to at most 8, where each node at a depth of 6 or less from the root. (Assume the tree has a binary branching structure, resulting from branches of the form "$x_j \leq v$ or $x_j \geq v + 1$".) Identify a leaf node at depth 6 as the node currently evaluated, and generate a concrete example of options for continuing the tree search by the following three choices of the next move to execute:

   (a) a restriction move that creates a new branch from the currently evaluated leaf node.
   (b) a reversal move, based on assuming the tree is generated by a LIFO (backtracking) rule. (All leaf nodes other than the currently evaluated node are terminal nodes, and only the single series of branches leading to the current leaf node is the surviving portion of the tree, as in Figure 6.1. The reversal must return to the most recent branch of the surviving tree whose alternative branch has not yet been generated, dropping the indicated branch and all branches following, and instituting the single alternative branch.)
   (c) a reversal move, based on assuming the tree is generated by a "jump tracking" or "best cost" rule, where some leaf nodes other than the current leaf node are not terminal nodes. (The reversal generates a restriction move consisting of an alternative branch at some node other than the leaf node currently examined. The associated original branch is not discarded unless its descendant leaf nodes are all terminal nodes.)

   Compare the outcomes of these options to outcomes that can be generated for this example by a tabu branching scheme, as follows. Identify the currently operative branches for the tabu branching approach to be those contained in the same sequence of branches that leads to the currently evaluated leaf node of the branch and bound tree. Then choose concrete examples of tabu branching that execute:

   (d) a restriction move
   (e) a relaxation move

(f) a reversal move

Identify how these options correspond to or differ from the options of (a), (b) and (c).

2. Indicate how the evaluation of the reversal move for tabu branching in Exercise 1(f) may differ from the evaluations for a branch and bound procedure, such as in Exercise 1(b) and 1(c). (In Exercise 1(b), the backtracking reversal is only triggered if the current leaf node is determined to be a terminal node. How does the nature of the reversal, which is compulsory, relate to the tabu branching possibilities in this case?)

3. Discuss how recency-based and frequency-based memory may be used to guide a tabu branching procedure. Identify a sequence of 3 consecutive reversal moves, by extension of Exercise 1(f), that may be consistent with a recency based memory design. Also, identify a simple aspiration criterion that may be used in such an approach to override tabu status.

   Compare the reliance on additional memory required by a jump-tracking branch and bound procedure to the use of additional memory required by a tabu search intensification strategy that recovers elite solutions. What further differences in strategic alternatives for branch and bound and tabu search arise in these circumstances? Indicate the differences that result when tabu search uses a path relinking strategy.

4. Discuss the advantages and disadvantages of a tabu branching framework in comparison with a branch and bound framework, based on the observations of Exercises 1 - 3. Do you think that a tabu branching approach would be easy to join with a branch and bound or a branch and cut scheme? In such a combined procedure, are there options for using tabu search other than to scout for good solutions subject to inheriting the restrictions transmitted by prior branches of the tree search framework?

Cut search and transformed coordinate systems (Exercises 5 - 16).

5. This exercise extends the example of Observation 6 to demonstrate the basic ideas for taking advantage of transformed coordinate systems. The LP cone illustrated in Figures 6.4 and 6.5 is derived from the basic (extreme point) representation of the following inequalities, for nonnegative integer variables x_1 and x_2:

$$5x_1 + 6x_2 \geq 30$$
$$7x_1 + 3x_2 \geq 25$$

A useful coordinate system for these inequalities results by a unimodular transformation that produces a "diagonal dominant" representation — one whose diagonal element in each row is somewhat larger in absolute value than other elements in the row. (This definition is purposely ambiguous to allow flexibility. Its significance will soon be clear.)

Consider the following sequence, which creates new variables by elementary integer column operations — operations that add or subtract integer multiples of one column to others, or multiply a column by -1. The variable whose column is chosen to modify others is the one whose associated variable changes, and the new variables that replace the old ones are designated y_1, y_2, etc., in the sequence that they are generated.

Step 0. (Initial)

	x_1	x_2
	5	6
	7	3
$x_1 =$	1	0
$x_2 =$	0	1

Step 1. (Subtract second column from first, replacing x_2 by y_1.)

	x_1	y_1
	-1	6
	4	3
$x_1 =$	1	0
$x_2 =$	-1	1

Step 2. (Subtract first column from second, replacing x_1 by y_2.)

	y_2	y_1
	-1	7
	4	-1
$x_1 =$	1	-1
$x_2 =$	-1	2

Now the transformed original matrix (the upper half of the representation) is diagonal dominant by reordering the columns so that 7 and 4 are on the diagonal. (Since we are not concerned about defining "diagonal dominant" strictly, we stop as soon as the matrix approximately has the desired form.)

Append an identity matrix to the bottom of the Step 2 table, to identify rows for y_2 and y_1 as shown below.

	y_2	y_1
	-1	7
	4	-1
$x_1 =$	1	-1
$x_2 =$	-1	2
$y_2 =$	1	0
$y_1 =$	0	1

Now reverse the steps so that the x_1 and x_2 rows return to the identity matrix. (It isn't necessary to carry the top two rows of the table at this stage. The rows for x_1 and x_2 provide enough "memory" to reverse the process — that is, any sequence of elementary column operations that returns these rows to an identity matrix will work.) Confirm that the resulting rows for y_1 and y_2 identify the transformed coordinate system displayed in Figure 6.5.

6. A result equivalent to that of Exercise 5 is obtained by the sequence that first subtracts column 1 from column 2, and then adds two times column 2 to column 1. (In this case, the diagonal elements are 7 and -4.) Show that one of the final integer variables is the negative of one in Exercise 5, and that when the representation is transformed back (to yield an identity matrix for x_1 and x_2) the row for this variable is the negative of the row for its counterpart in Exercise 5.

   The fact that different sequences of operations can lead to the same outcomes provides flexibility in designing rules to generate a diagonal dominant matrix. Can you identify a set of heuristic rules that will generate either the sequence of Exercise 5 or the sequence of present exercise? (These rules should allow the final result to be produced in more than two steps. For larger problems involving more than two integer variables, integer multiples of a given column can be used to transform a number of other columns in a single step. (The following exercises provide a basis for more rigorous rules.)

7. *Connections with the Euclidean algorithm* — The approach of Exercise 5 can also be applied to a problem that contains continuous as well as integer variables, simply by dropping the columns for the continuous variables (or introducing a variation discussed in Exercise 16). Under the special assumption that all variables are integer valued (not restricted to being nonnegative) and all data elements are integers, the Euclidean algorithm can be used to solve a corresponding system of linear equations.

   (a) The tabular representation of Exercise 5 can be used to execute a generalized application of the Euclidean algorithm, by performing elementary integer operations until yielding a lower diagonal matrix, with

0's above the main diagonal. Such a matrix can be generated by transforming a single row at a time, repeatedly choosing any two nonzero elements on or above the main diagonal (i.e., on the right of the diagonal in the given row), as a basis for a column operation that strictly reduces one of these elements. Specifically, the column operation subtracts an integer multiple of the element that has a smaller or equal absolute value from the other element to leave the smallest possible absolute value for this second element possible. Ultimately, the last nonzero element that remains on or above the main diagonal is moved to the main diagonal. Verify that, when a column of constants is appended to represent the right hand sides of the equations in integer variables, the lower diagonal matrix makes it possible to solve by successive substitution to obtain integer values for the transformed variables (and also, therefore, the original variables), provided an integer solution exists.

(b) Show that one last series of row operations with the generalized Euclidean algorithm of part (a) (disregarding the goal of solving a system of integer equations) can be used to create a diagonal dominant matrix by the following rule: select the final nonzero diagonal element in each row, and for each element of the row that lies below (to the left of) the diagonal, and which a larger or equal absolute value than the diagonal element, perform the same column operation indicated in part (a) to reduce the absolute value of this non-diagonal element as much as possible. Given that a random collection of integers has a fairly large probability of having a greatest common divisor of 1, and that this probability grows with the number of integers in the collection, can you argue that most of the diagonal elements will have an absolute value of 1, and hence the non-diagonal elements of most rows will be 0?

8. *Connections with bounding forms* — The representation of Exercise 5 additionally provides a connection to a method for finding integer solutions to inequality systems. The particular matrix obtained at the end of Step 2 of Exercise 5, with positive elements on the diagonal and nonpositive elements elsewhere, is a *bounding form* matrix that can be used to determine lower bounds for the transformed integer variables (y_1 and y_2) needed to satisfy the inequalities. Such a bounding form can be systematically generated for an IP problem in a manner to assure its inverse exists and has all nonnegative elements (Glover, 1964). (A square bounding form has a nonnegative inverse if there exists a positive linear combination of its columns (or rows) that yields a positive constant column (or row).)

We explore useful implications of this result as follows. A nonnegative inverse makes it possible to identify a lower bound for each transformed variable in the inequality system. (Explain why this is true.) Moreover, if the resulting bounds are not all integers, each non-integer bound can be "rounded up" to the next integer value and then a new constant column is generated by assigning each variable its resulting lower bound. If the inequalities are not satisfied by this assignment, the process is repeated, to identify new increments to some of the bounds until after some number of steps all inequalities are satisfied.

To see how this may be exploited, consider the system that appends the constant column to the Step 2 table of Exercise 5, as shown below.

	y_2	y_1	
	-1	7	≥ 30
	4	-1	≥ 25
$x_1 =$	1	-1	≥ 0
$x_2 =$	-1	2	≥ 0

(The constant column can be initially appended and carried forward from Step 0.) The bounding form in the upper half of the table has been generated under conditions that assure it has a nonnegative inverse, and hence the positive constant column (with entries 30 and 25) assures that each of y_1 and y_2 has a nonnegative (in fact positive) lower bound. This knowledge makes it possible to avoid doing a matrix inversion, if desired.

Specifically, show that the nonnegativity of the variables immediately implies bounds of $y_1 \geq 5$, $y_2 \geq 7$ for this system. In addition, show that plugging in these bound values discloses that further increments to the bounds are needed to satisfy the inequalities. How many rounds of incrementing are required in this example, and what are the final bounds for y_1 and y_2? What values result for x_1 and x_2 when y_1 and y_2 are assigned these bounds? (Assigning a variable y_j a lower bound L_j is the same as replacing y_j by the nonnegative variable $y'_j = y_j - L_j$. Show that the values for x_1 and x_2 that result by this replacement are the negatives of the values that appear to the right of the $\geq$ signs for x_1 and x_2, once these values are adjusted for the lower bound assignment.)

Finally, note in the present example a bounding form matrix also appears in the lower portion of the table. If the inequalities for x_1 and x_2 were not satisfied by assigning the other variables their lower bounds (replacing them by nonnegative variables), would this "double bounding form" structure assure that values can be found that ultimately satisfy all inequalities? (In general, a system of inequalities can have several rows that simultaneously share the structure to qualify for a given row of a bounding form, and these rows can be treated simultaneously in an incremental bounding process of the type illustrated.)

9. A somewhat different example from that of Exercise 5 is given by the system

$$7x_1 - 6x_2 \geq 0$$

$$-3x_1 + 7x_2 \geq 7$$

Show that this system can be readily transformed into the following diagonal dominant representation:

	y_1	y_2
	8	1
	1	4
$x_1 =$	2	1
$x_2 =$	1	1

Show also, that when transformed back, this yields

$$y_1 = x_1 - x_2$$
$$y_2 = -x_1 + 2x_2$$

Graph the cone for the original inequalities, and carry out a cut search process of probing with the edges, and alternately of probing with a ray formed from a convex combination of the edges. Show that probing with the edges will generate a number of infeasible points (relative to the cone) in the original (x_1, x_2) space, but that infeasible points will not be generated until going very much deeper in the transformed (y_1, y_2) space. Also, show that a ray through the approximate center of the cone generates a better set of points in the transformed space, but not with the same degree of difference as in the example of Figure 6.6.

The original inequality system of this exercise contains a bounding form. Would you anticipate that this might account for the fact that the ray through the middle of the cone was reasonably "productive" in the original space? Does the bounding form also make it possible to immediately generate lower bounds for x_1 and x_2 that go beyond the fractional LP vertex (which occurs at $x_1 = 1\ 11/31$, $x_2 = 1\ 18/31$)?

10. (*Creating a bounding form — preliminary*) Consider the single inequality

$$a_1 x_1 + \sum_{h \in Q} a_h x_h \geq a_0$$

where x_1 and x_h, $h \in Q = \{2, \ldots, q\}$ are integer-valued, and $x_h \geq 0$ for $h \in Q$. Suppose that $a_1 \neq 0$ and moreover $a_1 > 0$ (if necessary, by replacing x_1 by $-x_1$ and a_1 by $-a_1$). Create an integer variable

$$y_1 = x_1 + \sum_{h \in Q} k_h x_h - k_0$$

where the coefficients k_0 and k_h, $h \in Q$ are integers, and consider the inequality

$$a_1 y_1 + \sum_{h \in Q} a'_h x_h \geq a'_0$$

derived by substitution for x_1. Verify that $a'_h = a_h - k_h$ and $a'_0 = a_0 - k_0$, and show that if $a'_h \leq 0$ for $h \in Q$, then $y_1 \geq \left\lceil \frac{a'_0}{a_1} \right\rceil$. Show in general that this implies $y_1 \geq 0$ if $a'_0 > -a_1$.

Finally, show that the smallest values of k_h (largest values of a'_h) that will yield $a'_h \leq 0$, together with the largest k_0 (smallest a'_0) that will yield $a'_0 > -a_1$, are "best" in the following sense: $y_1 \geq 0$ for all larger values of k_h, $h \in Q$ and all smaller values of k_0.

11. *Creating a bounding form by a variant of the Euclidean algorithm* — A series of operations of the form indicated in the preceding exercise can be used to create a more general bounding form. Many different bounding forms may be produced, involving considerations suggested in later exercises, but here we consider a partial application of the Euclidean algorithm, coupled with the preceding transformation, to generate a bounding form as follows.

For concreteness, suppose we start with the following system of three inequalities in six nonnegative integer variables:

x_1	x_2	x_3	x_4	x_5	x_6		
*	*	*	*	*	*	$\geq$	a_{10}
*	*	*	*	*	*	$\geq$	a_{20}
*	*	*	*	*	*	$\geq$	a_{30}

where * denotes an arbitrary coefficient value. Assume the square submatrix associated with x_1, x_2 and x_3 is invertible. (Simple changes in the following description can be applied if not.) A bounding form is illustrated by the system

y_1	y_2	y_3	x_4	x_5	x_6		
+	$\ominus$	$\ominus$	$\ominus$	$\ominus$	$\ominus$	$\geq$	a'_{10}
$\ominus$	+	$\ominus$	$\ominus$	$\ominus$	$\ominus$	$\geq$	a'_{20}
$\ominus$	$\ominus$	+	$\ominus$	$\ominus$	$\ominus$	$\geq$	a'_{30}

where + denotes a positive coefficient and $\ominus$ denotes a negative or 0 coefficient. For the system to qualify as a bounding form, the transformation that yields must assure that y_1, y_2 and y_3 as well as x_4, x_5 and x_6 are nonnegative integer variables (and integer values of x_1, x_2 and x_3 can be recovered from the values of y_1, y_2 and y_3). Assume that all coefficients of the original system are integers. Applying the Euclidean algorithm to this system, replacing inequalities by equalities, as described in Exercise 7 (a), yields a system of the form

y_1	y_2	y_3	y_4	y_5	y_6		
*	0	0	0	0	0	=	a_{10}
*	*	0	0	0	0	=	a_{20}
*	*	*	0	0	0	=	a_{30}

where the * coefficients on the main diagonal are non-zero (and some of the y_j variables may possible correspond to original x_j variables). Consider instead a limited version of the Euclidean Algorithm, applied to the inequality system, which does not modify coefficients of x_4, x_5 and x_6, to yield

y_1	y_2	y_3	x_4	x_5	x_6		
*	0	0	*	*	*	$\geq$	a_{10}
*	*	0	*	*	*	$\geq$	a_{20}
*	*	*	*	*	*	$\geq$	a_{30}

where again the * coefficients on the main diagonal or non-zero. Show that the operation of Exercise 10 can now be used to transform this system into a bounding form (yielding new variables y_1, y_2 and y_3 that are nonnegative). By reference to the last part of Exercise 10, explain why all off-diagonal coefficients in each row of the bounding form should be smaller in absolute value than the coefficients on the main diagonal (and show this is easy to achieve). How might the process differ if the original square matrix for x_1, x_2 and x_3 were not invertible? What difference would result if the system did not also have full row rank?

12. *Complete tableau representations, including an objective function* — Bounding forms are particularly relevant for finding optimal solutions to systems of inequalities in nonnegative integer variables relative to a dual feasible objective function (which yields all coefficients nonnegative when expressed as a minimization objective). To illustrate, consider the following system, whose top row indicates the objective of minimizing $z = 6x_1 + 13x_2 + 7x_3$.

	x_1	x_2	x_3	
$z =$	6	13	7	
$x_4 =$	7	4	6	≥ 26
$x_5 =$	5	9	4	≥ 32
$x_6 =$	8	7	5	≥ 41
$x_1 =$	1	0	0	≥ 0
$x_2 =$	0	1	0	≥ 0
$x_3 =$	0	0	1	≥ 0

In addition to identifying z as associated with the top row, we have explicitly identified nonnegative slack variables, x_4, x_5 and x_6, associated with three row inequalities of the upper tableau. Just as the current values of the variables x_1, x_2 and x_3 in any tableau are given by the negative of the associated constant terms on the right hand side of the $\geq$ signs (see Exercise 8), the current values of the slack variables are also given by the negatives of their associated constant terms. For example, the x_4 row of the tableau identifies the equation $x_4 = 7x_1 + 4x_2 + 6x_3 - 26$, and $x_4 \geq 0$ corresponds to the inequality $7x_1 + 4x_2 + 6x_3 \geq 26$. Thus, as appropriate, a positive right hand side, which signals that a "$\geq$ constant" is not currently satisfied, assigns a negative value to the associated slack variable. (In order to give an intuitive rationale for understanding the bounding form constructions, the signs of all matrix coefficients are the negatives of those of customary LP expositions, as in the cut search development.) Within this tableau representation, consider the operations that keep the tableau dual feasible.

(a) Show that a column operation that subtracts positive multiples of any given column from various other columns automatically assures the new variable created is nonnegative, given that all current nonbasic variables (those attached to the columns) are nonnegative. Verify this both by direct algebraic argument (identifying the new variable as nonnegative combination of old ones) and by reference to a bounding form created by the transformation. (For example, if the x_3 column above is subtracted from each of the other two columns, show that the transformed row that expresses the inequality $x_3 \geq 0$ satisfies the bounding form structure to imply the new variable is nonnegative.)

(b) By the observation of part (a), show that it is always possible to maintain the objective function dual feasible, while keeping all variables nonnegative, by permitting one column to be subtracted from another only if the first column has a smaller or equal objective function coefficient. In particular, what multiples of the x_1 column can be subtracted from the other columns in the current example to keep a dual feasible objective?

(c) An important consequence of dual feasibility is as follows: if the outcome of assigning each current nonbasic variable its lower bound value satisfies a particular set of inequalities, then this lower bound assignment gives an optimal solution relative to these inequalities. (Explain why this is true.) Identify a transformation to subtract multiples of the x_1 column from the other columns in the current example problem to create a bounding form that will give an optimal solution relative to all of the problem inequalities. Show that this bounding form will give an optimal solution regardless of the initial "right hand sides" of the first three inequalities. What changes in the objective function coefficients of x_2 and x_3 in the original tableau will preserve the validity of this outcome? What changes in the objective function coefficients of x_1 and x_2 will allow a similar outcome in which only x_3 receives a positive value in an optimal solution?

13. *Alternative bounding forms* — Consider the problem summarized by

	x_1	x_2	x_3	
$z =$	8	9	7	
$x_4 =$	5	4	-6	≥ 17
$x_5 =$	-8	5	7	≥ 15
$x_1 =$	1	0	0	≥ 0
$x_2 =$	0	1	0	≥ 0
$x_3 =$	0	0	1	≥ 0

Two ways of creating a bounding form are as follows:

(a) Create a new nonnegative integer variable y_1 to replace x_1 by subtracting the x_1 column from the x_2 column and adding the x_1 column to the x_3 column. Why is the addition justified? (It is generally advantageous to use column additions since this yields y_1 by subtracting a nonnegative integer variable from x_1, which tends to reduce the implied lower bound for y_1.) Show that the bounding form that results for the first two row inequalities yields an optimal solution for the problem.

(b) Create a new nonnegative integer variable y_1 (different from that of part (a)) to replace x_3 by subtracting the x_3 column from the x_2 column and adding the x_3 column to the x_3 column to the x_1 column. Show in this case that the bounding form that results for the first two row inequalities does not yield an optimal solution for the problem, because the row inequality for x_3 becomes violated. However, show that a bounding form can be created from the x_3 row that allows the solution to the problem to be completed.

(c) What conditions make it legitimate to recover the original tableau (or some other earlier tableau) as a way of establishing a bounding form, as in the final step of part (b) above? What significance attaches to the fact that the recovery incorporates the constant column from the latest "current tableau," rather than from the tableau recovered? May there be advantages to such a recovery, as opposed to seeking new transformations to create bounding forms?

14. *Single-constraint bounding forms* — Single-constraint bounding forms that preserve dual feasibility can be created in one step by using very simple surrogate constraints that are always available. Consider the problem

	x_1	x_2	x_3	
$z =$	14	18	40	
$x_4 =$	8	10	21	≥ 66
$x_5 =$	-7	5	13	≥ 28
$x_1 =$	1	0	0	≥ 0
$x_2 =$	0	1	0	≥ 0
$x_3 =$	0	0	1	≥ 0

It is not possible to generate a single-constraint bounding form from either of the first two inequality constraints in one step, while preserving dual feasibility. However, such a bounding form can easily be generated from a "trivial" surrogate constraint that adds a selected multiple of a nonnegativity constraint $x_j \geq 0$. For example, adding a multiple of 3 times the inequality $x_1 \geq 0$ to the first inequality constraint yields

$$11x_1 + 10x_2 + 21x_3 \geq 66.$$

(a) Verify that this surrogate constraint has a large enough coefficient for x_1 to permit a single-constraint bounding form to be generated on one step (while preserving dual feasibility) to give the updated surrogate constraint

$$11y_1 - 1x_2 - 2x_3 \geq 66.$$

(b) How far should the coefficient of x_1 be increased in the first inequality to allow a one step bounding form to be created that subtracts only a unit multiple of the x_1 column from each of the other columns? What lower bound is implied for the resulting y_1 variable, and how does this lower bound compare to the one for the y_1 variable in part (a)?

(c) Show that the x_1 coefficient does not have to be increased as far as 11 (since it doesn't have to be given an integer value) in order to create a single-constraint bounding form that subtracts 2 times the x_1 column from the x_3 column, as in part (a). What value should the x_1 coefficient be given, and what is the implied lower bound for the resulting y_1 variable? (Is this better than increasing the x_1 coefficient to 11 as in part (a)?)

(d) Identify single-constraint bounding forms that can be created from the second inequality constraint of the example problem, and that will preserve dual feasibility. (Given the general desirability of adding a column to another where possible, does one of these surrogate constraints seem better than the rest? What other considerations may be relevant?

(e) Are there any conditions under which the surrogate constraint can actually replace the original constraint, instead of simply being used to determine an implied lower bound for the problem variable?

(f) As shown in Glover (1964), the transformations and bounds on variables when single-constraint bounding forms are created from trivial surrogate constraints are the same as those that result from the all-integer cutting planes of Gomory (1960). By contrast, the transformations and bounds that result from generating multi-constraint bounding forms may require very large numbers of all-integer cuts to achieve. (An example is cited of a 3 constraint bounding that can be generated to solve an IP problem in a small number of operations, but where over 1500 all-integer cuts are required to achieve the same outcome.) What advantages may derive from a combined approach, that seeks to generate multi-constraint bounding forms, while using trivial surrogate constraints in the process to identify inferred bounds on variables created?

(g) What advantages may result by first solving the LP problem associated with an IP problem to identify inequalities that are binding at the LP optimum as candidates to be transformed into a bounding form? (How are the observations of Exercises 11 and 12 relevant to such an approach?)

15. *Bounding forms, logical implications, and cut search* — The following problem has already been transformed to create a bounding from and to assign lower bound values to nonbasic variables, which now are currently nonnegative. (The problem contains additional inequalities not shown, including those for the original variables x_1, x_2 and x_3.)

	y_1	y_2	y_3	
$z =$	4	7	6	
$x_4 =$	12	-2	-4	≥ -3
$x_5 =$	-3	8	-2	≥ -2
$x_6 =$	-1	-3	9	≥ -4
$x_7 =$	3	4	2	≥ 3
$x_8 =$	4	-3	2	≥ -2

Observe that the fourth inequality constraint shown implies at least one of y_1, y_2 and y_3 must be positive. Also, verify that the bounding form of the first three inequalities yields the following direct conclusions: "$y_1 \geq 1$ implies $y_2 \geq 1$" and "$y_3 \geq 1$ implies $y_1 \geq 1$."

(a) Show that the preceding implications lead to the conclusion that $y_2 \geq 1$. Also, show that assigning y_2 this lower bound (replacing y_2 by $y_2' = y_2 - 1$) yields an optimal solution to the partial problem that includes the first four inequalities (associated with x_4 to x_7).

(b) Verify that after replacing y_2 by y_2' in (a), the fifth inequality (associated with x_8) implies y_1 or y_3 must be positive, and the bounding form further yields the conclusion "$y_3 \geq 1$ implies $y_1 \geq 1$." Show this in turn implies $y_1 \geq 1$, and that assigning y_1 its lower bound value (replacing y_1 by $y_1' = y_1 - 1$) yields an optimal solution to the problem that includes all five inequalities shown.

(c) Whenever a bounding form does not result in satisfying all problem inequalities, at least one current nonbasic variable must be positive. Moreover, if setting any single nonbasic variable equal to one does not yield a feasible solution, then the sum of the variables must be at least two. Can this observation be of any use in the type of analysis performed in parts (a) and (b)?

(d) Considerably stronger than the approach in part (c) is the following. Let y_h, $h \in \mathbf{NB}$, denote the current nonbasic variables (where some may correspond to original x_h variables), and define

$$R_k = \{h \in \mathbf{NB} : y_h \geq 1 \quad \text{implies} \quad y_k \geq 1\}$$

and

$$S_k = \mathbf{NB} - R_k$$

where the composition of R_k is determined by reference to the bounding form analysis, as illustrated in parts (a) and (b). (Note $k \notin S_k$, and S_k may be empty.) Show that

$$y_k + \sum_{h \in S_k} y_h \geq 1.$$

Moreover, if $y_h = 1$ does not by itself yield a feasible solution for each $h \in S_k$, show that

$$2y_k + \sum_{h \in S_k} y_h \geq 2.$$

Is such information useful for deriving further implications from the bounding form, or for generating new bounding forms?

(e) Once logical implications have extracted all the information that may be conveniently obtained from a current bounding form, describe how cut search may be used to generate trial solutions (and cutting planes) for the current problem representation. Discuss how cut search can be used to generate "conditional" logical implications in this setting, based on having already examined particular trial solutions.

16. In a mixed integer problem, which includes nonnegative continuous variables, it is possible to generate surrogate constraints from nonnegative linear combinations of the constraints, together with an appended objective function constraint, in which the coefficients of the continuous variables are nonpositive. (The objective function constraint simply requires the objective value to be better than any previously identified for a feasible MIP solution. Other constraints can be restricted to those that yield fractional values for integer variables in the LP basis.) What advantage or disadvantage might result from a process that generates a bounding form from such a set of surrogate constraints?

Star-Paths (Exercises 17 - 23).

17. In the 2-dimensional example of Figure 6.8, identify a single additional constraint hyperplane (line) for the zero-one IP problem that maintains $x(0)$ feasible (hence $x(0)$ lies within the half space determined by this constraint) and that also:

(a) eliminates the image of (0,1) on F^* (i.e., makes the region (F^*)-*Image*(0,1) infeasible) without eliminating the F^* image of (1,1) or (1,0). Similarly, consider cases to eliminate the F^* image of (1,1) and (0,1), respectively. Can the image of (1,1) be eliminated without eliminating the image of any other zero-one point?

(b) eliminates (1,1) without eliminating any portion of its F^* image and without eliminating any other zero-one point. Is it possible to find a hyperplane that eliminates either (1,0) or (0,1) and that also preserves its F^* image without eliminating any other zero-one point?

18. Construct a 3 dimensional example (using a visualized projection on the plane) whose alternatives correspond to those of Exercise 17 — that is, in which a single constraint (in addition to those defining the LP cone) will: (i) eliminate specific images on F^* without eliminating others, (ii) eliminate some images

only by eliminating others, (iii) eliminate particular zero-one solutions without eliminating their images. In the latter case, are there situations where the added constraint is also compelled to eliminate additional zero-one solutions?

19. Construct a 3 dimensional example of an LP cone, with vertex $x(0)$ in the interior of the unit cube, and a path P between two points on a face F^* so that the star-path $\delta(P)$ contains 3 zero-one points, all feasible. Can you structure the examples so that the second point on $\delta(P)$ is an optimal zero-one solution, but neither the first nor the last feasible point is optimal? (For help in identifying such a construction, see Exercise 20.)

    Granting the assumption that $x(0)$ is an optimal LP vertex and that F^* is obtained from a valid cutting plane, explain why the optimality of the second point on $\delta(P)$ in this construction is not inconsistent with the Optimal Focal Point Theorem.

20. Let $x(0) = (4/9, 3/9, 2/9)$ be an optimal LP vertex for a zero-one problem and suppose the two points $x' = (8/9, 0, 4/9)$ and $x'' = (0, 4/9, 4/9)$ are on the boundary of $F^* \cap X_{LP}$, for F^* given by a valid cutting plane.

    (a) Identify the zero-one points of the star-path $\delta(P)$, for $P = P(x', x'')$, using the Star-Path Generation Method.

    (b) Accompanying the identification performed in part (a), superimpose a set of coordinate axes, parallel to the original axes, at the point $x(0)$, for purposes of visualization. Verify that the point where the path P transitions from mapping into one zero-one point to mapping into the next occurs at the intersection with a hyperplane determined by this translated coordinate system.

    (c) Can you visualize an associated LP cone (and LP objective function hyperplane) such that all points of this star-path are feasible, and the second point (but no other) is optimal?

21. Suppose in Exercise 20 an additional boundary point of $F^* \cap X_{LP}$, given by (3/9, 5/9, 1), is selected to take the role of x''. Show that that the star-path $\delta(P(x', x''))$ now contains the same two zero-one endpoints, but a different zero-one middle point, than in Example 20.

    Consider the alternative boundary point $x'' = (0, 6/9, 8/9)$. Show that the star-path $\delta(P(x', x''))$ relative to this x'' has the same endpoints as in the preceding case, but contains a tie for the middle point. Confirm that resolving this tie arbitrarily results in generating either the star-path of Exercise 20 or the star-path previously generated for the current exercise.

22. The tie for generating the second point of $\delta(P)$ in the latter part of Exercise 21 occurs only between two variables (x_1 and x_2). Hence both of the possible "second points" of the star-path (that result by different ways of resolving the tie) can be examined while only adding one more element to the star-path. In general, when ties or "near-ties" occur only between two elements at a time, thereby producing only a single added zero-one solution that may optionally be examined, is it useful to include this added solution in an "enlarged" star-path? (When more than two variables tie simultaneously, would you consider it useful to examine a small number of related solutions instead of resolving the ties arbitrarily? What decision rule might be used to determine which of these solutions should be examined? Would you anticipate that choosing other points x' and x'' for generating star-paths could have greater value — for determining additional solutions to examine — than being concerned about tied elements?)

23. Identify an extended version of the Star-Path Generation Method that will produce the circled integer points of Figure 6.9 as members of a star-path $\delta(P)$, where P in this case corresponds to F^* (the line segment from A to E). Does your method apply to problems in more than two dimensions, where lines and hyperplanes do not correspond?

Branching on created variables (Exercises 24 - 28).

24. Consider the source equation $s - 2\frac{4}{5}x_1 + 3\frac{3}{5}x_2 = 6\frac{2}{5}$ where all variables are nonnegative and integer-valued. Graph and compare:

    (a) branches for the original variable s,
    (b) the P-Cut and the infeasible Counter P-Cut,
    (b) the N-Cut and the infeasible Counter N-Cut,
    (d) the G-MIP cut,
    (e) the S-D pair.

25. Multiply the equation of Exercise 24 by 2, and re-define the resulting integer variable $2s$ to be s (a new integer variable), thus giving $s - 5\frac{3}{5}x_1 + 7\frac{1}{5}x_2 = 12\frac{4}{5}$. Perform 24 (a) through (d) for this source equation. What do the outcomes disclose about the possibility of obtaining stronger cutting planes for pure IP problems? What do the outcomes disclose about special cases of S-D pairs?

26. Two dimensional examples, as in Exercises 24 and 25, can mislead the intuition about alternatives that may be anticipated in larger dimensions. Expand the source equation of Exercise 24 to become $s - 2\frac{4}{5}x_1 + 3\frac{3}{5}x_2 + 1\frac{4}{5}x_3 = 6\frac{2}{5}$, where x_3 is also a nonnegative integer variable. By comparing different 2-dimensional cross sections, and considering the result of multiplying the source equation by 2, how does this example reinforce or modify the perspective from the previous two exercises?

27. The pure IP case also does not give a clear appreciation of the mixed IP case. To illustrate, examine the consequences in Exercise 26 of assuming that x_3 is not integer-valued but continuous. Are these consequences essentially the same as instead assuming x_1 is continuous?

28. Consider the source equation $s - 1\frac{4}{5}x_1 + 4\frac{3}{5}x_2 + 2\frac{2}{5}x_3 = 13\frac{3}{5}$, where all variables are nonnegative and integer-valued.

    (a) Show that there are two different S-D pairs for this source equation, and show graphically by 2-dimensional cross sections that they give different ways to straddle-dominate the G-MIP cut.

    (b) Examine the Triple Branch that results by choosing one of the S-D pairs to be the "preferred" S-D pair, and letting the other take the role of the New branching pair. Identify the components of the Triple Branch, and compare the strength of these alternatives to the strength of either S-D pair taken separately? Does it matter which of the S-D pairs is chosen to be the "preferred" pair?

    (c) Consider the "preferred" S-D pair to be the one that contains the Down Branch $-\frac{1}{5}x_1 + \frac{3}{5}x_2 + \frac{2}{5}x_3 \geq \frac{3}{5}$. Create a New pair for the Triple Branch by the change that subtracts 1 from the coefficient $b_3' = \frac{2}{5}$, and compare the resulting Triple Branch alternatives to those examined in part (b). How do these alternatives compare to those that result instead by adding 1 to the coefficient $b_1' = -\frac{1}{5}$? Finally, consider the same respective changes in b_3' and b_1', applied to the other S-D pair, whose Down Branch is given by $-\frac{1}{5}x_1 - \frac{2}{5}x_2 + \frac{2}{5}x_3 \geq \frac{3}{5}$. What does this say about the different possibilities provided by different choices of a preferred S-D pair?

    (d) In general, do the outcomes of part (b) and (c) suggest rules for generating Triple Branches that are likely to have useful features?

7 SPECIAL TABU SEARCH TOPICS

In this chapter we begin by discussing probabilistic tabu search and the associated approach called tabu thresholding, which operates by reduced reliance on memory. Then we examine specialized memory mechanisms that have been used to enhance the performance of TS methods. We also discuss of approaches not yet widely considered. Finally, we examine basic strategies embodied in ejection chains, vocabulary building and parallel processing.

7.1 Probabilistic Tabu Search

In tabu search, randomization is de-emphasized, and generally is employed only in a highly constrained way, on the assumption that intelligent search should be based on more systematic forms of guidance. Randomization thus chiefly is assigned the role of facilitating operations that are otherwise cumbersome to implement or of assuming that relevant alternatives will not be persistently neglected (as a gamble to compensate for ignorance about a better way to identify and respond to such alternatives). Accordingly, many tabu search implementations are largely or wholly deterministic. However, as a potential hedge against the risks of strategic myopia, an exception occurs for the variant called *probabilistic tabu search*, which selects moves according to probabilities based on the status and evaluations assigned to these moves by the basic tabu search principles. The basic approach can be summarized as follows.

(a) Create move evaluations that include reference to tabu status and other relevant biases from TS strategies — using penalties and inducements to modify an underlying decision criterion.

(b) Map these evaluations into positive weights, to obtain probabilities by dividing by the sum of weights. The highest evaluations receive weights that disproportionately favor their selection.

Memory continues to exert a pivotal influence through its role in generating penalties and inducements. However, this influence is modified (and supplemented) by the incorporation of probabilities, in some cases allowing the degree of reliance on such memory to be reduced.

A simple form of this approach is to start by evaluating the neighborhood as customarily done in tabu search (including the use of candidate list strategies, special memory functions, and penalty and inducement values). Then, the best move can be selected by using a threshold so that probabilities are assigned to a subset of elements that consists of:

- the top k moves, or
- those moves having a value within α% of the best.

Choosing a restricted subset of highest evaluation moves for determining a probability assignment already creates a disproportionate bias in favor of such evaluations. To avoid completely excluding other options, that subset may be enlarged by random inclusion of a small number of additional elements that would be admitted by a less restrictive threshold. However, it should be kept in mind that diversification strategies already should be designed to provide evaluations that will bring elements into the subset that would not be included using a standard decision criterion, such as one based solely on an objective function change. Whether the subset is enlarged therefore depends to the extent that the method incorporates a reasonably effective diversification approach.

The probabilistic selection from the chosen subset can be *uniform* or can follow a distribution function empirically constructed from the evaluation associated with each move. The latter is considered preferable, provided the evaluation is correlated with effective quality (Chapter 5). An example is provided by the neighborhood evaluation given in Table 2.1 of Chapter 2. If k is set to 6, then only swaps (5,6), (2,4), (3,4), (3,6), (1,4) and (1,2) will be considered for selection, as shown in Table 7.1. Since the set of top moves contains the move (1,2), which was previously identified to be nonimproving (in Table 2.1), one alternative may be to adjust k so the selection is not made within a mixed set. This reinforces the aggressive orientation of TS and introduces a regional dependency.

Tabu status can be included in two forms. One option is to eliminate all tabu moves before the selection is made. The other option is to use the remaining tenure of tabu-active attributes to bias the selection (e.g., by decreasing the probability of selecting a move). In the case where relatively simple rules are used to generate tabu status, the bias against selecting tabu moves may periodically be decreased, to compensate for limitations in the basis for classifying a move tabu.

Table 7.1 Empirical probability distribution.

Swap	*Abs. Value*	*Cum. Value*	*Probability*	*Cum. Prob.*
(5, 6)	7	7	0.241	0.241
(2, 4)	6	13	0.207	0.448
(3, 4)	6	19	0.207	0.665
(3, 6)	6	25	0.207	0.862
(1, 4)	4	29	0.138	1.000
(1, 2)	—	—	—	—

An empirical probability distribution can be constructed using evaluations associated with each of the top moves. Suppose only improving moves are being considered for selection in Table 7.1, hence excluding the swap (1, 2). A cumulative probability function results from normalizing the cumulative absolute move values, as shown in the columns of Table 7.1 to the right of the *Swap* column. A uniform random number between zero and one is then generated to determine which move will be performed. For example, if 0.314 is generated, the swap (2,4) is selected.

In some situations, evaluations based on standard criteria do not sufficiently accentuate the relative differences in the effective quality of moves and it is preferable to modify quantities such as the absolute values in Table 7.1 by raising them to a power (e.g., by squaring or cubing) before generating the probability assignments (see Table 7.2). Such modification has been found particularly useful in scheduling applications where evaluations represent deviations from idealized "fits" or feasibility conditions (Glover and McMillan, 1986).

Table 7.2 Modified probability values.

	Probability	
Swap	*Squaring*	*Cubing*
(5, 6)	0.2832	0.3251
(2, 4)	0.2081	0.2047
(3, 4)	0.2081	0.2047
(3, 6)	0.2081	0.2047
(1, 4)	0.0925	0.0607

Table 7.2 shows the effect of squaring and cubing the absolute move values before calculating the empirical probability values for the swaps in Table 7.1. In this particular example, the only significant differences are observed in the probability values associated with the first and last swap. This is due to the fact that the three swaps between the first and the last all have the same absolute move value.

As in other applications of tabu search, the use of an intelligent candidate list strategy to isolate appropriate moves for consideration is particularly important in the probabilistic TS approach. Although a variety of ways of mapping TS evaluations into probabilities are possible (as illustrated above), an approach that sometimes performs well in the presence of "noisy" evaluations is as follows:

(1) Select the "*r* best" moves from the candidate list, for a chosen value of *r*, and order them from best to worst (where "evaluation ties" are broken randomly).
(2) Assign a probability *p* to selecting each move as it is encountered in the ordered sequence, stopping as soon as a move is chosen. (Thus, the first move is selected with probability *p*, the second best with probability (1-*p*)*p*, and so forth.) Finally, choose the first move if no other moves are chosen.

The effect of the approach can be illustrated for the choice $p = 1/3$. Except for the small additional probability for choosing move 1, the probabilities for choosing moves 1 through k are implicitly:

$$\frac{1}{3}, \frac{2}{9}, \frac{4}{27}, \frac{8}{81}, \ldots, \frac{2^{(k-1)}}{3^k}.$$

The probability of not choosing one of the first k moves is $(1 - p)^k$, and hence the value $p = 1/3$ gives a high probability of picking one of the top moves: about 0.87 for picking one of the top 5 moves, and about 0.98 for picking one of the top 10 moves.

Experimentation with a TS method for solving 0-1 mixed integer programming problems (Lokketangen and Glover, 1996) has found that values for *p* close to 1/3, in the range from 0.3 to 0.4, appear to be preferable. In this application, values less than 0.3 resulted in choosing "poorer" moves too often, while values greater than 0.4 resulted in concentrating too heavily on the moves with highest evaluations.

Basing probabilities on relative differences in evaluations can be important as a general rule, but the simplicity of the ranking approach, which does not depend on any "deep formula," is appealing. It still can be appropriate, however, to vary the value of *p*. For example, in procedures where the number of moves available to be evaluated may vary according to the stage of search, the value of *p* should typically grow as the alternatives diminish. In addition, making *p* a function of the proximity of an evaluation to a current ideal shows promise of being an effective variant (Xu, Chiu and Glover, 1996a).

The motivation for this type of approach, as already intimated, is that evaluations have a certain noise level that causes them to be imperfect — so that a "best evaluation" may not correspond to a "best move." Yet the imperfection is not complete, or else there would be no need to consider evaluations at all (except perhaps from a thoroughly local standpoint, keeping in mind that the use of memory takes the evaluations beyond the "local" context). The issue then is to find a way to assign probabilities that appropriately compensates for the noise level.

A potential germ for theory is suggested by the challenge of identifying an ideal assignment of probabilities for an assumed level of noise. Alternative assumptions about noise levels may then lead to predictions about expected numbers of evaluations (and moves) required to find an optimal solution under various response scenarios (e.g., as a basis for suggesting how long a method should be allowed to run).

7.2 Tabu Thresholding

There is an appeal to methods like simulated annealing and threshold acceptance that make no recourse to memory, but that operate simply by imposing a monotonically declining ceiling on objective function levels or degrees of disimprovement (treated probabilistically or deterministically). The framework embodied in tabu search instead advocates a nonmonotonic form of control, keyed not only to the objective function but to other elements such as values of variables, direction of search, and levels of feasibility and infeasibility. This creates a more flexible search behavior.

The question arises whether the nonmonotonic controls by themselves offer a sufficiently rich source of search trajectories to be relied upon as a primary guidance mechanism, with greatly reduced reliance on form of memory customarily used in tabu search.

Tabu thresholding was proposed to provide an easily implemented method of this type, which joins prescriptions of strategic oscillation with candidate list strategies. The candidate list and tabu search philosophies are mutually reinforcing, and the computational advantages contributed by these elements motivate a closer look at combining them. The result yields a method with a useful potential for variation and an ability to take advantage of special structure, yet which avoids some of the challenge of dealing with memory.

Tabu thresholding methods specifically embody the principle of aggressively exploring the search space in a nonmonotonic way. The motivation for using a nonmonotonic search strategy, in contrast to relying on a unidirectionally modified temperature parameter as in simulated annealing, derives from the following observation (Glover, 1986):

> ... the human fashion of converging upon a target is to proceed not so much by continuity as by thresholds. Upon reaching a destination that provides a potential "home base" (local optimum), a human maintains a certain threshold not a progressively vanishing probability for wandering in the vicinity of that base.

> Consequently, a higher chance is maintained of intersecting a path that leads in a new improving direction.
>
> Moreover, if time passes and no improvement is encountered, the human threshold for wandering is likely to be increased, the reverse of what happens to the probability of accepting a nonimproving move in simulated annealing over time. On the chance that humans may be better equipped for dealing with combinatorial complexity than particles wandering about in a material, it may be worth investigating whether an "adaptive threshold" strategy would prove a useful alternative to the strategy of simulated annealing.

Nonmonotonic guidance effects can be accomplished either deterministically or probabilistically, and in the following development we specifically invoke elements of probabilistic tabu search. That is, we use controlled randomization to fulfill certain functions otherwise provided by memory, assigning probabilities to reflect evaluations of attractiveness, dominantly weighted over near best intervals, and exerting additional control by judiciously selecting the subsets of moves from which these intervals are drawn. These linked processes create an implicit tabu threshold effect that emulates the interplay of tabu restrictions and aspiration criteria in TS procedures that are designed instead to rely more fully on memory.

In order to design a simple procedure that does not rely on memory, we start from the premise that a solution trajectory should be adaptively determined by reference to the regions it passes through. Thus, instead of obeying an externally imposed guidance criterion, such as a monotonically changing temperature, we seek to reinforce the tabu search strategy of favoring behavior that is sensitive to the current search state, accepting (or inducing) fluctuations while seeking best moves within the limitations imposed. Evidently, this must frequently exclude some subset of the most attractive moves (evaluated without reference to memory), because repeated selection of such moves otherwise may result in repeating the same solutions.

The probabilistic TS orientation suggests that choosing moves by reference to probabilities based on their evaluations (attaching high probabilities for selecting those that are near best) will cause the length of the path between duplicated solutions to grow, and this will provide the opportunity to find additional improved solutions, as with deterministic TS methods. By the same orientation, a probabilistic element can be injected into the manner of choosing move subsets in a candidate list approach, to create a reinforcing effect that leads to more varied selections.

The skeletal form of a tabu thresholding method is easy to describe, as shown in Figure 7.1. The method may be viewed as consisting of two alternating phases, an *Improving Phase* and a *Mixed Phase.* The Improving Phase allows only improving moves, and terminates with a local minimum, while the Mixed Phase accepts both non-improving and improving moves.

Fig. 7.1 Tabu thresholding procedure in overview.

Improving Phase

(a) Generate a subset S of currently available moves, and let S^* be the set of improving moves in S. (If S^* is empty and S does not consist of all available moves, expand S by adjoining new subsets of moves until either S^* is nonempty or all moves are included in S.)
(b) If S^* is nonempty, choose a probabilistic best move from S^* to generate a new solution, and return to (a).
(c) If S^* is empty, terminate the Improving Phase with a locally optimal solution.

Mixed Phase

(a) Select a tabu parameter t (randomly or pseudo randomly) between lower and upper bound t_{min} and t_{max}.
(b) Generate a subset S of currently available moves, and select a probabilistic best move from S to create a new solution.
(c) Continue step (b) for t iterations, or until an aspiration criterion is satisfied, and then return to the Improving Phase.

Choices of moves in the two phases are governed by employing candidate list strategies to isolate subsets of moves to be examined at each iteration, and by a probabilistic best criterion. The thresholding effect of the choice criterion is further influenced by a tabu timing parameter t which determines the number of iterations of the Mixed Phase, analogous to maintaining a tenure for tabu status in a more advanced system. Control over t is exerted by selecting lower and upper bounds, t_{min} and t_{max}, between which t is permitted to vary randomly, or according to a selected distribution. The form of the method given in Figure 7.1 assumes that a initial solution is constructed by an unspecified procedure and the best solution found is retained throughout the process.

Termination of the foregoing procedure occurs after a selected total number of iterations. The set S in these phases may consist of all available moves in the case of small problems, or where the moves can be generated and evaluated with low computational expense. In general, however, S will be selected by a candidate list strategy to assure that relevant subsets of moves are examined. Randomization is used to change the order in which the subsets are scanned.

The Mixed Phase may be expressed in a format that is more nearly symmetric to that of the Improving Phase, by identifying a subset S^* of S that consists of moves satisfying a specified level of quality. However, the determination of S^* can be implicit in the rules of the candidate list strategy for selecting S, and further can be controlled by the probabilistic best criterion, which itself is biased to favor choices from an elite subset of moves. We introduce S^* as distinct from S in the Improving Phase to emphasize the special role of improving moves in that phase.

As noted, this overview procedure is a direct embodiment of the strategic oscillation approach, keying on the movement of the objective function rather than that of other functional forms that combine cost and feasibility, or distances from boundaries or stages of construction. In spite of this narrowed focus, it is apparent that a significant range of strategic possibilities present themselves, drawing on standard ideas from the tabu search framework. For example, we observe that the method offers immediate strategic variability by permitting the Improving Phase to terminate at various levels other than local optimality, passing directly to the Mixed Phase at each such point. In this instance, the Improving Phase need not rigidly adhere to a policy of expanding S to include all moves, when no improving moves can be found. At an extreme, by suitably controlling nonimproving moves, the method can operate entirely in the Mixed Phase, or alternately, the Improving Phase can be given a more dominant role and the Mixed Phase can be replaced by a Nonimproving Phase.

Temporarily disregarding these supplementary considerations, the simple tabu thresholding approach outlined in Figure 7.1 is governed by three critical features: determining the subset S of the candidate moves to be considered, defining the probabilistic best criterion for choosing among them, and selecting the bounds t_{min} and t_{max} that affect the duration of the Mixed Phase.

7.2.1 Candidate Lists and Scanning

Studies of linear and continuous optimization problems sometimes contain useful implications for solving nonlinear and discrete optimization problems. An instance of this is an investigation of linear network optimization approaches for selecting pivot moves in basis exchange algorithms (Glover, et al. 1974). Two findings emerged that are relevant for our present purposes: first, a best evaluation rule, which always selects a move with the highest evaluation, produced the fewest total pivot steps (from a wide range of procedures examined); and second, a straightforward candidate list strategy proved notably superior in overall efficiency for problems of practical size (in spite of requiring more iterations). The first finding was consistent with outcomes of related studies. The second finding, however, was contrary to the accepted folklore of the time.

The basis of the network candidate list strategy was to subdivide the moves to be examined into subsets, one associated with each node of the network. Let **MS**(x) denote the set of all moves associated with a given solution x, i.e., the set of moves that can transform x into $x' \in$ **N**(x). For example, in the network setting, elements of **MS**(x) can represent pivot moves that generate a new basis from the current one denoted by x. Each move has an evaluation, where a higher evaluation corresponds to a more attractive alternative.

To create a candidate list, divide **MS**(x) into indexed subsets **MS**(1, x), **MS**(2, x), ..., **MS**(m, x). For the network problem **MS**(i, x) identifies all the moves associated with a given node $i \in \mathbf{M} = \{1, 2, ..., m\}$; specifically, all the pivot nodes that introduce one of the arcs meeting i into the network basis. The candidate list strategy (CLS) operates as outlined in Figure 7.2.

Fig. 7.2 Candidate list strategy for tabu thresholding.

Step 0 (Initialization.)
Start with the set **R** (rejected) empty and set $i = 0$ (to anticipate examination of the first subset).

Step 1 (Choose a candidate move from the current subset.)
Increment i by setting $i = i + 1$. (If i becomes greater than m, set $i = 1$.) Select candidate move by a predefined selection rule.

Step 2 (Execute the move or proceed to next subset.)
(a) If the candidate move is acceptable, execute the move to produce a new solution. Reset **R** to be empty and return to Step 1.
(b) If the candidate move is unacceptable, add i to **R**. If $\mathbf{R} = \mathbf{M}$, stop (all subsets have been examined and none have acceptable candidate moves). Otherwise, return to Step 1.

The success of this simple strategy soon led to a number of variations. It was observed, for example, that defining **MS**(i, x) relative to nodes of the network could cause these sets to vary somewhat in size (in non-dense networks), thus making them unequally representative. Improvement were gained by redefining these sets to refer to equal sized blocks of moves (in this case, blocks of arcs, since each nonbasic arc defines a move). This change also made it possible to choose the number of subsets m as a parameter, rather than compelling it to equal the number of nodes. Good values for m in the network setting were found to range robustly over an interval from 40 to 120, although it appeared to be better to put each successive block of m "adjacent" arcs into m different subsets. In each instance, it also appeared preferable to retain the policy of examining all subsets before returning to the first.

In the original CLS and its variants, all scanning processes involve sampling without replacement; i.e., randomization is constrained to examine all elements of a set, in contrast to performing sampling that may revisit some elements before scanning others. (Strategies for revisiting elements with tabu search are allowed only by restricting the frequencies of such visits.) One way to apply the CLS prescription for scanning the indexes of **M**, i.e., for examining the associated subsets **MS**(i, x), $i \in$ **M**, is to start completely fresh after each full scan, to examine the elements in a new order. This approach has the shortcoming that some subsets can become reexamined in close succession and others only after a long delay. It is usually preferable for each subset to be re-examined approximately m iterations after its previous examination. This leads to the examination sequence of the Improving Phase.

Block-Random Order Scan. Divide **M** into successive blocks, each containing a small number of indexes relative to m (e.g., Min(m, 5) if $m < 100$, and otherwise no more than $m/20$). As a given block is encountered, reorder its elements randomly before examining them, thus changing the sequence of this portion of **M**. The index i of the CLS strategy thus identifies positions in **M**, rather than elements of **M**.

Unless the number of elements in a block divides m, which can be countered by varying the block sizes, the block-random order scan permits the resequenced elements gradually to migrate. On any two successive scans of **M**, however, a given element will be scanned approximately m moves after its previous scan.

The scanning procedure for the Mixed Phase is similar and determines the initial conditions for the scan of the next improving phase.

Full-Random Order Scan. A full-random order scan corresponds approximately to selecting a block of size m; all elements are potentially re-ordered. (Once a local optimum is reached, such a re-ordering of **M** is conceived appropriate.) The simplest version of this scan is to randomly re-order **M**, subject to placing the last element that yielded an improving move in the Improving Phase at the end of **M**, and then scan the elements in succession. If $t > m$, the process reverts to a block-random order scan after all of **M** is examined.

This scan is resumed in the subsequent Improving Phase at the point where it is discontinued in the Mixed Phase. A preferable size for **MS**(i, x) in the Improving Phase may not be ideal for the Mixed Phase, and in general several such subsets may be combined at each iteration of the Mixed Phase as if they composed a single subset, for the purpose of selecting a current move. If the value $m - t$ is large, time may be saved by randomly extracting and re-ordering only the sets actually selected from **M** during the Mixed Phase, without randomly re-ordering all of **M**.

The combined effect of randomization and non-replacement sampling in the preceding scanning processes approximates the imposition of using tabu restrictions recency-based memory. The use of the block-random order scan in the Improving Phase (and in the Mixed Phase, once all elements of **M** have been examined), succeeds in preventing the selection of moves from sets recently examined, starting from each new local optimum. This will not necessarily prevent reversals. However, randomization makes it unlikely to generate a sequence consisting of moves that invariably correspond to reversals of previous moves, and hence acts as a substitute for memory in approximating the desired effect. If moves can be efficiently isolated and reassigned upon initiation of the full-random order scan, the composition of moves within subsets also may be randomly changed.

7.2.2 Candidate Selection

The method can be implemented at two different levels. The first level employs an *absolute best* candidate selection while the second uses a *probabilistic best* function. Absolute best refers to selecting a move in **MS**(i, x) that shares the highest evaluation in this subset. This implementation level is a simplified form of the method that is relevant only for larger problems. We include it because it is fast and because it can be viewed as a component of the approach at the next level.

A *probabilistic best* function, represented notationally as a set containing the proposed candidate, is created as follows. A subset of k best (highest evaluation) moves is extracted from **MS**(i, x) (e.g., for $k = 10$ or $k = 20$). The current evaluations of these moves are then used a second time to generate probabilities for their selection. A small probability may be retained for choosing an element outside this set.

Once a set **K** of k best moves has been identified, a simple way to determine the probabilities for selecting an element from this set is as follows. Let k_{min} equal the smallest evaluation over **K**, and create a normalized evaluation for each move in **MS**(i, x).

$$NormValue = MoveValue + \varepsilon + k_{min},$$

where ε is given a value representing a "meaningful separation" between different evaluations (or is determined by more refined scaling considerations). Further, let *NormSum* denote the sum of the normed evaluations over **K**. Then we may specify the probability for choosing a given move to be

$$Probability = \frac{NormValue}{NormSum}.$$

Note that the evaluation function itself may be modified (as by raising a simple function to different power) to accentuate or diminish the differences among its assigned values, and thus to produce different probability distributions.

7.2.3 Specialized Tabu Thresholding Procedure

As noted in the method outlined in Figure 7.1, we seek to achieve an aggressive search of the solution space by diving the solution process into two phases. During the Improving Phase we employ the CLS in exactly the form specified in Section

7.2.1 along with the special procedure for scanning **M**. Once a local optimum is reached by this phase, the Mixed Phase is activated, again applying CLS by reference to a different approach for scanning **M**. The two phases are alternated, retaining the best solution obtained, until a selected cutoff point is reached for termination. The resulting method is sketched in an outline form in Figure 7.3.

The bounds t_{min} and t_{max} in the Mixed Phase can be set to values customarily used to bound tabu list sizes in standard tabu search applications. For example, setting t_{min} and t_{max} to bracket a simple function of the problem dimension, or in some instances setting t_{min} and t_{max} to constants, may be expected to suffice for many problems in the short term. The same type of rules used to set tabu tenures in tabu search may give reasonable guidelines for the magnitude of these values (see Chapter 2).

In the mixed Phase, a candidate move selected by the probabilistic criterion is automatically deemed acceptable for selection. Automatic acceptance also results for the Improving Phase, since only improving moves are allowed to compose the alternations considered. A special case occurs for the transition from the Improving Phase to the Mixed Phase, since the final step of the Improving Phase performs no moves but simply verifies the current solution is locally optimal. To avoid wasting the effort of the scanning operation of this step, an option is to retain a few best moves (e.g., the best from each of the k most recently examined **MS**(i, x)'s), and use these to select a move by the probabilistic best criterion, thus giving a first move to initiate the Mixed Phase.

Fig. 7.3 Specialized tabu thresholding procedure.

Improving Phase

(a) Apply CLS by a Random-Block Order Scan of **M**, accepting a candidate move if it is improving
(b) Terminate with a local optimum.

Mixed Phase

(a) Select the tabu timing parameter t randomly or pseudo randomly between t_{min} and t_{max}.
(b) Apply CLS by a Full-Random Order Scan of **M**, automatically accepting a candidate move generated for each given $i \in$ **M** examined
(c) Continue for t iterations, or until an aspiration criterion is satisfied, and then return to the Improving Phase.

Ways to enhance this tabu thresholding method are presented in Glover (1995c). These enhancements maintain the feature of convenient implementation, considering the dimensions of efficiency and solution quality.

7.3 Special Dynamic Tabu Tenure Strategies

In Chapters 2 and 3 we introduced static and dynamic structures to manage the tabu tenure in attributive TS memory. The following sections offer alternative dynamic strategies to explicitly or implicitly manage tabu tenures.

7.3.1 The Reverse Elimination Method

The reverse elimination method (REM) is a dynamic strategy that determines tabu status without relying on a tabu tenure at all, but by accounting for logical relationships in the sequence of attribute changes. Appropriate reference to these relationships makes it possible to determine in advance if a particular current change can produce cycling, and thus to generate tabu restrictions that are both necessary and sufficient to keep from returning to previously visited solutions. A small tabu tenure introduces extra vigor into the search, since the avoidance of cycling is not the only goal of recency-based memory. In addition, a bounded memory span reduces overhead and provides increased flexibility, as where it may sometimes be preferable to revisit solutions previously encountered. This means of exploiting logical interdependencies also provides information that is useful for diversification strategies.

The REM operates as follows. In each iteration, a running list (RL) that contains information about the search history, is traced back to determine all moves which must be given a tabu status to avoid leading the search to an already explored solution. When tracing RL, a set called the residual cancellation sequence (RCS) is built to determine the difference between the current solution and all those solutions included in the running list. An outline of the REM for single-attribute moves is given in Figure 7.4.

Fig. 7.4 REM for single-attribute moves.

for (all a) *TabuStatus*(a) $\leftarrow$ inactive;
RCS $\leftarrow \varnothing$;
for (i = *CurrentIteration*, …, 1) {
    if ($\overline{\text{RL}(i)} \in$ RCS) {
        RCS $\leftarrow$ RCS $\setminus \overline{\text{RL}(i)}$;
    } **else** {
        RCS $\leftarrow$ RCS $\cup \overline{\text{RL}(i)}$;
    }
    if ($|\text{RCS}| \leq t$) {
        for (all $a \in$ RCS) *TabuStatus*(a) $\leftarrow$ active;
    }
}

Starting with an empty RCS, in each tracing step two things can happen. Either an attribute is processed whose complement is not in RCS and the difference between the current solution and those previously visited increases, or the attribute cancels its complement in RCS. When an attribute cancels its complement in RCS, the current solution is closer to one of the solutions visited in the search history contained in RL. If the set RCS contains only t attributes, the complement of these attributes must become tabu-active to avoid re-visiting an already explored solution. Although it is sufficient for t to be equal to one in order to avoid re-visiting a solution, the t value may be increased to achieve a larger degree of diversification (defined as the difference between the current solution and those visited in the past).

Consider a RL that consists of all moves that either add (a) or drop ($\overline{a}$) an attribute. The attributes may represent items in a knapsack, edges or nodes in a graph, etc. The initial solution is (1, 2, 5, 6) and the RL after 10 iterations is ($\overline{1}$, $\overline{6}$, 7, 3, 1, $\overline{5}$, 6, 4, $\overline{6}$, 5). The tracing steps in the REM determine which attributes must be made tabu-active to prevent the search from going back to one of the solutions visited during the first 10 iterations. Table 7.3 illustrates the mechanics of the REM, by showing the membership of RCS in each tracing step. If t is equal to 1, then $\overline{5}$ and $\overline{4}$ are tabu-active in iteration 11.

Table 7.3 REM illustration.

Tracing step	*RCS*	*Tabu-active*
1	5	$\overline{5}$
2	$\overline{6}$ 5	-
3	4 $\overline{6}$ 5	-
4	4 5	-
5	4	$\overline{4}$
6	1 4	-
7	3 1 4	-
8	7 3 1 4	-
9	$\overline{6}$ 7 3 1 4	-
10	$\overline{6}$ 7 3 4	-

The effort required by the reverse elimination method to assign tabu status to appropriate attributes clearly grows as the number of iterations increases, therefore, the main research in this area is devoted to developing mechanisms to reduce the number of computations associated with this activity. Another important consideration is the adjustment of the parameter t, to achieve a balance between intensification and diversification. Useful applications and special elaboration of the REM approach have been made by Voss (1993), Dammeyer and Voss (1993), and Sondergeld and Voss (1996a, 1996b).

7.3.2 Moving Gaps

Another form of systematic dynamic tabu tenure is provided by a *moving gap* approach, which subdivides tabu tenure into a static part and a dynamic part. The configuration of the dynamic part is changed so that shorter tenure values are followed by longer tabu tenure values. This is done by moving a gap in the tabu tenure structure. The movement is generally controlled by particular strategies for implementing intensification and diversification. To understand how moving gaps work, it is useful to think about tabu tenure in terms of a list that records attributes of the most recent moves. Consider, for example, a list consisting of 8 time slots, where four of those slots are assigned to the static part of the list (see Figure 7.5).

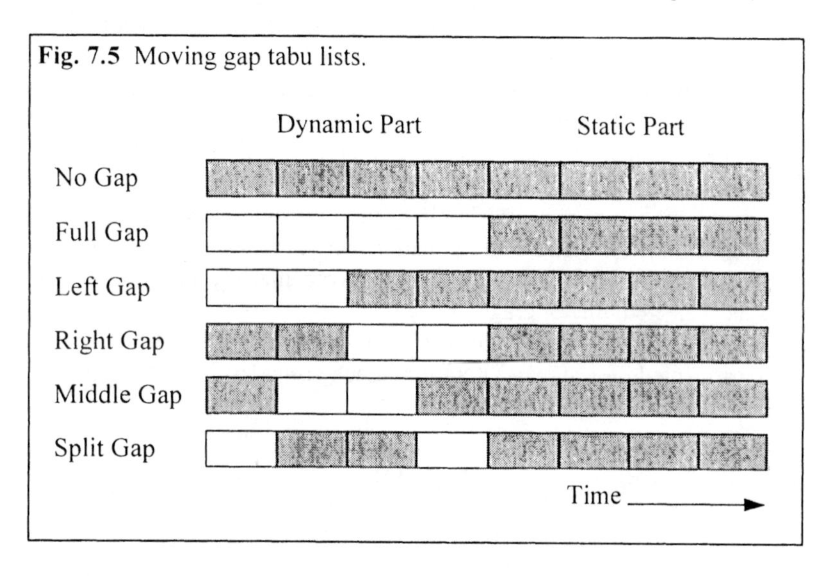

Fig. 7.5 Moving gap tabu lists.

Figure 7.5 shows that the "No Gap" list records the attributes of the last 8 moves, the "Full Gap" remembers attributes from the last 4 moves, the "Left Gap" stores attributes from the last 6 moves, and the "Right Gap" uses attributes from the last 4 moves plus the moves performed 7 and 8 iterations earlier to enforce a given tabu activation rule. "Middle Gap" and "Split Gap" have similar interpretations. Effective applications of the moving gap approach have been made by Skorin-Kapov (1994) and by Hubscher and Glover (1994).

7.3.3 The Tabu Cycle Method

The tabu cycle method and its probabilistic variant that follows represent dynamic approaches which, although proposed some time ago, have not been studied empirically. We include them in our catalog because they contain interesting features that achieve an implicit thresholding effect across different spans of iterations.

The tabu cycle method is based on creating a set of groups that consist of elements added between given iteration ranges. For example, a group may consist of elements (move attributes) that were added between h and k iterations ago. A common approach for accomplishing a progressive relaxation of the tabu status is to apply successively smaller penalties to the elements in older groups. By contrast, the tabu cycle method allows the elements of the groups to fully escape their tabu status according to certain frequencies that increase with the age of the groups. The method is based on the use of iteration intervals called *tabu cycles*, which are made smaller for older groups than for younger groups. Specifically, if Group k has a tabu cycle of $TC(k)$ iterations, then at each occurrence of this many iterations, on average, the elements of Group k escape their tabu status and are free to be chosen. To illustrate, suppose that there are three groups (each older than the preceding), whose tabu cycles are 4, 3, and 2. Then, roughly speaking, an element could be selected from the first group once in every 4 iterations, from the second group once in every 3 iterations, and the third group once in every 2 iterations. (A buffer group whose elements never escape tabu status has an implicit tabu cycle of infinity.)

However, the process is not quite as simple as the illustration suggests. There is no clear provision of how to handle the situation where no element is chosen from Group k for a duration of several tabu cycles, given that the goal is to allow an element to be selected, on average, once every cycle. To take care of this consideration, we introduce a *cycle count*, $CC(k)$, for Group k. Initially, $CC(k)$ starts at 1 and is incremented by 1 at every iteration. Each group has three states, OFF, ON and FREE. We define Group k to be:

OFF	if $CC(k) < TC(k)$
ON	if $CC(k) \geq TC(k)$
FREE	if Group h is ON for all $h \geq k$.

An element is allowed to be chosen from Group k only if it is FREE, hence only if its cycle count equals or exceeds the tabu cycle value (qualifying the group as ON), and only if this same condition holds for older groups. The ON and FREE states are equivalent for the oldest group. As implied by our earlier discussion, we assume $TC(k) < TC(k\text{-}1)$ for all $k > 1$.

The definition of the FREE state derives from the fact that each $CC(k)$ value should appropriately be interpreted as applying to the union of Group k with all groups younger than itself. Accordingly, once an element is selected from Group k, the cycle count $CC(k)$ is re-set by the operation:

$$CC(h) = CC(h) - TC(h) \text{ for all } h \geq k,$$

whereupon the cycle counts are incremented again by 1 at each succeeding iteration. By this rule, if Group k becomes FREE as soon as it is ON, and if an element is chosen from Group k at that point, then $CC(k)$ is re-set to 0 (which gives it a value 1 the iteration after it receives the value $TC(k)$). On the other hand, if no element is selected from Group k (or any younger group) until $CC(k)$ is somewhere between $TC(k)$ and $2TC(k)$ iterations, the rule for re-setting $CC(k)$ assures that Group k will again become ON when the original (unadjusted) cycle count reaches $2TC(k)$ iterations. Thus, on the average, this allows the possibility of choosing an element from Group k once during every $TC(k)$ iterations.

This process is illustrated in Table 7.4. An additional buffer group ("Group 0") may be assumed to be included, although not shown, whose elements are never allowed to escape tabu status, hence which is always OFF. For convenience, a sequence of iterations is shown starting from a point where all cycle counts have been re-set to 1 (iteration 61 in the illustration). Since a group is OFF until its cycle count reaches its tabu cycle value, each group begins in the OFF state. A group that is both ON and FREE is shown as FREE, and a FREE group from which an element is selected is indicated by an asterisk.

Table 7.4 Tabu cycle method illustrated.

	Group 1: $TC(1) = 4$		*Group* 2: $TC(2) = 3$		*Group* 3: $TC(3) = 2$	
Iteration	$CC(1)$	*State*	$CC(2)$	*State*	$CC(3)$	*State*
61	1	OFF	1	OFF	1	OFF
62	2	OFF	2	OFF	2	FREE*
63	3	OFF	3	ON	1	OFF
64	4	FREE	4	FREE*	2	FREE
65	5	ON	2	OFF	1	OFF
66	6	FREE	3	FREE*	2	FREE
67	7	ON	1	OFF	1	OFF
68	8	ON	2	OFF	2	FREE
69	9	FREE*	3	FREE	3	FREE
70	6	ON	1	OFF	2	FREE*

Table 7.4 discloses how the choice of an element from a FREE group affects the cycle counts, and hence the states, of each group. Thus, for example, on iteration 64 the choice of an element from Group 2 reduces the cycle counts of both Group 2 and Group 3 by the rule:

$$CC(2) = CC(2) - TC(2) = 4 - 3 = 1,$$
$$CC(3) = CC(3) - TC(3) = 2 - 2 = 0.$$

Hence on iteration 65, where these counts are again incremented by 1, their values are shown as $CC(2) = 2$ and $CC(3) = 1$.

An additional feature of this method is that the cycles do not have to be integer. A value such as TC(k) = 3.5 can be selected to allow the method, on average, to choose an element from Group k once every 3.5 iterations (hence twice every 7 iterations). The rules remain exactly as specified.

A slight elaboration of the rules is required, however, to handle a situation that may occur if no element is selected from a Group k or any younger group for a relatively large number of iterations. In this case, CC(k) may attain a value which is several times that of TC(k), causing Group k to remain continuously ON, and hence potentially FREE, until a sufficient number of its elements (or of younger groups) are chosen to bring its value back to below TC(k). This leads to the possibility that a series of iterations will occur where elements are repeatedly selected from Group k. (The frequency of selection will be limited however, by the cycle count values of older groups. Hence the greatest risk of inappropriate behavior occurs when elements are selected entirely outside of the tabu list for a fairly high number of iterations.) Such "statistically exceptional" outcome can be guarded against by bounding the value of CC(k) by specified multiple of TC(k).

7.3.4 Conditional Probability Method

A further interesting variant of the tabu cycle method chooses elements by establishing the probability that a group will be FREE on a given iteration. The probability assigned to Group k may be viewed conceptually as the inverse of the tabu cycle value TC(k), i.e., $P(k) = \frac{1}{TC(k)}$. As with the treatment of cycle counts in the Tabu Cycle Method, the appropriate treatment of probabilities in the Conditional Probability Method is based on interpreting each P(k) as applicable to the union of Group k with all groups younger than itself. Also, Group k can only be FREE by implicitly requiring all other groups likewise to be FREE.

Under the assumption P(k) > P(k-1) for all k > 1, we generate a *conditional probability,* CP(k), as a means of determining whether Group k can be designated as FREE. (We may again suppose the existence of a buffer, Group 0, which admits no choice of its elements, hence for which P(0) = 0.) The rule for generating CP(k) is as follows:

$$CP(1) = P(1)$$

$$CP(k) = \frac{P(k) - P(k-1)}{1 - P(k-1)} \quad \text{for } k > 1.$$

Then at each iteration, the process for determining the state of each group starts at $k = 1$, and proceeds to larger values in succession, designating Group k to be FREE with probability CP(k). If Group k is designated FREE, then all groups with larger k values are also designated FREE and the process stops. Otherwise, the next larger k is examined until all groups have been considered.

The derivation of CP(k) is based on the following argument. By interpretation, P(k) represents the probability that Group k or some younger group is the first FREE group, while CP(k) designates the probability that Group k is the first FREE group but no younger group is FREE. The assignment CP(1) = P(1) is appropriate since it is not possible for any group younger than Group 1 to be FREE. In general. For larger values of k, the event that Group k or a younger group is FREE derives from the two exclusive events (assumed independent) where either (1) Group k-1 or some younger group is the first FREE group, which occurs with probability P(k-1), or (2) this is not the case and Group k is the first FREE group. This gives rise to the formula

$$\mathrm{P}(k\text{-}1) + (1 - \mathrm{P}(k\text{-}1))\,\mathrm{CP}(k) = \mathrm{P}(k)$$

and solving for CP(k) gives the value specified.

Consideration of the rationale underlying the Tabu Cycle Method described earlier, however, shows that an appropriate characterization of the CP(k) probability is not yet complete. Designating Group k to be FREE does not imply an element will be selected from this group. For some number of iterations, elements outside of the tabu list may be accepted regardless of the FREE state of the groups of elements within the list. When this occurs, the expected number of elements per iteration chosen from groups no older than any given Group k will generally fall below P(k), without any compensating increase in freedom to choose elements from these groups (which would potentially allow the average per iteration to come closer to P(k)). Moreover, the same result will occur if for some number of iterations no element is selected from a group as young as Group k.

To handle this, the original P(k) values may be replaced by "substitute probabilities" $\mathrm{P}^*(k)$ in the determination of CP(k). This substitute probabilities make use of the same cycle count values CC(k) used in the Tabu Cycle Method, invoking the relationship $\mathrm{TC}(k) = \frac{1}{\mathrm{P}(k)}$. To begin, $\mathrm{P}^*(k)$ = P(k) until CC(k) exceeds TC(k) (bounding CC(k) in early iterations as specified previously). Then $\mathrm{P}^*(k)$ is allowed to exceed P(k) as an increasing function of the quantity $\frac{\mathrm{CC}(k)}{\mathrm{TC}(k)}$. In contrast to the Tabu Cycle Method, a negative value can result for CC(k) in this approach, as a result of the update CC(k) = CC(k) - TC(k) (which occurs whenever an element is selected from a Group h, for $h \le k$).

A variation with an interesting interpretation resets the cycle count CC(k) to 0 instead of subtracting TC(k). Then CC(k) counts consecutive iterations where no element is chosen from any Group h, $h \le k$, an event which may be (loosely) constructed occurring with probability $(1-\mathrm{P}(k))^{\mathrm{CC}(k)}$. Then the substitute probability $\mathrm{P}^*(k)$ on the following iteration may be established by setting

$$P^*(k) = 1 - (1 - \mathrm{P}(k))^{\mathrm{CC}(k)+1}$$

(Note this gives $\mathrm{P}^*(k) = \mathrm{P}(k)$ when $\mathrm{CC}(k) = 0$.)

Finally, to provide a valid basis for computing CP(k), we require $\mathrm{P}^*(k) \geq \mathrm{P}^*(k\text{-}1)$ for $k > 1$. Hence, beginning with the largest k and working backward, we set

$$\mathrm{P}^*(k\text{-}1) = Min\{\mathrm{P}^*(k\text{-}1), \mathrm{P}^*(k)\}.$$

The substitute probability approach has the advantage of increasing the probability of choosing elements of neglected groups as long as they remain neglected. The variation that resets $\mathrm{CC}(k) = 0$ on the update step also automatically avoids the creation of inappropriately high probabilities for the type of situation handled in the Tabu Cycle Method by capping CC(k).

By assigning numerical values to tabu attributes, with an associated limit on the sum (or other function) of these values to determine that tabu status of a move, both the Tabu Cycle Method and the Conditional Probability Method can be applied to multiple attribute moves as readily as to single move attributes.

7.4 Hash Functions

Hash functions have been used within tabu search with the purpose of incorporating a mechanism to avoid cycling in a way that is computationally inexpensive (Woodruff and Zemel, 1993; Carlton and Barnes, 1995). As with other forms of memory used in TS, the tabu activation rules and the tenure of attribute-based memory do not need to be designed with the sole purpose of cycling avoidance. The simplest (but computationally very expensive) scheme for mitigating cycling would be to keep a list of the last s solutions visited during the search and to compare all trial solutions during the current iteration with those in the list. For even moderate values of s the storage requirements and the effort to search the list become excessive. Hashing proposes that the solution vectors be mapped to integers that can be stored for a large number of recent iterations. This process is considered practical, since for most problems the computational effort associated with keeping and searching a list of integers is negligible compared with the evaluation of the neighborhood. The *hash functions* have the role of mapping a solution vector to an integer and a *hash list* contains the function values for recent solutions.

Since hashing is employed as an algorithmic device to avoid cycling, an effective hash function h must meet the following criteria:

- Computation and update of h should be as easy as possible. This means that the structure of h should reflect the neighborhood structure.

- The integer value generated by h should be in a range that results in reasonable storage requirements and comparison effort (e.g., an integer requiring two or four bytes).
- The probability of *collision* should be low. A collision occurs when two different vectors are encountered with the same hash function value.

The simplest type of hashing, for problems where the decision variables are bounded integers (i.e., $a \le x_i \le b$), can be achieved by a function of the following form

$$h_0 = \sum_{i=1}^{n} z_i x_i ,$$

where z is a pre-computed vector of pseudo-random integers in the range 1, ..., m. This function is particularly easy to evaluate when the neighborhood structure is defined so that very few elements of the solution vector change after each iteration. For example, if x_i and x_j are swapped to form y, then

$$h_0(y) = h_0(x) + \left(z_i\left(x_j - x_i\right) + z_j\left(x_i - x_j\right)\right).$$

When the function h_0 is used, the parameters that affect the collision probabilities are a, b, m, and n. Unless m is very small, the function h_0 may take values in excess of the maximum integer that can be represented by a computer (MAXINT). In most machines, this will result in an overflow, so the function will effectively be

$$h_1 = \left(\sum_{i=1}^{n} z_i x_i\right) \mathbf{mod}(\text{MAXINT} + 1).$$

In the usual case where m is not small compared to MAXINT, the distribution of h_1 values will be approximately uniform on the interval from zero to MAXINT for all but the smallest values of $n(b - a)$. If the elements in x and y are sufficiently random, the probability of collision using h_1 is approximately 1/MAXINT. If MAXINT is 65,535 (two bytes for unsigned integers), then the probability of collision is about 0.0000153 for all n, b, a, such that $n(b - a)$ and m are sufficiently large and the solution vectors are sufficiently random. For a hash list of length 100, the probability of seeing a collision at any iteration is still only 0.00153, suggesting that a two byte h_1 function will be adequate for many applications.

Hash functions constitute a special case of a design that maps standard types of solution attributes into additional (composite) attributes. One appeal of hash functions is that the resulting attributes, which consists simply of integers, can be referenced and accessed by the most elementary form of memory. Hash functions can be used either to replace or complement other attribute-based memory within tabu search implementations. When used to supplement another attribute-based memory function, hashing may be implemented as follows:

- The hash list can be circular and checked at every iteration or sampled with a frequency that increases if cycling is detected.
- The hash list can be maintained as a binary tree, using the function value as the key. Since the hash values are fairly random, the tree should remain balanced without any special effort.

In some cases where the search trajectory of a TS procedure is guided by longer term memory and an evaluation function that includes penalties and incentives, hashing can provide a useful mechanism to avoid cycling without becoming overly restrictive.

7.5 Ejection Chains

The rudiments of ejection chains were examined in Chapter 5. We now elaborate by describing additional features of these special types of compound moves. Neighborhoods defined by ejection chains provide a foundation for more advanced levels of solution methods based on local search. Such neighborhoods are designed to produce moves of greater power with an efficient investment of computer effort.

An ejection chain is an embedded neighborhood construction that compounds the neighborhoods of simple moves to create more complex and powerful moves. In rough overview, an ejection chain is initiated by selecting a set of elements to undergo a change of state (e.g., to occupy new positions or receive new values). The result of this change leads to identifying a collection of other sets, with the property that the elements of at least one must be "ejected from" their current states. State-change steps and ejection steps typically alternate, and the options for each depend on the cumulative effect of previous steps (usually, but not necessarily, being influenced by the step immediately preceding). In some cases, a cascading sequence of operations may be triggered representing a domino effect. The ejection chain terminology is intended to be suggestive rather than restrictive, providing a unifying thread that links a collection of useful procedures for exploiting structure, without establishing a narrow membership that excludes other forms of classifications.

As an illustration, consider creating a neighborhood structure for the generalized assignment problem (see Laguna, et al. 1995). In general, the generalized assignment problem consists of assigning tasks to agents. Each task must be assigned to one and only one agent. Each agent has a limited amount of a single resource. An agent may have more than one task assigned to it, but the sum of the resource requirements for these tasks must not exceed the agent's budget. The resource requirements of a particular task and the assignment cost depend on the agent to which the task is assigned. The objective of this combinatorial optimization problem is to minimize the total assignment cost.

A very simple move definition for the generalized assignment problem changes the assignment of task i from agent j to agent k. In terms of a netform (network related formulation), this simple move is equivalent to deleting the arc going from node i to node j, and inserting the arc that goes from node i to node k. An ejection chain results by allowing this move to eject an element (arc i') at node k, which must then be repositioned (replaced by another arc from the same source) to lead to some new node l (see Figure 7.6). The destination node l is given in the form of a pointer for each task, to avoid the computational cost of searching for a new agent to assign the task ejected by i. The process may continue through additional nodes until a suitable termination criterion is met. The adaptation of the ejection chain idea to this setting may be viewed as a heuristic generalization of a weighted alternating path approach, as applied in the solution of matching problems.

Fig. 7.6 Ejection chain illustration.

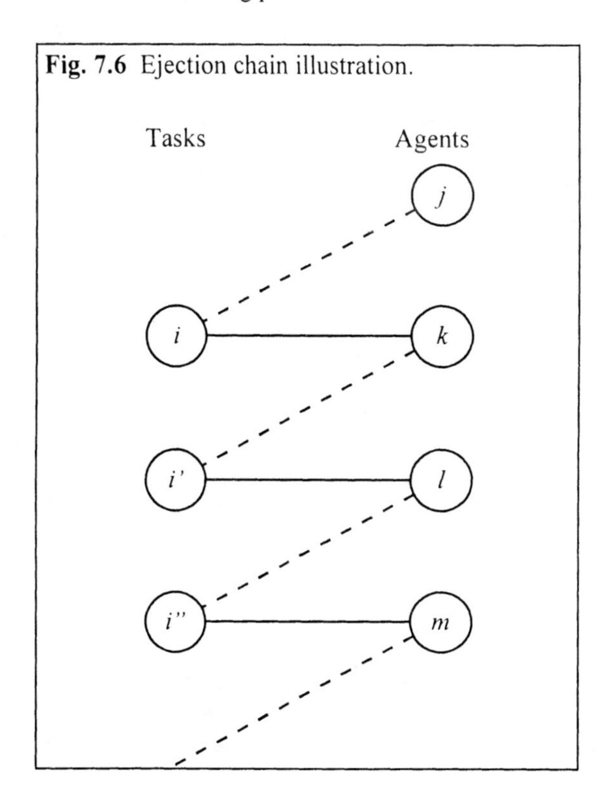

One possibility is to continue the ejection process until the chain reaches the agent in the first link (i.e., agent j). Another option is to continue the chain until no ejection occurs, usually because an agent is found with enough spare capacity to perform the last ejected task. Regardless of the length of the chain, the most important aspect of this neighborhood generation procedure is to maintain appropriate information to minimize the effort associated with evaluating each chain. In this context, after an iteration is completed, a pointer to a preferred agent must be maintained for each task. This pointer indicates the assignment to be made in case a task is ejected. The pointers can be found by using the cost and resource requirement information, as well as the total slack for each agent. A design consideration is to either restrict the process to create only feasible chains or allow infeasible assignments. If infeasible assignments are considered, a strategic oscillation process can be implemented.

7.5.1 Reference Structures and Traveling Salesman Applications

As pointed out in Chapter 5, the use of reference structures provides an important way to increase the effectiveness of ejection chains in a variety of settings. The first application of reference structures was introduced in the context of the traveling salesman problem (Glover, 1992), and we sketch some of the basic ideas of this application as a foundation for understanding the relevance of reference structures more generally.

The purpose of reference structures is to guide the generation of acceptable moves. This is done in a manner that effectively enlarges the alternatives that might otherwise be considered, as by freeing an ejection chain construction from the necessity to operate directly within the space of feasible solutions. Instead, by means of a reference structure, a succession of "partial moves" is concatenated that allows a separation from feasibility, which is bridged by trial solutions whose form depends on the reference structure considered.

Fig. 7.7 Stem-and-cycle reference structure.

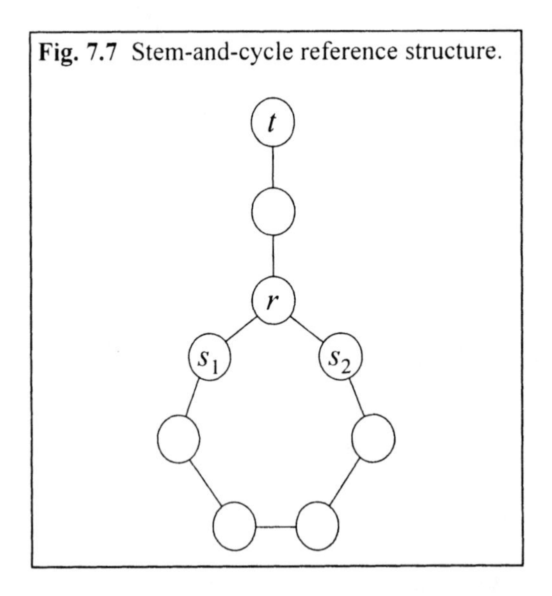

In the TSP setting such a structure can be controlled to produce transitions between tours with desirable properties, in particular generating alternating paths (or collections of such paths) of a non-standard yet advantageous type. A simple type of reference structure for the TSP is illustrated in Figure 7.7.

The *stem-and-cycle* reference structure is a spanning subgraph that consists of a node simple cycle attached to a path, called a stem. The node that represents the intersection of the stem and the cycle is called the root node, denoted by r, and the two nodes adjacent to the root are called the subroots. The other end of the stem is called the tip of the stem, denoted by t.

The stem can be degenerate, consisting of a single node, in which case $r = t$ and the stem-and-cycle structure corresponds to a tour. Two trial solutions are available for creating a tour when the stem is non-degenerate, each obtained by adding an edge (t,s) from the tip to one of the subroots s_1 or s_2, and deleting the edge (r,s) between this subroot and the root. (When the stem is degenerate, this operation adds and deletes the same edge, leaving the tour unaffected.)

An ejection chain construction can be developed to operate by cutting out and relinking tour subpaths. The process generates an evolving stem-and-cycle configuration which is initiated by selecting a node of the current tour to be the root node r, and hence also the initial tip t. The stem is then extended by a series of step that consist of attaching a non-tour edge (t,j) to a chain of tour nodes $(j, ..., k)$ which is thereby ejected from the tour.

Fig. 7.8 An ejected subpath tour construction.

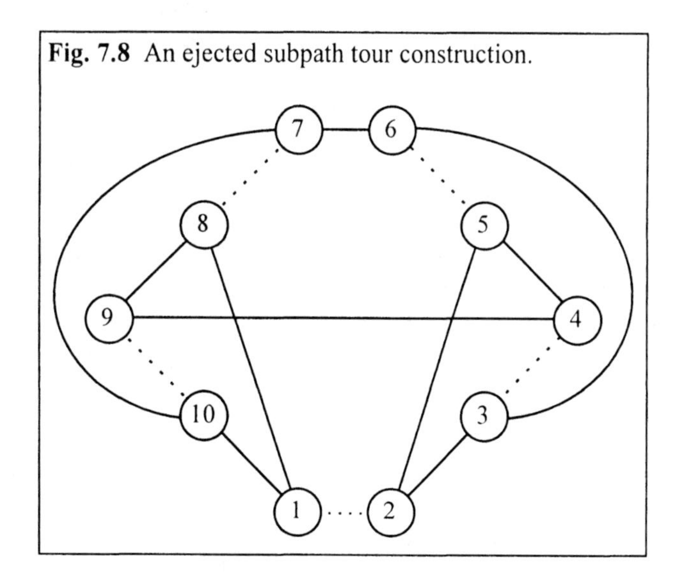

An example that illustrates the kind of outcomes a simple stem-and-cycle approach can generate, establishing a basis that leads to more advanced considerations, is based on the 10-node graph shown in Figure 7.8. The current tour is assumed to visit the nodes in their numerical order. The illustrated process ejects two subpaths, each consisting of a single edge, and then selects a trial solution. In Figure 7.8, the ejection chain starts with node 1 as the root *r*, and then adds edge (1,8) to attach to and eject ("cut free") the subpath (8,9) (which in this case consists of an edge), followed by adding edge (9,4) to attach to and eject the subpath (4,5) (likewise in this case consisting of an edge). Thus the capping edges added by these two ejections are (7,10) and (3,6) while the edges deleted are (7,8), (9,10), (3,4) and (5,6). Of the two trial solutions available at this point, we select the one associated with subroot 2 to conclude the process, thus adding edge (5,2) and deleting edge (1,2).

The transformation of the current tour into the new one of this example produces an outcome that cannot be achieved by the popular heuristic of Lin and Kernighan (1973), or by a heuristic that is designed similarly to maintain an underlying feasible tour construction at each step. Specifically, we see that the two deleted edges adjacent to the first added edge destroy the connectivity of the graph. That is, the initial addition of edge (1,8) cannot be accompanied by the deletion of both (1,2) and (7,8), since this divides the graph into a disjoint path and subtour. Moreover, the same sort of infeasibility occurs in this example regardless of which edge is added first. Thus, the resulting tour cannot be obtained by a "feasibility preserving" approach regardless of which node of the graph is selected to initiate the process.

Additional somewhat different but equally simple examples using the stem-and-cycle reference structure yield the same type of outcome. For example, an instance of such a construction occurs by replacing the three interior edges of Figure 7.8 by the three edges (1,4), (5,8), and (9,2).

The concept of using reference structures to permit the creation of ejection chains that can pass through infeasible regions is extended in Glover (1992) to provide additional types of reference structures with special properties. The use of these structures can be controlled in an entirely natural way by TS memory mechanisms, often very simple ones. Highly effective uses of the stem-and-cycle reference structure with simple TS memory have been developed for traveling salesman and vehicle routing problems by Rego (1996a, 1996b). Effective applications of other types of reference structures are provided by Punnen and Glover (1997).

7.6 Vocabulary Building

A frontier area of tabu search involves the creation of new attributes out of others, by an approach that imparts meaning or structure to the resulting attributes which can be exploited by the search process. In this section, we focus on creating new attributes by reference to a process called vocabulary building, related to concept formation. The basic idea of this approach is to identify meaningful fragments of solutions, rather than focusing solely on full vectors, as a basis for generating combinations. A pool of such fragments is progressively enriched and assembled to create larger fragments, until ultimately producing complete trial solutions. In some settings these fragments can be integrated into full solutions by means of optimization models

(Glover, 1992; Glover and Laguna, 1993). Procedures using this design have been developed with highly successful outcomes by Rochat and Taillard (1995) and Kelly and Xu (1995). Approaches that heuristically assemble fragments have also been successfully implemented by Taillard, et al. (1995) and Lopez, Carter and Gendreau (1996).

Vocabulary building effectively may be conceived as an instance of path relinking. There are two key objectives: (1) to identify a good collection of reference points, in this case consisting of "partial solutions" (which include the fragments), and (2) to identify paths in one or more neighborhood spaces that will unite components of these partial solutions, with suitable attendant modifications, to produce complete solutions.

The neighborhood spaces for vocabulary building need not be the same as those used by tabu search processes that introduce vocabulary building to generate solutions as candidates for subsequent refinement. For example, the threshold based form of logical restructuring examined in Chapter 3 constitutes a means for creating elements of vocabulary, and the recourse to neighborhoods based on shortest path concepts to connect such elements illustrates the shift in neighborhoods (contrasting with swap neighborhoods used in the Chapter 3 examples) in order to execute a vocabulary building function. Ejection chains are also relevant to vocabulary building, since attributes of different partial solutions may be imperfectly compatible, and hence the synthesis of such partial solutions can benefit from compound transformations to create effective linkages. As a special instance, solution fragments may be united by linear combinations, following a scatter search design. A formalism for expressing (and exploiting) several key forms of vocabulary building is as follows (Glover and Laguna, 1993).

A chosen set S of solutions may be viewed as a text to be analyzed, by undertaking to discover attribute combinations shared in common by various solutions $x \in \mathbf{X}$. Attribute combinations that emerge as significant enough to qualify as units of vocabulary, by a process to be described below, are treated as new attributes capable of being incorporated into tabu restrictions and aspiration conditions. In addition, they can be directly assembled into larger units as a basis for constructing new solutions.

We represent collections of attributes by encoding them as assignments of values to variables, which we denote by $y_j = p$, to differentiate the vector y from the vector x which possibly may have a different dimension and encoding. Normally we suppose a y vector contains enough information to be transformed into a unique x, to which it corresponds, but this assumption can be relaxed to allow more than one x to yield the same y. (A specified range of different assignments for a given attribute can be expressed as a single assignment for another, which is relevant to creating vocabulary of additional utility.)

Let $Y(S)$ denote the collection of y vectors corresponding to the chosen set S of x vectors. In addition to assignments of the form $y_j = p$ which define attributes, we allow each y_j to receive the value $y_j = *$, in order to generate subvectors that identify specific attribute combinations. In particular, an attribute combination will be implicitly determined by the non-$*$ values of y.

To generate vocabulary units we compare vectors y' and y'' by an intersection operator, $Int(y', y'')$ to yield a vector $z = Int(y', y'')$ by the rule: $z_j = y'_j$ if $y'_j = y''_j$ and $z_j = *$ if $y'_j \neq y''_j$. By this definition we also obtain $z_j = *$ if either y'_j or $y''_j = *$. *Int* is associative, and the intersection $Int(y\colon y \in \mathbf{Y})$, for an arbitrary $\mathbf{Y}$, yields a z in which $z_j = y_j$ if all y_j have the same value for $y \in \mathbf{Y}$, and $z_j = *$ otherwise.

Accompanying the intersection operator, we also define a relation of containment, by the stipulation that y'' *contains* y' if $y'_j = *$ for all j such that $y'_j \neq y''_j$. Associated with this relation, we identify the *enclosure* of y' (relative to S) to be the set $Y(S{:}\, y') = \{y \in Y(S){:}\ y \text{ contains } y'\}$, and define the enclosure value of y', $EncValue(y')$, to be the number of elements in this set, i.e., the value $|Y(S{:}y')|$. Finally, we refer to the number of non-$*$ components of y' as the *size* of the vector, denoted $size(y')$. (If $y \in Y(S)$, the size of y is the same as its dimension.)

Clearly the greater $Size(y')$ tends to become, the smaller $EncValue(y')$ tends to become. Thus for a given size s, we seek to identify vectors y' with $Size(y') \geq s$ that maximize $EncValue(y')$, and for a given enclosure value v to identify vectors y' with $EncValue(y') \geq v$ that maximize $Size(y')$. Such vectors are included among those regarded to qualify as vocabulary units.

Similarly we include reference to weighted enclosure values, where each $y \in Y(S)$ is weighted by a measure of attractiveness (such as the value $c(x)$ of an associated solution $x \in S$), to yield $EncValue(y')$ as a sum of the weights over $Y(S{:}y')$. Particular attribute values likewise may be weighted, as by a measure of influence, to yield a weighted value for $Size(y')$, equal to the sum of weights over non-$*$ components of y'.

From a broader perspective, we seek vectors as vocabulary units that give rise to aggregate units called *phrases* and *sentences* with certain properties of consistency and meaning, characterized as follows. Each y_j is allowed to receive one additional value, y_j = *blank*, which may be interpreted as an empty space free to be filled by another value (in contrast to $y_j = *$, which may be interpreted as a space occupied by two conflicting values). We begin with the collection of vectors created by the intersection operator *Int*, and replace the $*$ values with *blank* values in these vectors. We then define an extended intersection operator *EInt*, where $z = EInt(y', y'')$ is given by the rules defining *Int* if y'_j and y''_j are not *blank*. Otherwise $z_j = y'_j$ if

$y''_j = blank$, and $z_j = y''_j$ if $y'_j = blank$. *EInt* likewise is associative. The vector $z = EInt(y: y \in Y)$ yields $z_j = *$ if any two $y \in Y$ have different non-*blank* values y_j, or if some y has $y_j = *$. Otherwise z_j is the common y_j value for all y with y_j non-*blank* (where $z_j = blank$ if $y_j = blank$ for all y).

The y vectors created by *EInt* are those we call *phrases.* A *sentence* (implicitly, a complete sentence) is a phrase that has no *blank* values. We call a phrase or sentence *grammatical* (logically consistent) if it has no $*$ values. Grammatical sentences thus are y vectors lacking both *blank* values and $*$ values, constructed from attribute combinations (subvectors) derived from the original elements of $Y(S)$. Finally we call a grammatical sentence y *meaningful* if it corresponds to, or maps into a feasible solution x. (Sentences that are not grammatical do not have a form that permits them to be translated into an x vector, and hence cannot be meaningful.)

The elements of $Y(S)$ are all meaningful sentences, assuming they are obtained from feasible x vectors, and the goal is to find other meaningful sentences obtained from grammatical phrases and sentences constructed as indicated. More precisely, we are interested in generating meaningful sentences (hence feasible solutions) that are not limited to those that can be obtained from $Y(S)$, but that also can be obtained by one of the following strategies:

- (S1) Translate a grammatical phrase into a sentence by filling in the blanks (by the use of neighborhoods that incorporate constructive moves or that algorithmically solve subproblems for generating complete assignments);
- (S2) Identify some set of existing meaningful sentences (e.g., derived from current feasible x vectors not in S), and identify one or more phrases, generated by *EInt* over S, that lie in each of these sentences. Then, by a succession of moves from neighborhoods that preserve feasibility, transform each of these sentences into new meaningful sentences that retain as much of the identified phrases as possible;
- (S3) Identify portions of existing meaningful sentences that are contained in grammatical phrases, and transform these sentences into new meaningful sentences (using feasibility preserving neighborhoods) by seeking to incorporate additional components of the indicated phrases.

The foregoing strategies can be implemented by incorporating the same tabu search incentive and penalty mechanisms for choosing moves indicated in previous sections. We assume in these strategies that neighborhood operations on x vectors are directly translated into associated changes in y vectors. In the case of (S1) there is no assurance that a meaningful sentence can be achieved unless the initial phrase itself is meaningful (i.e., is contained in at least one meaningful sentence) and the constructive process is capable of generating an appropriate completion. Also, in (S3) more than one grammatical phrase can contain a given part (subvector) of a meaningful sentence, and it may be appropriate to allow the targeted phrase to change according to possibilities consistent with available moves.

Although we have described vocabulary building processes in somewhat general form to make their range of application visible, specific instances can profit from special algorithms for linking vocabulary units into sentences that are both meaningful and attractive, in the sense of creating good $c(x)$ values. A number of combinatorial optimization problems are implicit in generating good sentences by these approaches, and the derivation of effective methods for handling these problems in various settings may provide a valuable contribution to search procedures generally.

7.7 Nonlinear and Parametric Optimization

This section identifies ways to apply tabu search to nonlinear and parametric optimization problems. The basic ideas for applying tabu search to discrete linear and nonlinear settings can readily be adapted to provide methods for continuous and mixed-discrete nonlinear settings. Such nonlinear methods can also be applied to certain combinatorial problems, such as those arising in production, manufacturing and resource planning, where frameworks have evolved for obtaining discrete solutions by reference to nonlinear continuous equivalents, or by reference to special parameterizations.

Parametric optimization problems, as referred to here, involve searching an auxiliary space in order to obtain vectors that map into solutions in a primary space. The auxiliary space may be a superset of the primary space, or may be defined over and entirely separate domain. The goal is to generate parameters or coefficient values that map into solutions in primary space by means of algorithmic processes that become well defined once these values are specified.

Among the more conspicuous parametric optimization problems are those arising from Lagrangean and surrogate constraint relaxations. But there are also a number of other types, whose susceptibility to treatment by nonlinear search strategies is sometimes overlooked. Examples of such "problems within problems," where continuous trial solutions map into solutions in a different space, are as follows.

Example 1. The auxiliary space is defined over vectors of objective function or constraint coefficients, which modify original problem coefficients for strategic purposes. Typical applications include project selection problems, capital budgeting problems and multiconstraint knapsack problems. Coefficient manipulations induce the generation of alternative solutions by standard heuristics, and the solutions are used to gauge the merit of the modified coefficients. (An important special instance is the strategy of searching for modified objective function coefficients for a zero-one mixed integer programming problem, in order to cause the optimal LP solution to correspond to an optimal MIP solution to the original problem.)

Example 2. Auxiliary spaces based on modified coefficients as in Example 1 are linked with structural modifications in order to adapt network models to problems with more complex constraints. In applications such as resource distribution and telecommunications, selected constraints are replaced by associated simpler restrictions that impose bounds on newly created variables, giving rise to a pure or generalized network problem. The bounds are systematically adjusted, and the solutions to the resulting network problems give trial solutions for the original (primary) problem.

Example 3. Weights attached to decision rules in job shop scheduling problems serve as parameters for defining auxiliary space. These parameters are used either probabilistically or deterministically to generate schedules for the primary job shop problem.

Example 4. An auxiliary space based on a more rigorous treatment of embedded network structures occurs in layering strategies for linear and integer programs. Linking constraints are replaced by parameterized bounds to create decompositions into collections of problems with special structure such as generalized network problems. Trial solutions establish a feedback mechanism to evaluate the parameterized bounds.

Example 5. Applications of target analysis create an auxiliary space where historical search information is parameterized to generate decision rules. The quality of solutions derived from the decision rules measures the efficacy of the parameters. (A standard search process at one level thereby is used to create an improved search process at another.)

In each of these instances, a successful problem solving effort depends on the search of a nonlinear continuous space, using algorithmic mappings to link the outcomes to trial solutions in another space.

To apply tabu search to nonlinear problems, arising both from such parameterized applications and from standard nonlinear models, the types of memory structures described in Chapter 3 for handling increase/decrease changes in values of variables, and their matrix based extensions for handling more complex assignment changes, are directly relevant. We have already noted that such structures have been used with variable scaling strategies (which dynamically subdivide the ranges of values for the variables into subintervals of different sizes) as a basis for solving nonlinear continuous problems. In the following section we discuss how related ideas can be used to control *directional search*, where a transition from one point to another occurs by reference to gradients, subgradients, feasible directions, and projections.

7.7.1 Directional Search Strategies

We begin by focusing on ways to treat directional search strategies with the short term component of tabu search. By the standard TS approach, a memory is maintained of selected attributes of recent moves and their associated solutions, where a current move is classified tabu if it reinstates an attribute (or attribute set) that was changed by earlier moves. In more refined implementations the tabu status, or degree to which a move is classified tabu, depends on the quality of the move and the contribution of its component attributes to the quality of past moves. In turn, this status determines how long the move remains tabu (or the penalty attached to its selection).

Tabu search strategies involving recency-based memory can be adapted to the nonlinear context as follows. Let x' and x'' denote a pair of successive solutions where x'' was generated from x' by a recent move. Following the approach standardly used in discrete settings, we seek to prohibit a current move that would create a solution x containing selected attributes of x'. For the present purpose, we will select problem variables as a basis for defining move attributes, specifically identifying a subset of variables that change values in going from one solution to another. Then a move from a current solution to a new solution is classified tabu if the designated variables return to the values they had in x'.

Applying this approach to spaces with continuous as well as discrete properties, tabu restrictions may be imposed to prevent moves that bring the values of selected variables "too close" to values they held previously. (For an integer programming problem, "too close" may mean "less than a unit away.")

Specifically, let t denote a selected tabu tenure value. Then two kinds of tabu restrictions for nonlinear problems are the following.

(R1) A move is tabu if it creates a solution x which lies closer than a specified tabu distance d to any solution visited during the preceding t iterations. (The distance d can be a function of x and also of t.)

(R2) A move is tabu if it employs a direction vector contained in a *tabu region* created on one of the preceding t iterations. (Such a region created by a move from x' to x'' can be defined, for example, to be a set of direction vectors that lie in a cone with vertex x'' and with center line passing through x', for a selected angle, e.g., 30 degrees or 45 degrees.)

Focusing on components of vectors rather than on complete vectors provides a useful way to make (R1) and (R2) easier to implement. This may be accomplished by the following special variants of (R1).

(V1) Identify k components of x whose values change the most in the move from x' to x''. Create a new variable y which equals a weighted linear combination of the chosen components; e.g., choosing a weight for the component x_j equal to its change, $x''_j - x'_j$, or selecting all weights to be 1 or -1, according to the direction of change. For a duration of t iterations, prevent this variable from receiving a value less than $\frac{(y' + y'')}{2}$, where y' and y'' respectively denote the values that result for y when x equals x' and x''. More generally, the variable y may be prevented from receiving a value less than $y' + w(y'' - y')$, or excluded from falling inside the line interval bounded by this value and $y' - w(y'' - y')$, where $1 \geq w \geq 0$.

If k is selected equal to n, the dimension of x, then precisely the components of x that change their values will receive non-zero weights in defining y. At the other extreme, choosing k equal to 1 corresponds to selecting a single component x_j of x, which offers the advantage of making it unnecessary to keep track of definitions of t additional variables. In this case an alternative form of elaboration is the following.

(V2) Choose two components of x whose values change the most from x' to x''. Impose a tabu restriction that prevents at least one of these two components from changing in the reverse direction by an amount greater than half of its current change. If one (or both) of these components is already subject to such a restriction, then impose the stronger restriction of (V1). (The remaining component can be paired with the former partner of the first component to create a weaker restriction, preventing at least one of the two from reversing direction.)

Implementation of the preceding rules requires consideration of three strategic elements. (1) The x values should be normalized so that a given change in one component is comparable to the same change in another. If x is a vector of weights for a Lagrangean or surrogate relaxation method, for example, then the problem constraints should be normalized to assure that the change in one constraint has the same relative effect as that of another (since the weight changes are typically derived as a function of changes in constraint violations). (2) The "largest change" choice rule should be amended in a diversification strategy designed for the long term to favor the choice of components not yet subject to tabu restrictions. (This may be a self-regulating consequence of handling other considerations appropriately.) (3) Derivative factors can be relevant in choosing a component of x. Specifically, in the context of Lagrangean and surrogate relaxations, a choice based on the magnitude of a change in weight (dual variable) may alternately be replaced by one based on the magnitude of a change produced in a corresponding problem constraint, giving priority to constraints that undergo the transition from being violated to satisfied.

Applying the basic ideas indicated in Chapter 4, frequency-based memory can be used to track attributes derived from components of x (and their values) that change according to the formats of variants (V1) and (V2). Nonlinear TS approaches based on these types of rules provide a natural way to exploit problems where gradient information is available, or can be approximated by standard nonlinear optimization approaches. Unlike the usual application of such approaches, however, these tabu search mechanisms can be applied directly to the solution of nonconvex problems. Additional considerations relevant to intensification and diversification in these settings are treated in Chapter 9, in the discussion of scatter search processes for nonlinear problems.

7.8 Parallel Processing

Parallel implementations of tabu search have been proposed to deal with problems of realistic size within reasonable computing times. Toulouse, et. al (1996) identify three strategies often used to create parallel meta-heuristic procedures:

- *Functional decomposition* — In this strategy, some or all operations within an iteration of the corresponding sequential solution method are executed in parallel. This strategy is also known as low level parallelism.
- *Domain decomposition* — The problem domain or search space is decomposed. Typically in this strategy, a master-slave framework is used, where the master creates partitions and assigns them to the slave processors.
- *Multi-search threads* — The strategy consists of coordinating several concurrent searches. These searches may start from the same or different initial solutions and may or may not communicate during the search. Communication may be performed synchronously or asynchronously, and may be event driven or executed at predetermined or dynamically determined times.

Low level parallelism has been extensively applied to tabu search. The typical implementation of this type consists of a master thread that executes a sequential TS procedure and uses multiple processors to evaluate a predefined neighborhood at each iteration. The slave processes evaluate a set of moves assigned by the master. The master process collects the evaluations sent by the slaves and applies the corresponding tabu rules to select the next move to be executed. The master is also in charge of updating tabu search memories and applying any additional strategies implemented. This type of parallel strategy is employed either to reduce the computational time of a sequential TS implementation or increase the ability to search a larger fraction of a neighborhood in every iteration. The strategy works well when the neighborhood tends to be extremely large (e.g., a swap neighborhood in the quadratic assignment problem), but the time to evaluate each move in the set is relatively small (compared to the execution of the rest of the procedure).

Tabu search implementations of a domain decomposition parallel strategy partition the set of decision variables and perform concurrent searches on each subset. In the context of vehicle routing, Rochat and Taillard (1995) implemented in a sequential procedure the principles of this kind of parallelism. Cities are partitioned into several subsets on which slave processes operate using tabu search. The master process then combines the routes to construct a complete solution. Cities are again partitioned into subsets and the process repeats. The domain decomposition strategy is based on *speculative computation*, because the resulting procedure performs some steps (i.e., a speculative domain decomposition) without knowing if the results of these steps are needed in order to directly reach a solution to the problem. The speculative domain decomposition strategy and the corresponding procedure to create complete solutions from partial ones has elements in common with the vocabulary building ideas described in the preceding section.

Multi-thread parallelism in tabu search has been implemented using a framework in which a number of independent threads operate on the same solution space. The search threads may or may not start from the same initial solution and generally employ different TS strategies. If the search threads are totally independent and the best solution is captured at the end, the approach is identical to a parallel restarting procedure. Alternatively, the process may involve cooperation among the TS threads by exchanging information during the search. Information can be exchanged employing a synchronous or an asynchronous process. Synchronized exchange of information requires procedural parameters that determine the intervals or stages at which information will be exchanged. Asynchronous processes, on the other hand, do not have to deal with the question of "when" to exchange information; however, they still have to determine "what" information to exchange.

The exchange of information in memoryless methods such as genetic algorithms and simulated annealing is usually confined to the objective value of the best solution found by the different search threads. In tabu search, however, the information exchange may include search history stored in the TS memory of each thread. One approach is to create a candidate list of elite solutions found by the different search threads. The elite solutions can then be used by each search thread to implement diversification and intensification strategies in the same way as in the sequential TS procedures.

An interesting variant of multi-thread parallelism is the so-called asynchronous team or A-Team (Talukdar, et al. 1996). An A-Team consists of an evolving sequence of strongly cyclic data flows. A data flow is a directed hypergraph, where nodes represent memory structures (or the objects that these structures can contain) and arcs represent the autonomous agents that read from (tail) and write to (head) these structures. A data flow is strongly cyclic when most of the arcs form closed loops. An A-Team is developed in a network of computers to solve a problem, where a basic data flow can be made more complex by adding memory structures and agents. Trial solutions accumulate in the nodes of the A-Team to form populations (like those in scatter search or genetic algorithms). Two types of agents operate on the nodes: *construction* agents that add members to the population and *destruction* agents that delete members from the population. The adding and deleting of population members is controlled by tabu search strategies embedded in short and long term memory functions.

Consider the data flow in Figure 7.9. The data flow consists of one memory structure (Solutions), three constructive agents (C1, C2, and C3) and one destructive algorithm (D1). The three constructive agents add solutions to the population. The destructive agent may delete solutions based on quality and TS strategies.

Fig. 7.9 Simple data flow.

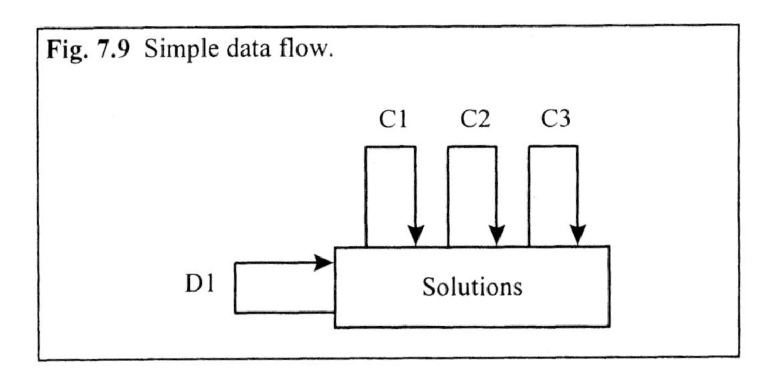

Fig. 7.10 Data flow with two nodes.

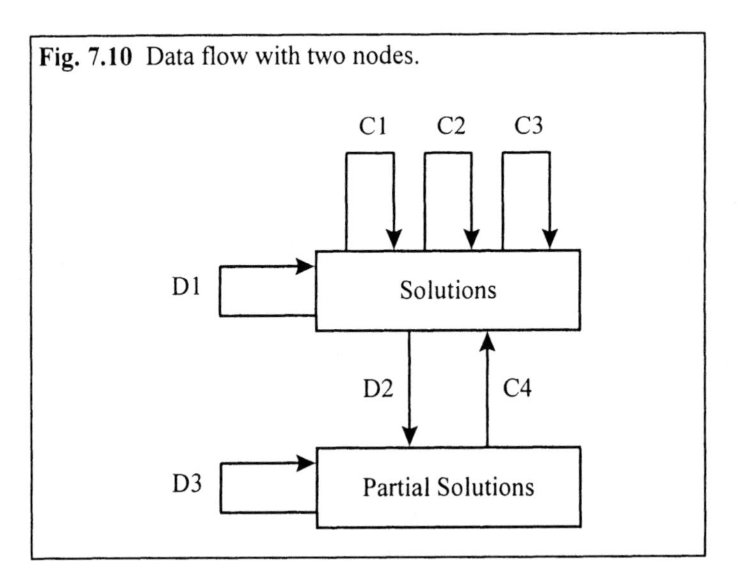

A more complex data flow can be developed by adding a memory structure that contains partial solutions (see Fig. 7.10). A destructive agent is added to create partial solutions from two complete solutions (D2). Another destructive agent (D3) is also added to control the size of the population of partial solutions. Finally, a constructive agent (C4) produces complete solutions from the population of partial solutions. We not that relationship of this design to that of vocabulary building processes as described in the preceding section.

Increasingly complex data flow designs such as those incorporating vocabulary building considerations may be necessary to improve performance. (A number of more complex data flow structures are suggested by Talukdar, et al. (1996).) Since the agents are autonomous, they decide what do next and when to do it (without the assistance of a centralized control mechanism). This feature allows A-Teams to be quite flexible while adding and deleting different types of agents before settling on a final design.

7.8.1 Taxonomy of Parallel Tabu Search Procedures

A taxonomy for the classification of parallel tabu search procedures has been recently suggested by Cranic, Toulouse and Gendreau (1997). The taxonomy is based on three dimensions: *control cardinality, control and communication type,* and *search differentiation.* The first two relate to the control of the search trajectory and the communication approach, while the third classifies the strategies to partition the domain and to specify the parameters of each search. Table 7.5 summarizes the dimensions and their possible cases.

A parallel search can be controlled by one or several processors. Typically, in the 1-*control* case one processor executes the algorithm and distributes numerically intensive tasks (e.g., the move evaluations in a large neighborhood) to other processors in the system. This case corresponds to the functional decomposition or low level parallelism referred to at the beginning of Section 7.7. In the *p-control* case, each processor is in charge of both its own search and the communications with other processors.

Table 7.5 Taxonomy dimensions.

Dimension	*Case*
Control cardinality	1-control *p*-control
Control and communication type	Rigid synchronization Knowledge synchronization Collegial Knowledge Collegial
Search Differentiation	SPSS and SPDS MPSS and MPDS

The *rigid synchronization* case in Table 7.5 refers to a synchronous operational mode where processes are instructed to stop and exchange limited amount of information at predetermined points (e.g., number of iterations or time intervals). *Knowledge synchronization* is also a synchronous form of communication that forces processes to stop at predetermined points. The main difference between rigid and knowledge synchronization is the amount of information exchanged every time the processes stop performing their individual tasks. The last two forms of communication are asynchronous. *Collegial* refers to a design where each process executes a different search and broadcasts improving solutions to all or some of the other processes. Communications in the collegial case are therefore simple. In the *knowledge collegial* design, however, communications are more complex, since findings from the processors are analyzed to modify search strategies being employed by each process. In this operational mode, the information that a process sends and the one that it receives may not necessarily be the same.

The search differentiation dimension considers the number of initial solutions and the search strategies used in the parallel design. The first two letters of each case refer to the number of initial points, thus SP = single point and MP = multiple points. The last two letters of each case refer to the number of strategies, thus SS = single strategy and DS = different strategies. While SPSS allows only for the implementation of low level parallelism, MPDS is the most comprehensive and general case for parallel implementations.

Table 7.6 Parallel TS methods.

Reference	*Application*	*Classification*
Battiti and Tecchiolli (1992)	Quadratic assignment	*p*-RS MPSS
Chakrapani and Skorin-Kapov (1992, 1993a and 1995)	Quadratic assignment	1-RS SPSS
Fiechter (1994)	Traveling salesman	*p*-RS MPSS
Malek, et al. (1989)	Traveling salesman	1-KS SPDS
Taillard (1991)	Quadratic assignment	1-RS SPSS
Taillard (1994)	Job shop scheduling	*p*-RS MPSS
Taillard (1993)	Vehicle routing	*p*-KS MPSS

RS = rigid synchronization
KS = knowledge synchronization

The taxonomy described above can be used to classify a few tabu search procedures that have been reported in the literature (see Table 7.6). Note that none of the procedures in Table 7.6 use an asynchronous type of communication nor do they use a MPDS search differentiation. This indicates that parallel tabu search implementations are still in their infancy and higher level of parallelism is a goal still to be achieved.

7.8.2 Parallelism and Probabilistic Tabu Search

Probabilistic TS has several potential roles in parallel solution approaches, which may be briefly sketched as follows.

(1) The use of probabilities can produce a situation where one processor may get a good solution somewhat earlier than other processors. The information from this solution can be used at once to implement intensification strategies at various levels on other processors to improve their performance. Simple examples consist of taking the solution as a new starting solution for other processors, and of biasing moves of other processors to favor attributes of this solution. (This is a situation where parallel approaches can get better than linear improvement over serial approaches.) In general, just as multiple good solutions can give valuable information for intensification strategies in serial implementations, pools of such solutions assembled from different processors can likewise be taken as the basis for these strategies in parallel environments.

(2) Different processors can apply different probability assignments to embody different types of strategies — as where some processors are "highly aggressive" (with probabilities that strongly favor the best evaluations), some are more moderate, and some use varying probabilities. (Probabilities can also be assigned to different choice rules, as in some variants of strategic oscillation.) A solution created from one strategy may be expected to have a somewhat different "structure" than a solution created from another strategy. Thus, allowing a processor to work on a solution created by the contrasting strategy of another processor may yield an implicit diversifying feature that leads to robust outcomes.

(3) Solution efforts are sometimes influenced materially by the initial solutions used to launch them. Embedding probabilistic TS within methods for generating starting solutions allows a range of initial solutions to be created, and the probabilistic TS choice rules may have a beneficial influence on this range. Similarly, using such rules to generate starting solutions can be the basis for diversification strategies based on restarting, where given processors are allowed to restart after an unproductive period.

At present, only the simplest instances of such ideas have been examined, and many potential applications of either deterministic or probabilistic TS in parallel processing remain to be explored.

7.9 Discussion Questions and Exercises

1. Replace the deterministic selection rule in the TS procedure for the Min k-Tree problem described in Chapter 2 by a probabilistic rule based on move values. Discuss the potential advantages and disadvantages of the new method.

2. Design a tabu thresholding method for the Min k-Tree Problem. Do you anticipate this memoryless process to be sufficient to provide high quality solutions to instances of the Min k-Tree problem?

3. Design a TS procedure that uses REM as its only memory mechanism in the solution of 0-1 knapsack problems. Compare this design with a short term memory TS procedure. Discuss the implementation of strategic oscillation in both TS variants, where the oscillation is around the feasibility boundary.

4. Indicate how a hash function related to h_0 might be designed for swap moves in contexts other than that of a sequencing problem.

5. Propose an ejection chain mechanism and a reference structure for the Min k-Tree Problem.

6. There are problems that are not readily exploitable by reference structures, but that can be exploited by compound moves derived from the ejection chain perspective. Illustrate this by identifying ejection chain moves for clustering and multiple set partitioning problems, where the objective is to maximize some measure of cohesiveness among elements assigned to a particular set (and/or minimize some measure of repulsion among elements assignments to different sets). Is it relevant in this setting to consider ejection chains with loops, where an element that ejects another may in turn be ejected at a later step of the chain? What relevance do you envision for chains that allow "nonadjacent ejections" and "ejections by attraction," where a series of ejections generates a fertile condition for dislodging an element from a set other than the one that most recently receives a new element (in order to enter a set from which another element has departed)? In general, what types of complexities are reasonable to allow in creating such ejection chains?

7. Indicate how vocabulary building ideas suggest parallel processing data flows with features beyond those illustrated in Figure 7.10.

8 TABU SEARCH APPLICATIONS

This chapter provides a collection of "vignettes" that briefly summarize applications of tabu search in a variety of settings. These vignettes are edited versions of reports by researchers and practitioners who are responsible for the applications The TS research summarized in this chapter was contributed by academicians and practitioners from all over the world. A debt of gratitude is owed to the individuals whose contributions have made this summary possible. (Deficiencies in describing their work are solely due to the current editing, and should not be interpreted to reflect shortcomings of the original reports.) In a number of cases, the work cited presents only a small sampling of significant contributions by the authors referenced.

Because of the rapid pace of innovation in applying tabu search — as by taking advantage of incompletely explored areas we have undertaken to disclose in this book — it is likely that a number of the "best outcomes" identified in the following applications will soon be superseded. It is quite conceivable that some of these advances will be initiated by the same individuals who are responsible for the applications described in this chapter (or cited elsewhere in this book). By the same token, innovations are also continuously emerging from researchers who are examining tabu search for the first time, and "new records" may be soon expected from this source as well.

There are of course many applications that space has not permitted us to include. Undoubtedly, a portion of these will occur in areas that some will find at least as important as those represented. Our apologies both to authors and to readers whose favorite topics are not among those we have incorporated. (Researchers or practitioners who may be interested in establishing new beachheads, however, may find the absence of a topic worth investigating, as a possible indicator that it has not yet been broached, and therefore as a potential opportunity to provide advances in additional terrain.)

We emphasize there is no implied priority of importance or general interest that is intended by the order in which the following applications appear. Readers are invited to choose their own sequence for visiting the topics listed.

8.1 Planning and Scheduling

8.1.1 Scheduling in Manufacturing Systems

Effective scheduling in manufacturing systems leads to the reduction of manufacturing costs (inventory costs, labor costs, etc.) and improves the operational efficiency of management. The most frequently used and extensively studied problems in the literature are (A) the job shop problem and (B) the flow shop problem. In addition, a basic model for a broad family of cases called flexible flow line scheduling problems is given by the problem known as (C) the flow shop problem with parallel machines. Industrial applications arise in computer systems, telecommunication networks, and the chemical and polymer industries.

Nowicki and Smutnicki (1993, 1994, 1995) have developed effective tabu search methods for problems A, B, and C to optimize the makespan criterion. These algorithms employ a classical insertion neighborhood which is significantly reduced by a candidate list strategy for removing useless moves, in order to concentrate on "the most promising part" of the neighborhood.

The proposed algorithms employ a short-term memory tabu list which stores attributes of visited solutions, represented by selected pairs of adjacent jobs on a machine. Linked intensification and diversification occurs by storing the best solutions collected during the search on a list of limited length. An extended sequence of unproductive steps triggers a 'back jump' on the search trajectory to the nearest elite solution, which is recovered together with its associated search history as a basis for re-initiating the search.

Implementations made on a PC are able to improve significantly the best known solution found by other algorithms. Computation times are only a few minutes for instances of A&B problems containing 10,000 operations, and for instances of C problems containing 3,000 operations. An extensive comparative study shows the significant superiority of TS over other approaches including iterative improvement, genetic search, simulated annealing, threshold accepting, constraint satisfaction, neural networks, and other local search methods (see Vaessens, Aarts and Lenstra (1995)).

8.1.2 Audit Scheduling

Scheduling problems often become increasingly difficult as they acquire greater realism. Audit scheduling adds complexities to traditional scheduling. Characteristically, in such problems the processing units (the auditors):

1) have unequal processing times,

2) are not all fit (educated) for all jobs,

3) are not always available (yet must work at least a minimum amount, and at most a maximum amount of time), and

4) are movable (at a cost), and can transfer between various projects: hence introducing sequence dependent setup costs and times.

Dodin, Elimam, and Rolland (1995) develop a tabu search procedure that utilizes traditional dispatching rules (such as forward loading) with the short- and long-term memory structures of tabu search. The tabu search intensifies the search using short-term memory, and diversifies the search using controlled dispatching, long-term memory and candidate lists. Computational tests show the tabu search has approach produces schedules superior to those obtained by heuristics traditionally applied to these problems.

8.1.3 Scheduling a Flow-Line Manufacturing Cell

Effective scheduling of flow-lines for manufacturing cells improves the operational efficiency of manufacturing processes, leading to reductions in setup costs, labor costs, tooling and inventory costs. This leads to further reductions in throughput times and a corresponding increase in the shipment of on-time deliveries. A tabu search method for this problem has been proposed and successfully implemented by Skorin-Kapov and Vakharia (1993).

A manufacturing cell consists of a group of similar machines located in close proximity to one another and dedicated to the manufacture of a specific number of part families. Part families consist of a set of jobs with similar processing requirements. In this context, a feasible schedule S consists of a sequence of part families and a sequence of jobs within each family in a manufacturing cell. The tabu search heuristic of Skorin-Kapov and Vakharia efficiently schedules a pure flow-line manufacturing cell under varying parameter conditions (given F families, M machines and N_f jobs in family f).

A collection of alternative tabu search strategies (designed to test different aspects of tabu search) was compared against state-of-the-art simulated annealing heuristic that was tailored to solve this problem. Results from testing multiple data sets with alternative ratios of family set up times to job processing times showed the clear superiority of tabu search for these scheduling problems.

8.1.4 Generalized Capacity Requirements in Production Planning

Production planning problems that arise in real world applications are typically attended by capacity requirements. The incorporation of generalized capacity effects such as economies and diseconomies of scope and the learning-curve effect gives rise to a capacity-consumption function that is nonlinear in the tasks assigned to each facility. The resulting models for facility planning and loading decisions often involve nonlinear optimization problems in which some or all of the decision variables are integer-valued. The combined conditions of nonlinearity and discreteness makes these problems exceedingly difficult to solve.

Mazzola and Schantz (1995a) consider the resource allocation of a single facility under capacity-based economies and diseconomies of scope, and develop two models for this problem. In the first (more general) model the capacity of a single facility is considered to be a general function of the *subset* of tasks selected to be produced. In the second model the capacity is assumed to be consumed as a function of the *number* of tasks assigned to the facility. These can be viewed as single-facility production loading models that capture economies and diseconomies of scope within the production planning framework. The problems arising within each of these models generalize both the 0-1 knapsack problem and the 0-1 collapsing knapsack problem. Tabu search heuristics and branch-and-bound algorithms are defined for each model. Computational testing shows the tabu search heuristics are capable of obtaining exceedingly high-quality solutions to these problems, including the more difficult problems that exhibit a high degree of nonlinear behavior.

This study is extended in Mazzola and Schantz (1995b) to the multiple-facility setting, focusing on under capacity-based economies (and diseconomies) of scope (MFLS), including applications of MFLS in hierarchical production planning, group technology, and professional services. MFLS is formulated as a nonlinear 0-1 mixed-integer programming problem which generalizes many well known and widely applicable optimization problems, such as the generalized assignment problem and the capacitated facility location problem. A tabu-search heuristic and a branch-and-bound algorithm are developed for MFLS, and computational testing of the solution procedures is discussed. Once again, tabu search proves to be a highly effective approach for heuristically solving MFLS, and is reported to be a powerful tool for capturing complex capacity requirements.

8.1.5 Production Planning with Workforce Learning

Mazzola, Neebe, and Rump (1995) consider a production planning problem in which the work-force productivity for each product depends on the amount of previous production of the product. This change in productivity is captured in the corresponding resource coefficients that occur in the work-force requirement constraint for each time period.

The model includes a learning effect in each period that depends on the level of production of each product in the preceding period. The corresponding work-force coefficient can increase, decrease, or remain the same, representing a forgetting, learning, or status-quo production effect. The resulting problem is modeled as a mixed-integer programming problem. In addition to establishing problem complexity and defining a branch-and-bound algorithm for solving the problem to optimality, this paper also examines heuristics for the problem. A forward pass, linear programming-based heuristic previously defined in the literature is examined and shown to produce arbitrarily bad solutions for this problem. The paper then proposes a new tabu search heuristic for the problem. Extensive computational experiments with the solution procedures establishes the effectiveness of the tabu-search approach in this problem setting.

The TS approaches of this study are concluded to provide an ability to handle new levels of modeling complexity, allowing for closer approximation of real-world phenomena. The findings indicate that this is particularly true for problems involving complex, nonlinear behavior involving discrete decision variables.

8.1.6 Process Plan Optimization for Multi-Spindle, Multi-Turret CNC Turning Centers

Veeramani and Stinnes (1996), Veeramani, et al. (1996) and Veeramani and Brown (1996) have developed a tabu search based system that generates optimal process plans for machining parts in minimum cycle time on multi-spindle, multi-turret CNC turning centers. Due to the presence of multiple spindles (maximum two) and turrets (one to three) and live tooling, this new generation of machines have enhanced processing capabilities such as simultaneous machining (two turrets working simultaneously on one part), parallel machining (two turrets working on two parts on the two spindles) and mill-turn machining. However, these capabilities also imply that the number of alternative process plans for a given part can be very large resulting in a vast solution space, and the process plan optimization problem can, therefore, be complex. Further, due to difficulties in representing simultaneous machining actions and the cycle time objective function, it is not convenient to employ traditional optimization methods that are typically used to solve similar problems such as the job shop scheduling problem, the timetable problem, and the assembly line balancing problem.

In the tabu search procedure for process plan optimization, candidate solutions are generated by one of the following moves: (i) changing the spindle assignment of an operation; (ii) application/dis-application of simultaneous machining by changing the turret assignment; and (iii) changing the sequence of operations at a spindle. The first two types of moves can significantly change the structure of the process plan (and hence the quality of the solution) in comparison to the third type of move. This realization has enabled a geometric characterization of the solution space as a collection of "patches". The first two types of moves result in jumping from a solution on one patch to another solution on an adjacent patch, and the third type of move leads to different solutions within the same patch. Another novel aspect of this Tabu Search application is the use of an auxiliary lower bound performance measure, called the *projected cycle time*, to guide the search.

The tabu search based process plan optimization system has been implemented in C/C++ on MS Windows. Experiments with a set of sample parts have demonstrated that this approach can find quickly (typically within seconds) process plans having a high quality, e.g., within 2% to 7% of the theoretical (possibly unattainable) minimum cycle time for the part.

8.1.7 Sustainable Forest Harvesting

The long-term planning of sustainable forest treatment at the landscape level is a complex task that is becoming increasingly more challenging as forest resources face increasing demand. Local treatment schedules, pertaining to homogeneous sub-areas called stands, must be developed over a time horizon of a few centuries. Thousands of local schedules must be coordinated to satisfy hard constraints, and balance soft constraints and optimization criteria. Constraints and objectives are defined in terms of economical, recreational, and environmental effect. The aim of the forest treatment schedule is twofold. First, it must provide clear instructions for forest treatment over the near time horizon. Second, it must demonstrate sustainability over the full horizon.

Lokketangen and Hasle (1997) show how this problem can be solved using hybrid, opportunistic, tabu search based techniques. The search neighborhood consists of two levels where the top level selects a forest stand to move in, and the second level selects an alternative schedule for this stand. The concept of localized tabu tenure (for loosely coupled domains) is applied to the stand selection.

The solution system relies heavily on close interaction with a sophisticated stand simulator, which provides forestry knowledge necessary to guide the scheduling process, including the definition of abstract forest treatment actions. The system also incorporates mechanisms for dynamically changing the view of the various constraints (switching them between the objective function and the constraint set), to be able to easily change focus during the search. This application is carried out by SINTEF Applied Mathematics, part of the largest research establishment in Norway. The clients are the major Norwegian forest owners, and the system is designed for their long-term forest management planning.

8.1.8 Forest Harvest Scheduling to Satisfy Green-Up Constraints

A distinct change in public attitude toward the environment has led to demands for a forest management paradigm shift from one of a dominant timber use to one in which forests are managed for multiple values. Among the most important implications of such a shift in policy on forest management was the introduction of adjacency or green-up constraints. Green-up constraints are imposed to avoid large clear cut areas. For example, in British Columbia legislation requires that a block of forest which has been harvested must reach a mean tree height of three meters before any adjacent block can be harvested.

One result of these green-up constraints is that the forest harvesting scheduling problem, which previously was often formulated as a linear program, becomes combinatorial in nature. Current harvest scheduling codes, like FORPLAN, Timber RAM and others, are unable to generate harvest schedules satisfying the green-up constraints. Brumelle, et al. (1996) formulate forest harvesting problems with green-up constraints arising in the Tangier watershed in British Columbia as multicriteria discrete optimization problems.

The study considers two harvest scheduling problems associated with the Tangier watershed. The small problem focuses on a 219 cut-block subset of the watershed located at the southern end. The northern boundary coincides with that of the proposed Serenity Peaks Wilderness Area. The larger problem included all 491 cut-blocks comprising the entire watershed, which will be appropriate should the proposed park not eventuate.

Tabu search is used to investigate the tradeoffs between different criteria, which were chosen as the total volume of lumber cut, the period to period deviation from even-flow of lumber during a harvest rotation and adjacency violations. The tabu search methodology easily obtained good solutions to these problems, and was shown to be much superior to a biased random search method which is cited as one of the most effective methods to obtain good schedules satisfying green-up constraints. In fact, the tabu search method generates schedules which harvest more timber than the upper bound of the confidence interval suggested by previous empirical and algorithmic analysis.

8.2 Telecommunications

8.2.1 Hub-and-Spoke Communication Networks

The location of hub facilities is an important issue arising in the design of communication networks. Applications include: traffic networks (airline passengers flow and parcel delivery networks), as well as telecommunication networks (location of digital switching offices for Digital Data Service (DDS) networks, location of base stations for wireless networks).

Determining optimal locations of hub nodes and allocations of non-hub nodes to those hubs is an NP-hard combinatorial problem. For a widely used benchmark set of problems (the Civil Aeronautics Board (CAB) data set) efficient tabu search algorithms and lower bounds for a class of uncapacitated multiple and single allocation p-hub median problems have recently been developed which notably improve on results previously obtained. (Attempts to solve the problems of the CAB data set have been undertaken in more than 70 research papers.)

Skorin-Kapov and Skorin-Kapov (1994) provide an efficient tabu search heuristic for the single allocation p-hub median problem, which models the situation when n nodes can interact only via a set of fully interconnected hubs. The goal is to determine the location of hubs and the allocation of each non-hub node to one of the hubs. The new tabu search approach, in addition to being efficient, obtains a number of new best solutions for the CAB data set.

Skorin-Kapov et al. (1995) provide a novel way to further take advantage of these improved heuristic outcomes by using high quality heuristic solutions to derive lower bounds. Accompanying this, they provide tight linear programming relaxations for the hub location and some other relevant uncapacitated p-hub median problems that allow the CAB data set to be solved to optimality for the first time. By this means, they verified that the solutions obtained by the tabu search approach were in fact optimal for all of the test problems.

8.2.2 Optimization of B-ISDN Telecommunication Networks

Costamagna, Fanni and Giacinto (1995) develop a tabu search algorithm for topological optimization of broad band communication networks whose structure is based on a single exchange and on a number of multiplexing centers.

The topology of the network is represented by an undirected graph $G(N,A)$. The set of nodes N represents locations of both existing or possible multiplexers, and location of users. The set of arcs A represents the possible communication links that may be used to connect the users, the multiplexers and the exchanges among them (constituting a cable conduit graph). This graph contains all the information about the area in which the network must be built.

The design problem consists of choosing a spanning tree T of G that connects all the users to the multiplexers through the distribution network, and the multiplexers to the exchange through the transport network, allowing the overall cost of the plant be minimized.

An empirical study was performed comparing TS with three other methods: a Simulated Annealing method, a genetic algorithm and a heuristic "Add & Drop" procedure, in terms of computational time and cost. The TS approach reached better configurations in a time equal to or lower than that required by other algorithms. The work has been supported by Marconi, Genova, Italy. Thesis awards have been also granted by S.I.P. S.p.A. (now Telecom Italia), and the Italian public telephone company, which also provided cost data and evaluation of results.

8.2.3 *Design of Optical Telecommunication Networks*

The increased complexity and globalization of today's world has been accompanied by the emergence of optical networks as flexible, fast, efficient, and reliable media for transferring information. Congestion minimization presents one of the main challenges of telecommunication network design. The problem often includes the goals of devising efficient routing and management techniques in case of network failures. An improved approach for minimizing congestion in optical networks, based on tabu search, has been developed by Skorin-Kapov and Labourdette (1995).

Changing traffic conditions create a need for fast algorithms to re-arrange logical connections. Algorithms that quickly obtain very high quality solutions are mandatory in order to optimize the use of network relative to a given criterion. The goal is to find the logical connection diagram and routing of flow which minimizes the maximum congestion on a link. (This goal also effectively increases the relative capacity of the network.)

The tabu search approach for this problem has generated improved solutions for data sets established to provide comparative benchmarks. The approach not only improves on previous results, but has been calibrated by Skorin-Kapov and Labourdette to identify performance characteristics for different parameter values, and on different patterns of input data. The outcomes also yield guidelines for solving larger problems.

8.3 Parallel Computing

8.3.1 *Mapping Tasks to Processors to Minimize Communication Time in a Multiprocessor System*

Connectionist machines are attracting widespread attention for their value as an embodiment of massively parallel computer architecture. This is particularly true for solving combinatorial optimization problems arising in a variety of engineering applications. At the same time, the goal of designing and implementing a connectionist machine as effectively as possible introduces challenging optimization problems.

An important problem is to minimize the communication time required by a connectionist machine. Communication time often is a substantial determinant of overall cost and efficiency. In a significant class of applications, such as finite element analysis, the communication pattern is static. The memory locations defining the source and destinations of messages do not change in these applications, but only the communicated data varies. Improved designs for allocating processors to chips according to the structure of their communication pattern offer considerable potential for savings in cost and time.

Chakrapani and Skorin-Kapov (1995) have developed an effective tabu search method for the problem of mapping tasks to processors to minimize communication time in a multiprocessor system. The method incorporates a parallel processing implementation which includes tabu search memory and guidance mechanisms for iteratively selecting pairs of tasks and swapping their processor assignments. The implementation employs two levels of parallelism. First, the candidate tasks to be swapped are identified in parallel. Second, more than one pair of tasks are swapped in a single iteration. This strategy is designed to operate with efficient approximations that allow inaccurate (incomplete) information for evaluating moves. The authors propose a diversification strategy which makes the search robust under these circumstances. Due to its robust parallel implementation, the algorithm can be used to develop heuristics for "quasi-dynamic" communication patterns, in which the task graph changes slowly with time.

8.3.2 Multiprocessor Task Scheduling in Parallel Programs

When parallel application programs are executed on MIMD machines, the parallel portion of the application can be speeded up according to the number of processors allocated to it. In a homogeneous architecture, where all processors are identical, the sequential portion of the application will have to be executed in one of the processors, considerably degrading the execution time. In a heterogeneous structure, where a faster processor, responsible for executing the serial portion of the parallel application and is tightly coupled to other smaller processors, higher performance may be achieved. The procedure of assigning tasks to processors (task scheduling) is more complex in the heterogeneous case, where the processors have distinct processing speeds.

Porto and Ribeiro (1995a) have applied the tabu search metaheuristic to the task scheduling problem in a heterogeneous multiprocessor environment under precedence constraints. A series of different tabu search parameters and strategies were studied side-by-side with a variety of task precedence graphs (topology, number of tasks, serial fraction, service demand of each task) and system configurations (number of processors, architecture heterogeneity measured by the processor power ratio). The algorithm showed itself to be very robust and effective, systematically improving by approximately 25% the makespan of the solutions obtained by the best greedy algorithm used to provide an initial solution.

8.3.3 Multiprocessor Task Scheduling Using Parallel Tabu Search

Porto and Ribeiro (1995b) have designed and implemented parallelization tabu search strategies for the multiprocessor task scheduling problem. Parallelization relies exclusively on the decomposition of the solution space exploration. Four different parallel strategies were proposed and implemented on a 32-processor IBM SP1 parallel machine running PVM for varying problem sizes and number of processors: the master-slave model, with two different schemes for improved load balancing, and the single-program-multiple-data model (SPMD), with single-token and multiple-token message passing schemes. These two basic models mainly differ in the way information is exchanged between parallel tasks at the end of each iteration of the tabu search. The computational results confirmed the high adaptability of the TS algorithm to parallelization, showing that communication is not a burden to achieving almost ideal efficiency in the majority of the test problems. The task scheduling problem considered in this study is characterized by very large neighborhood structures that are costly to explore. However, the speedup achieved through simple parallelization techniques made possible a less restricted neighborhood search, which not only reduced computation time but produced better solutions for several test problems.

8.3.4 Quadratic Assignment Problem on a Connection Machine

The Quadratic Assignment Problem (QAP) is a classical NP-hard problem arising in many applications involving, for example, facility layout or VLSI design. Skorin-Kapov (1990, 1994) solves the quadratic assignment problem suboptimally using the so-called Tabu-Navigation procedure, obtaining improved outcomes over previous results. In the process of designing an efficient massively parallel algorithm for the QAP, Chakrapani and Skorin-Kapov (1992) first generalized the connectionist model proposed by Aarts and Korst (1989a, 1989b) for the Traveling Salesman Problem (TSP) to solve the QAP. This was the first study replacing simulated annealing with (deterministic) tabu search in a connectionist model. This was also the first study involving dynamically changing connection strengths for such problems. In a subsequent paper Chakrapani and Skorin-Kapov (1993a) developed a massively parallel tabu search algorithm and implemented it on the Connection Machine. The careful implementation on a Connection Machine, a massively parallel computer architecture, proved to be extremely suitable: provided enough processors, the computational time grows with $O(\log n)$.

8.3.5 Asynchronous Parallel Tabu Search for Integrated Circuit Design

The logical test of integrated circuits is one of the main phases of their design and fabrication. The pseudo-exhaustive approach for the logical test of integrated circuits consists in partitioning the original circuit to be tested into non-overlapping subcircuits with a small, bounded number of input gates, which are then exhaustively tested in parallel. Andreatta and Ribeiro (1994) developed an approximate algorithm for the problem of partitioning integrated combinational circuits, based on the tabu search metaheuristic. The circuits are modeled as directed acyclic graphs. The proposed algorithm contains several original features, including reduced neighborhoods; complex moves (similar to an ejection chain strategy); a multicriteria cost function and the use of a bin-packing heuristic as a post-optimization step. Computational results were compared with those obtained by the best algorithm previously published in the literature, with significant improvements. The average reduction rates have been on the order of 30% for the number of subcircuits in the partition, and of the order of 40% for the number of cuts required.

8.3.6 Asynchronous Multithread Tabu Search Variants

The use of alternative types of move attributes for the formation of the tabu lists, and multiple strategies for obtaining initial solutions, can very often enhance the quality of solutions obtained in TS approaches. As shown by Aiex, Martins, Ribeiro and Rodriguez (1996) the combination of different initial solution procedures with different types of move attributes may be usefully integrated in an asynchronous multithread procedure, in which each processor runs a tabu search algorithm with a different pair of initial solution procedure and move attributes.

Each time one of the search threads finds a new local optimum, it writes this solution in a pool of elite solutions, which is kept by a master processor in charge of search coordination. When a search thread is not able to improve its local best solution, it accesses the pool of solutions and randomly chooses one of them to restart the search. Global stopping criteria are used. Aiex et al. implement this asynchronous strategy for the circuit partitioning problem, using 10 processors, where one of them is the master in charge of the search coordination, and the other processors run nine different strategy combinations. The results obtained by this multithread version of the tabu search algorithm yield much better solutions than those obtained by the sequential tabu search algorithm, within very reasonable computational times. This multithread search procedure was implemented using two different parallel programming tools, PVM and Linda, also leading to comparative results concerning these tools. (Reviews of the parallel TS literature and evaluations of several parallel programming tools have been recently published by Martins, Ribeiro and Rodriguez (1996) and Toulouse, Crainic and Gendreau (1996). See also Chapter 3 for a discussion of strategies for exploiting multiple choice rules and neighborhoods.)

8.4 Transportation, Routing and Network Design

8.4.1 The Fixed Charge Transportation Problem

In a fixed charge transportation problem, a fixed ("all or none") cost is incurred whenever a route in the transportation network is used. Goods transported along that route are additionally subject to a unit variable cost. The underlying model is also applicable to plant and warehouse location problems, purchase/lease problems, and personnel hiring problems.

A tabu search approach was developed for this problem by Sun et al. (1995) using recency based and frequency based memories, two strategies for each of the intermediate and long term memory processes, and a network based implementation of the simplex method as the local search method. A computational comparison was performed to evaluate the performance of this approach on randomly generated problems of different sizes and of different ranges of magnitude of fixed costs relative to variable costs. Objective function values and CPU time were used as criteria to compare the performance of this procedure with that previously proposed methods consisting of an exact solution algorithm and a heuristic procedure.

The tabu search procedure obtained optimal and near-optimal solutions much faster than the exact solution algorithm for simple problems, and thoroughly dominated the exact algorithm for more complex problems. In a set of 15 randomly generated problems in the class studied, restricting the problem size, the tabu search procedure found the optimal solutions for 12 problems, and the objective function value of the worst solution to the remaining 3 problems was less than 0.06% higher than that of the optimal solution. The exact solution procedure used an average of 5888 CPU seconds for these problems, while the tabu search procedure used an average of 1.63 seconds. As problem size increased or as fixed costs became high relative to variable costs, the solution time for the exact algorithm became inordinate. Compared to the alternative heuristic approach, statistical results showed that the tabu search procedure found comparable solutions at least as good for very small and easy test problems, and found significantly better solutions for all other problems. For the small problems, the solution times used by both heuristics were similar, while for larger problems and for problems with higher fixed relative to variable costs, the tabu search procedure was 3 to 4 times faster than the competing heuristic.

8.4.2 The Transportation Problem with Exclusionary Side Constraints

The transportation problem with exclusionary side constraints represents a class of practical distribution problems where goods at a set of sources need to be shipped to a set of destinations, but the simultaneous shipment of goods from some pairs of sources to the same destination is prohibited. For example, some chemicals cannot be stored in the same storage room, and food items and poison products cannot be stored together, although they may be handled through the same distribution system. The objective of a transportation problem with exclusionary side constraints is to determine which routes to use and how much to ship along each route so as to minimize the total shipping and handling costs, while satisfying the source and destination requirements without shipping goods from any pair of prohibited sources simultaneously to the same destination.

To solve the transportation problem with exclusionary side constraints, a new heuristic procedure based on tabu search is developed and implemented by Sun (1996). Net changes in total cost and in total infeasibility are used to evaluate the attractiveness of a move, and strategic oscillation is used to implement the intensification and diversification functions. To take advantage of the network structure inherent in the problem, the simplex method on a graph is used as a local search method to lead the search from one trial solution to an adjacent trial solution.

Computational experiments tested the performance of the tabu search procedure, both in terms of solution quality and CPU time, compared to two procedures, one heuristic and one exact, which are the only special purpose solution methods that are in established use for this problem. The study examines the effects of changes in problem size, density of exclusionary side constraints, problem shape and the range of cost coefficients. The tabu search procedure finds optimal or near optimal solutions by expending only a small fraction of the CPU time required by the exact method, and can effectively handle problems much larger (of more realistic sizes) than the exact method can treat. Compared with the heuristic approach, the tabu search procedure not only requires less CPU time but finds better solutions, and can handle more complicated and more difficult problems.

8.4.3 The Vehicle Routing Problem

Gendreau, Hertz and Laporte (1994) develop a tabu search heuristic called TABUROUTE for the vehicle routing problem with capacity and route length restrictions. The algorithm considers a sequence of adjacent solutions obtained by repeatedly removing a vertex from its current route, and reinserting it into another route. This is done by means of a generalized insertion procedure developed by the authors.

Results obtained on a series of benchmark problems, including a comparison of twelve alternative procedures, clearly indicate that tabu search outperforms the best existing heuristics, and among TS approaches TABUROUTE often produces the best known solutions. The success of TABUROUTE can be attributed to two implementation devices. First, the search is allowed to visit infeasible solutions, which are managed by means of an objective function with penalty terms. Second, a highly effective insertion mechanism is employed. The insertion procedure has the additional feature of periodically perturbing the solutions and thus reducing the risk of being trapped at a local optimum.

A major advantage of TABUROUTE is its flexibility. The procedure may start from either a feasible or an infeasible solution. It can be adapted to contexts where the number of vehicles is fixed or bounded, or where vehicle are not homogeneous. Finally, the procedure can easily handle additional problem characteristics such as assigning particular cities to specified vehicles, using several depots, and allowing for primary and secondary routes.

8.4.4 Vehicle Routing Problem with Time Windows

In the face of today's global competition, the need for more efficient logistical planning has become a pressing issue for most manufacturing and distribution concerns. Barnes and Carlton (1995) present a reactive tabu search (RTS) approach to the vehicle routing problem with time windows (VRPTW). The VRPTW considered has available, at a single depot, m identical vehicles with a specified cargo capacity. Each of n nonidentical customers require a specified volume of cargo which must be delivered within a specified contiguous interval of time. Each customer must be visited exactly once and the objective is to find the feasible set of vehicle routes that minimize the total travel time.

The Barnes and Carlton study furnishes a brief review of the most recent literature associated with the VRPTW, presents their RTS algorithm and gives computational results for the algorithm when applied to a widely used benchmark test set of vehicle routing problems with time windows due to Solomon. The results were produced without any attempt at "tuning," and were obtained in a small fraction of the time required by current exact techniques. The proposed algorithm does not suffer from the computational limitations of exact approaches, which are unable to successfully attack larger problems. The algorithm experienced no difficulty in obtaining solutions to all 56 Solomon problems for the 100-customer sets.

The vehicle routing problem with time windows can be used to model many real-world problems and has recently been the subject of intensive research. Applications of the VRPTW include bank deliveries, postal deliveries, industrial refuse collection, national franchise restaurant deliveries, school bus routing, and security patrol services.

Chiang and Russell (1995) have developed a reactive tabu search method for the VRPTW that dynamically varies the size of the list of forbidden moves (in order to avoid cycles as well as an overly constrained search path). The method incorporates intensification and diversification strategies to achieve higher quality solutions. Chiang and Russell also developed simulated annealing procedures which achieve solutions that are generally superior to previously reported results for this approach. Computational tests of problems from the literature as well as of large-scale real-world problems, which obtain several new best known solutions, show that tabu search outperforms simulated annealing in solution quality. The tabu search approach is especially effective in reducing fleet size requirements for routing problems constrained by time windows.

8.4.5 Routing and Distribution

Rochat and Semet (1994) consider a real-life vehicle routing and distribution problem that occurs in a major Swiss company producing pet food and flour. In contrast with usual hypothetical problems, a large variety of restrictions must be considered. The main constraints involve the accessibility and the time windows at customers, the carrying capacities of vehicles, the total duration of routes and the drivers' breaks. The optimization problem for the transport plan consists in elaborating a set of routes that minimizes the total travel distance while satisfying the indicated constraints.

The general scheme used to solve this real-life VRP first applies a straightforward construction procedure to generate an initial solution which provides a starting point for the tabu search procedure. The key features of the TS approach consist of a constraint relaxation strategy for diversification and an intensification strategy. The relaxation of constraints makes it possible to expand the solution space, diversifying the search by examining infeasible solutions as well as feasible ones. The intensification strategy plays a complementary role, and leads the search to visit solutions close to the best solution found so far by rendering some routes tabu. Computational results show that the TS method yields solutions dominating those of the constructive heuristic even when the total number of iterations is small. Thus, good solutions are obtained in a reasonable amount of CPU time. Moreover, the study shows that embedding the TS algorithm in decision support software can be particularly useful. The fact that the TS approach generates multiple solutions that have approximately the same length of routes as the best makes it possible to propose several transport plans to the user. Comparisons of the solutions produced with the routes actually covered by the company disclose that the total distance traveled is reduced significantly by these solutions.

8.4.6 Probabilistic Diversification and Intensification in Local Search for Vehicle Routing

Rochat and Taillard (1995) develop a probabilistic TS technique to diversify, intensify and parallelize almost any local search for almost any VRP. This technique makes the local search more robust since it converges more often solutions whose quality is close to that of the best known solution. This technique has several advantages: First, it is relatively easy to design a local search that locally finds good tours, but it is hard to design a search that finds good tours for all customers simultaneously; the proposed technique makes it possible to overcome this difficulty and to design a fairly robust method more easily. Second, this technique may be applied to several types of VRPs, for example those including the following constraints:

1) Time windows for the customer deliveries.
2) Differentiated vehicles (cost of use per kilometer, volume capacity, carrying capacity).
3) Constraints on the tours (maximum length, driver breaks, customers that cannot be reached by any vehicle).
4) Backhauls.
5) Multiple depots.

Third, this technique may easily be parallelized with an arbitrary number of processors (not depending on problem size). The Rochat and Taillard approach exploits two primary perspectives, as follows.

The first comes from probabilistic tabu search, which is founded on the idea of translating information generated by the search history, coupled with current measures of attractiveness, into evaluations that are monotonically mapped into probabilities of selection. Operating in a neighborhood framework, the approach then successively selects among available alternatives according to a probability assignment that is strongly biased to favor the choice of higher evaluations.

The second main perspective that underlies the R&T approach derives from one of the most basic types of intensification strategies. The heart of this approach lies in generating solutions by reference to the notions of strongly determined and consistent variables.

It is shown that efficient first level tabu searches for vehicle routing problems may be significantly improved with this technique. Moreover, the solutions produced by this technique may often be improved by a post-optimization technique developed by the authors, which embodies an effective means for applying a vocabulary building strategy in this context. The solutions of nearly 40 problem instances of the literature have been improved. This technique may also be applied to other local searches or other VRPs.

8.4.7 Quadratic Semi-Assignment and Mass Transit Applications

The quadratic semi-assignment problem (QSAP) is related to the quadratic assignment problem by the requirement of assigning a set of n objects to any of m locations. The QSAP differs by allowing each location to be assigned none, one, or even more than one object, whereas the QAP requires a one-one mapping of objects to locations ($m = n$).

Voss (1992), Domschke et al. (1992) and Daduna and Voss (1995) develop dynamic tabu search approaches for the QSAP, in a series of applications for modeling and solving a schedule synchronization problem in a mass transit system, where the goal is to minimize the total transfer waiting times of passengers. Both from an economic and a social standpoint, reducing passenger waiting time is a major issue in the operation of mass transit systems.

The outcomes of the tabu search applications show that better schedules are produced than those obtained by previous approaches, which were based on simulated annealing. The Daduna and Voss study also reports the successful incorporation of the tabu search schedule synchronization procedures into an overall solution approach for changing a public mass transit system by introducing new bus lines. The resulting advances include the ability to perform sensitivity analysis more effectively, disclosing that small changes, appropriately determined, can create large improvements in both cost and quality of service.

8.4.8 Automated Guided Vehicle Systems Flowpath Design Applications

Automated Guided Vehicle Systems (AGVSs) have been of great interest to industry for the last two decades. The number of applications of these systems has increased to a point where AGVSs are considered to be a basic concept in material handling. Although initial applications of AGVSs were generally limited to warehouses, in recent years an increasing number of applications in manufacturing systems have been reported.

Chiang and Kouvelis (1994a) address the flowpath design issue of AGVSs. The authors concentrate on the design of unidirectional flowpaths (where vehicles are restricted to travel only on one direction along a given segment of the flowpath), and develop different versions of simulated annealing and tabu search algorithms for the design of unidirectional AGVSs. Extensive computational results indicate that a tabu search implementation with the use of a frequency based memory structure dominates all tested heuristics in terms of solution quality, with an impressive average performance over 45 test problems of less than 0.85% deviation from optimality.

8.4.9 Multicommodity Capacitated Fixed Charge Network Design

The fixed charge capacitated multicommodity network design problem (CMND) is a fundamental model for addressing issues in infrastructure construction or improvement, service network design, telecommunication network design, power system design, etc. The problem arises in settings where several "commodities" (goods, data packets, people,...) have to be moved, from their respective origins to their respective destinations, over the links of a network with limited capacities. In addition to the usual transportation costs related to the volume of each commodity flowing through a given link, a "fixed" (construction or utilization) cost is incurred as soon as a linked is used. The trade-off between the variable and fixed costs, compounded by the competition among the various commodities for limited capacity, creates a model with many practical applications. At the same time, however, the problem poses various obstacles to existing solution methods, especially for applications of realistic size.

Although exact methods have been proposed, only specially tailored heuristics have proved of help for capacitated problems of practical dimensions. These local search heuristics find only a local optimum, however, and generally require significant computation times.

Crainic, Gendreau and Fardolven (1996) develop an efficient Tabu Search procedure to find good feasible solutions to realistically sized capacitated multicommodity fixed cost network design problems. The procedure combines simplex pivot type moves to column generation to produce a search that explores the space of the continuous path flow variables, while evaluating the actual mixed integer objective of the CMND. The method represents one of the first attempts to devise an efficient Tabu Search-based procedure for a mixed integer programming problem that displays both a very strong combinatorial nature and a difficult linear programming component. The ideas of pivoting exclusively in the space of the path variables and of combining local tabu search and column generation are also applied for the first time in the context of Tabu Search.

Experimental results show that the procedure is robust with respect to the type of problem, in terms of relative importance of the fixed costs and capacities, size and, especially, the number of commodities. In addition, it successfully obtains exceedingly good solutions to large problems with many commodities. The outcomes which include comparisons with the leading heuristic and exact methods, indicate this approach is the best currently existing procedure for finding good feasible solutions for the CMND.

8.5 Optimization on Structures

8.5.1 Protein Conformation Lattice Model

The determination of the three-dimensional structure of a protein from a given sequence of amino acids is one of the most challenging unsolved problems in the science of molecular biology. There have been many computer models designed to solve the protein folding problem. All computer models, though employing different types of energy minimization, can be expressed as the global optimization of a non-convex potential energy function. The basic difficult in solving these models is the existence of multiple local minimizers. Recently, there have been various approaches used to solve these models arising from protein folding.

The determination of the three-dimensional structure of a protein from a given sequence of amino acids is one of the most challenging unsolved problems in the science of molecular biology. There have been many computer models designed to solve the protein folding problem. All computer models, though employing different types of energy minimization, can be expressed as the global optimization of a non-convex potential energy function. The basic difficult in solving these models is the existence of multiple local minimizers. Recently, there have been various approaches used to solve these models arising from protein folding.

Pardalos, Liu and Xue (1995) design an algorithm for a class of lattice models using tabu search and test their approach with a chain of 27 monomers. The algorithm was developed in C and tested using the same data for a fundamental test set published by the American Mathematical Society. Among the protein sequences tested, only a few sequences fail to match the best results previously reported. In all other cases (from a set of 200 examined) the tabu search method obtains results as good or better than the previous best.

8.5.2 Optimization of Electromagnetic Structures

Fanni, Giacinto and Marchesi (1996) develop a Tabu Search strategy to optimize the design of a magnet for Magnetic Resonance Imaging (MRI). This is an important biomedical device whose optimal design is sought by many companies, such as general Electric, Siemens and Oxford Instruments. Among different magnetic structures, MRI magnet systems are a tough benchmark for optimization procedures.

The goal of the problem considered was to design coils to yield a 'homogeneous' magnetic field in a fixed region, according to an appropriate function. For each coil the position and thickness have to be determined. To apply a TS based method,

Fanni, Giacinto and Marchesi discretize the range of variation of each variable dividing it in sub-ranges, yielding a finite alphabet. The neighborhood of a solution consists of all the configurations obtained by considering all the possible symbols for each variable, keeping the others constant. Short term memory prevents repetitions of configurations of the coils (in terms of symbols of the finite alphabet). In non improving phases, frequency based memory also penalizes choices of moves that drive toward configurations often visited. Finally, local minimization using golden search is applied after the choice of the move. Computational tests show the TS approach performs more effectively than an SA and a GA approach, and requires less than half as much solution time.

8.5.3 The Damper Placement Problem on Space Truss Structures

NASA has conducted a set of laboratory experiments investigating the control of space structures. To facilitate these experiments, a large, flexible structure was assembled from truss elements and antenna support members and dubbed the Controls-structures-interactions Evolutionary Model (CEM). The CEM was designed to simulate characteristics of a large earth-observation platform and was dynamically tested in the NASA Langley Space Structures laboratory.

The overall structural motion of a flexible truss structure can be reduced by the use of structural dampers that both sense and dissipate vibrations. Kincaid and Berger (1993) develop a tabu search method for the problem of locating these dampers. The goal is to assure that vibrations arising from the control or operation of the structure and its payloads, or by cyclic thermal expansion and contraction of the space structure, can be damped as effectively as possible.

Given a strain energy matrix with rows indexed on the modes and the columns indexed on the truss members, the problem can be expressed as that of finding a set of p columns such that the smallest row sum, over the p columns, is maximized. The TS approach obtained high quality solutions, as verified by comparisons with designs previously proposed and also with upper bounds provided by the optimum value of an LP relaxation. Outcomes from the study led NASA engineers to reconsider their design assumptions, and in consequence to change the rigidity of support arms for the truss structure.

8.5.4 The Polymer Straightening Problem

Polymer chemists at NASA-Langley Research Center are interested in the crystallization of high-performance aromatic polyimides. The value of these polymers lies in their thermal stability, strength, and toughness. Aromatic polyimides are used to build high-performance carbon fiber composites for structural components in aircraft and spacecraft, which depends on their crystallization. A key problem is to determine a priori if there exists a conformation for which a given aromatic polyimide crystallizes.

The role of the optimizer in this application is to determine if a straight line conformation for a given polyimide exists among all possible combinations of allowable (minimum energy) torsion angles for the rotable bonds. The total number of combinations may be quite large, easily containing as many as 100 million possible conformations.

Kincaid, Martin and Hinkley (1995) develop a simple tabu search procedure to find a conformation that maximizes the cosine of the angle between the first bond and the projection of the last bond over all allowable conformations. The method was applied to the analysis of three polyimides of interest to NASA Langley Research Center, and succeeded in finding the optimal conformation in all three cases. (Normally, a research chemist could require as much as three years to perform such an analysis.) The ultimate goal of this research is to provide a technique that will serve as an aid to chemists in deciding what conformations are most likely to result in crystallizable structures when produced in a laboratory.

8.5.5 Active Structural Acoustic Control

Active structural acoustic control is a method in which the control inputs used to reduce interior noise are applied directly to a vibrating structural acoustic system. The ultimate goal is to use active acoustic control to decrease the interior noise in propeller driven jet aircraft. Kincaid (1995) studies the instance of this problem in which the objective consists of damping noise generated by a single exterior source in the interior of a cylinder.

The model requires a determination of the force inputs and sites for piezoelectric actuators so that (1) the interior noise is effectively damped; (2) the level of vibration of the cylinder shell is not increased; and (3) the power requirements needed to drive the actuators are not excessive.

A tabu search approach was developed to determine the best set of actuator sites to meet the three specified objectives. Experiments confirmed that the TS procedure is able to uncover better solutions than those selected based upon the judgment of engineers. In addition, the high quality solutions generated by tabu search, when minimizing interior noise, do not further excite the cylinder shell. Thus, it was possible to meet objective (2) without imposing an additional constraint or forming a multi-objective performance measure. The TS solutions also led to identifying natural groupings that require fewer control channels and that permit a simpler control system.

8.6 Optimization on Graphs

8.6.1 The P-Median Problem

Many location analysis problems are prime targets for tabu search techniques. One such binary decision problem, the p-median problem, can be stated as follows: given a graph $G = (V,E)$, the goal to find a set of nodes, S, of size p, such that the weighted sum of the distances from the remaining nodes (those of V-S) to the set S is minimized. Rolland, Schilling and Current (1996) provide a tabu search algorithm that utilizes single node transfers with a simple short-term memory structure, and a strategic oscillation scheme that allows the procedure to search through an infeasible solution space. Long-term memory structures are used to penalize moves that occur inordinately often. The resulting tabu search procedure outperformed all known heuristics both with respect to solution quality and computational effort.

8.6.2 Graph Partitioning

The uniform graph partitioning, or equicut, problem may be described as follows. Given a graph $G = (V,E)$, where $|V| = 2n$, we seek a partition of V into two node sets V_1 and V_2 such that $V = V_1 \cup V_2$ and $|V_1| = |V_2| = n$. The goal is to identify such a partition that minimizes the sum of the cost of edges (i,j) where $i \in V_1$ and $j \in V_2$. This problem has several applications. In VLSI design it models the problem of optimally placing "standard cells" in order to minimize the routing area required to connect the cells. It also models the problem of minimizing the number of holes on a circuit board, subject to pin pre-assignment and layer preferences and has applications in Physics, where it models the problem of finding the ground state magnetization of spin glasses having zero magnetic field. Other applications arise in layout and floor planning, in computer memory paging and in Group Technology.

This partitioning problem has been investigated for more than twenty five years. With the exception of a classical deterministic heuristic due to Kernighan and Lin (the first approach presented) and an exact branch-and-cut algorithm (one of the last approaches), all the other algorithms from the literature use metaheuristic techniques. This is probably due to the simple structure of the problem which makes it possible to represent a solution with compact data structures, to have simple and natural neighborhoods and to have immediate feasibility tests. Exact approaches, however, cannot solve instances with more than 100 vertices and classes of instance exist for which the performances of the various metaheuristics differ greatly.

Rolland, Pirkul and Glover (1995) developed a tabu search algorithm for this problem that outperformed all other heuristics tested (including the reported best versions of simulated annealing and the Kernighan-Lin approach), both with respect to solution quality and computational requirements. A key element of the tabu search was a strategic oscillation that drove the search through infeasible solution configurations, causing the cardinality of the node sets to grow and shrink in coordinated waves.

More recently Dell'Amico and Maffioli (1996) and Dell'Amico and Trubian (1997) investigated the application of tabu search to the problem with 0-1 weights and with general weights, respectively. The main idea in both algorithms is to use very simple attributes and a straightforward short term memory improved by adding a dynamic updating of the tabu tenure (in order to locally intensify or diversity the search) and an aspiration criterion based on a restricted short term memory. In contrast to the simplicity of the short term memory functions, strong emphasis is given to the long term memory and to associated longer term strategies.

The weighted and unweighted problems differ significantly both in the implementation details and in the choice of the various strategies and parameters. Dell'Amico and Maffioli (1996) show that the new tabu search is superior to all other algorithms on all classes of instances of the unweighted problem, with the exception of very sparse ones. For the weighted case, Dell'Amico and Trubian (1997) compare their tabu search with exiting algorithms on several classes of graphs, with up to 4,000 nodes and 320,000 edges. The computational results show that the new approach easily determines the optimal solution for small graphs and its average performance is notably superior to that of the other heuristic algorithms for all classes.

8.6.3 Determining New Ramsey Graphs

A $(k,l;n)$-*Ramsey graph* is a graph on n vertices which does not contain a clique of order k or an independent set of order l. The Ramsey number $R(k,l)$ is defined to be the smallest integer $n > 0$ such that there is no $(k,l;n)$-*Ramsey graph*. Since 1930 when F. P. Ramsey proved the existence of $R(k,l)$ for any integer parameters k,l only a few (non trivial: $k,l \geq 3$) exact values have been established: $R(3,3)$ ... $R(3,9)$, $R(4,4)$, $R(4,5)$.

Piwakowski (1996) provides a tabu search algorithm for finding Ramsey graphs that establishes seven new lower bounds for classical Ramsey numbers: $R(3, 13) \geq 59$, $R(4,10) \geq 80$, $R(4, 11) \geq 96$, $R(4,12) \geq 106$, $R(4, 13) \geq 118$, $R(4,14) \geq 129$, and $R(5,8) \geq 95$. The approach searches for a $(k,l;n-1)$- Ramsey graph in a space of highly regular graphs on n - 1 vertices. Two decisions improved the efficiency of the approach and contributed to its ability to obtain new lower bounds: a) The family of graphs searched is restricted to highly symmetrical graphs (mostly cyclic); b) The attributes that are used to define tabu status consist of the indexes of subgraphs recently removed or added to the actual graph. (All these subgraphs are generated at the beginning as a given partition of a complete graph. No subgraph on the tabu list can play role in forming a new graph.) Opportunities to incorporate additional longer term tabu search strategies afford a basis for future research to discover additional new Ramsey Graphs and lower bounds.

8.6.4 The Maximum Clique Problem

Given a simple undirected graph $G = (V, E)$, a *complete subgraph* of G is one in which all vertices are pairwise adjacent and a *clique* denotes the vertex-set associated with any such subgraph of G. The maximum clique problem (MCP) consists in finding a clique whose cardinality is *maximum* among all subgraphs of G. This problem is equivalent to finding a *maximum vertex packing* (i.e., a set of vertices where no two vertices are adjacent, also referred to as a *stable* or *independent* set) or a *minimum vertex cover* in the complement of G, the graph $\overline{G} = (V, \overline{E})$, where $\overline{E}$ is the complement of E relative to $V \times V$.

The MCP and its equivalents are important combinatorial optimization problems that arise in several application contexts, such as classification theory, signal transmission theory, fault tolerance diagnosis, and timetabling among others. These problems belong to the NP-hard class. Several exact algorithms, mostly of the branch-and-bound type, have been proposed to solve them, as well as several heuristics.

Gendreau, Soriano and Salvail (1993) and Soriano and Gendreau (1996a) develop a tabu search approach that explores the set X of vertex-sets of complete subgraphs of G. The neighborhood structure is based on augmenting and decreasing moves that respectively add a vertex adjacent to all elements of the current solution S and remove vertices from S.

Three variants of TS were developed and analyzed, and extensive tests were conducted which showed these methods to be very efficient for solving randomly-generated problems. In addition to these outcomes, the methods were tested by Soriano and Gendreau (1996b) on the set of MCP benchmarks chosen for the *2nd DIMACS Challenge.* In comparison to other tested procedures, these methods performed extremely well, finding optimal or quasi-optimal solutions for the majority of the instances consisting of problems designed to be particularly hard to solve. They also required far less computing time than most other approximate methods tested. Further enhancements of these tabu search approaches are currently being investigated, and preliminary outcomes disclose that they offer promise of additional success.

8.6.5 Graph Drawing

Graphs are commonly used as a basic modeling tool in areas such as project management, production scheduling, line balancing, business process reengineering, and software visualization. Drawings of graphs are called maps and their value for modeling and analysis is widely heralded in various fields of the economic, social and computational sciences.

The main quality desired for maps is readability: A map is readable if its meaning is easily captured by the way it is drawn. It is extremely difficult to make a readable map by hand of a graph that represents a real system, even when the graph size is relatively small. Therefore, an automatic procedure for drawing graphs by computer is indispensable for generating readable maps quickly.

An important problem in the area of graph drawing is to minimize arc crossings in a hierarchical digraph. It is customary to draw a hierarchical digraph by placing the vertices on a set of equally spaced horizontal or vertical lines called layers and then drawing the arcs as straight-line segments.

Bipartite Diagraph. Valls, Marti and Lino (1995) provide a Tabu Thresholding (TT) approach for the problem of minimizing the number of arc crossings in a 2-layer hierarchical digraph (a bipartite digraph). The procedure combines elements of probabilistic tabu search, candidate list strategies and thresholds, yielding a simplified implementation of basic TS ideas that does not make explicit use of memory structures.

The computational study has been carried out on a set of 250 randomly generated problems of varying sizes and densities. The TT procedure has been compared with the Greedy Switching and the Splitting methods, which are reported to be the best methods in the literature previously available for graph drawing. Outcomes from the TT methods are also compared with the optimal solutions for the test problems, in those cases where the problems are small enough to permit them to be solved by state-of-the-art exact methods. Results show that in the 130 test problems where an optimum is available, the TT solution is optimal. Moreover in each of the 250 examples tested, the TT solution and generally is superior (and never inferior) to the best of those given by the Greedy Switching and Splitting heuristics.

Marti (1995) has developed a TS heuristic for the problem of minimizing the number of arc crossings in a bipartite graph. To perform an aggressive search for the global optimum, the author has considered intensification, diversification, influential moves and strategic oscillation elements of tabu search.

The procedure has three different search states: normal, influential and opposite, and oscillates among them according to the search history. In each state there are two alternately applied phases, an intensification phase and a diversification phase. The moves defined in the intensification are based on a positioning function while those defined in the diversification are based on permuting consecutive vertices. The use of different moves reinforces the non-monotonic search strategy. The criteria for differentiating between "improving" and "disimproving" moves within the oscillation strategy are not limited to the objective function evaluation, but consider factors of move influence, as determined by context and the search history.

Two variants of the general TS procedure are developed and compared with the Greedy Switching, Splitting, Barycentric, and Semi-Median methods and with the Tabu Thresholding heuristic. The computational results show that both of these TS variants perform better than the other methods, closely followed by the TT procedure, which in turn also performs better than the methods remaining.

Multi-Layered Diagraph. Laguna, Marti and Valls (1995) propose a TS algorithm for the general k-layer graph drawing problem ($k > 2$). Existing solution methods for this problem are based on simple ordering rules for single layers that may lead to inferior drawings. The Tabu Search implementation consists of an intensification phase that seeks local optimal orderings of layers using an insertion move, and two levels of diversification. The first level of diversification is a strategy for selecting layers for intensification, while the second one escapes local optimality by means of switching moves. The authors utilize two different termination criteria (TABU1 and TABU2).

Computational testing was performed on a set of 200 randomly generated instances, including graphs with up to 571 vertices and 2,241 arcs. Comparisons were performed with procedures that have shown to be effective for arc crossing minimization, i.e., the barycentric and the semi-median methods with switching (BC+SW, SM+SW).

The results of the experiments show that in terms of solution quality the procedures are ranked in the order TABU2, TABU1, BC+SW, and SM+SW. In terms of computational time, the tabu search version TABU1 is quite competitive with the procedures based on simple ordering rules plus switching, in spite of yielding significantly better outcomes. This allows TABU1 to be considered as a powerful procedure for real-time drawing (e.g., drawing on a computer screen). When still higher quality drawings are important, at the cost of additional computational time, TABU2 obtains solutions with fewer arc crossings with a maximum running time of 209 seconds.

8.7 Neural Networks and Learning

8.7.1 Sub-Symbolic Machine Learning (Neural Networks)

While derivative-based methods for training from examples have been used with success in many contexts (error backpropagation is an example in the field of neural networks), they are applicable only to differentiable performance functions and are not always appropriate in the presence of local minima. In addition, the calculation of derivatives is expensive and error-prone, especially if special-purpose VLSI hardware is used. Battiti and Tecchiolli (1995b) use a significantly different approach: the task is transformed into a combinatorial optimization problem (the points of the search space are binary strings), and solved with a reactive tabu search algorithm. To speed up the neighborhood evaluation phase a stochastic sampling of the neighborhood is adopted and a "smart" iterative scheme is used to compute the changes in the performance function caused by changing a single weight. The RTS approach escapes rapidly from local minima, it is applicable to non-differentiable and even discontinuous functions and it is very robust with respect to the choice of the initial configuration. In addition, by fine-tuning the number of bits for each parameter one can decrease the size of the search space, increase the expected generalization and realize cost-effective VLSI.

8.7.2 Global Optimization for Artificial Neural Networks

The ability of neural networks to closely approximate unknown functions to any degree of desired accuracy has generated considerable demand for neural network research in business. The attractiveness of neural network research stems from researchers' need to approximate models within the business environment without having a priori knowledge about the true underlying function. Gradient techniques, such as backpropagation, are currently the most widely used methods for neural network optimization. Since these techniques search for local solutions, they are subject to local convergence and thus can perform poorly even on simple problems when forecasting out-of-sample. Consequently, a global search algorithm is warranted. Sexton, et al. (1997) examine tabu search as a possible alternative to the problematic backpropagation approach.

A Monte Carlo study was conducted to test the appropriateness of TS as a global search technique for optimizing neural networks. Holding the neural network architecture constant, 530 independent runs were conducted for each of seven test functions, including a production function that exhibits both increasing and diminishing marginal returns and the Mackey-Glass chaotic time series. In the resulting comparison, tabu search derived solutions that were significantly superior to those of backpropagation for in-sample, interpolation, and extrapolation test data for all seven test functions. It was also shown that the TS approach required fewer function evaluations and only a fraction of the replications to find these optimal values.

8.7.3 Neural Networks for Combinatorial Optimization

When artificial neural networks are used to solve combinatorial optimization problems, two phases are involved. In the first phase, the problem is mapped onto the structure of the neural network in such a way that the constraints of the combinatorial optimization problem are represented by the connection weights and node biases, and the objective function is represented by the energy function of the neural network. An appropriate mapping usually allows a local minimum energy state of the neural network to correspond to a local optimal solution and a global minimum energy state to correspond to a global optimal solution of the combinatorial optimization problem. In the second phase, a state transition mechanism is employed to guide the time evolution of the states of the neural network in order to find a global or close to global minimum energy state. Consequently, the state transition mechanism determines the effectiveness and the efficiency of the neural network in solving combinatorial optimization problems.

With an appropriate mapping of the problem structure in the first phase, Aarts and Korst (1989a; 1989b) used the Boltzmann Machine, whose state transition mechanism is governed by simulated annealing, to solve many combinatorial optimization problems. However, to find high quality solutions, the computational time required by the Boltzmann Machine is prohibitively long.

The Tabu Machine (Nemati and Sun 1994; Sun and Nemati 1994), provides an alternative to and an improvement over the Boltzmann Machine, embodied in a new form of artificial neural network whose state transition mechanism is governed by tabu search.

To empirically compare the performance of the Tabu Machine against that of the Boltzmann Machine, a rigorous computational experiment was designed involving two types of combinatorial optimization problems, the maximum cut problem and the independent node set problem. In both problem instances, identical mappings were used and three dimensions were considered: size of the problem, arc density of problem, and the size of tabu list. Performance was measured in terms of the quality of the solution obtained and the CPU time used. Although the Tabu Machine was organized to use only short term memory and elementary tabu search techniques, the results of the computational experiment showed that for all problem types and classes, as compared to the Boltzmann Machine, the Tabu Machine finds higher quality solutions within less CPU time.

8.7.4 VLSI Systems with Learning Capabilities

In contrast to the exhaustive design of systems for pattern recognition, control, and vector quantization, an appealing possibility consists of specifying a general architecture, whose parameters are then tuned through Machine Learning (ML). ML becomes a combinatorial task if the parameters assume a discrete set of values. A reactive tabu search algorithm developed by Battiti et al. (1994a, 1994b) permits the training of these systems with low number of bits per weight, low computational accuracy, no local minima "trapping", and limited sensitivity to the initial conditions.

A board with the TOTEM chip used for Machine Learning applications — a collaboration between the Department of Mathematics at the University of Trento and IRST — aims at developing special-purpose VLSI modules to be used as components of fully autonomous massively-parallel systems for real-time adaptive applications. Because of the intense use of parallelism at the chip and system level and the limited precision used, the obtained performance is competitive with that of state-of-the-art supercomputers (at a much lower cost), while a high degree of flexibility is maintained through the use of combinatorial algorithms. In particular, neural nets can be realized. In contrast to many "emulation" approaches, the developed VLSI completely reflects the combinatorial structure used in the learning algorithms. The first chip of the project (TOTEM, partially funded by INFN, the Department of Mathematics at the University of Trento and EU (Esprit project MInOSS), and designed at IRST) achieves a performance of more than one billion multiply-accumulate operations. Applications considered are in the area of pattern recognition (Optical Character Recognition), events "triggering" in High Energy Physics [A+] (Anzellotti, et al. 1995), control of non-linear systems (Battiti and Tecchiolli (1995c), compression of EEG signals [B+] (Battiti, et al. 1995). Test boards for ISA, PCI or VME buses with software and technical documentation are available at IRST.

8.8 Continuous and Stochastic Optimization

8.8.1 Continuous Optimization

A simple benchmark on a function with many suboptimal local minima is considered in Battiti and Tecchiolli (1994a), where a straightforward discretization of the domain is used. A novel algorithm for the global optimization of functions (C-RTS) is presented in Battiti and Tecchiolli (1995a), in which a combinatorial optimization method cooperates with a stochastic local minimizer. The combinatorial optimization component, based on reactive tabu search, locates the most promising boxes, where starting points for the local minimizer are generated. In order to cover a wide spectrum of possible applications with no user intervention, the method is designed with adaptive mechanisms: in addition to the reactive adaptation of the prohibition period, the box size is adapted to the local structure of the function to be optimized (boxes are larger in "flat" regions, smaller in regions with a "rough" structure).

A hybrid scatter genetic tabu search approach (HSGT) is proposed by Trafalis and Al-Harkan (1995) to solve an unconstrained continuous nonlinear global optimization problem. This approach combines the characteristics of the following metaheuristics: scatter search, genetic algorithms, and tabu search. The proposed approach has been tested against a simulated annealing algorithm and a modified version of a hybrid scatter genetic search approach (HSG) by optimizing twenty-one well known test functions. From the computational results, the HSGT approach proved to be quite effective in identifying the global optimum solution which makes the HSGT approach a promising approach to solve the general nonlinear optimization problem. In the hundred runs performed for each of the twenty-one functions, the HSGT approach performed better than the HSG and the simulated annealing approaches, except for one function. Also, the HSGT approach converged to a near global optimum in CPU times ranged between 1.3 seconds and 19.55 seconds. The algorithm was implemented in a GATEWAY 2000 (Pentium, P5-90) computer using the Microsoft FORTRAN PowerStation version 4.

8.8.2 Mixed Integer, Multi-Stage Stochastic Programming

A very large class of problems is characterized by a multi-stage decision process where the future is uncertain and some decisions are constrained to take on values of either zero or one (as in the decision of whether or not to open a facility at a particular location). Although some mathematical theory exists for such problems, no general purpose algorithms have been available to address them. Lokketangen and Woodruff (1996) introduce the notion of integer convergence for progressive hedging, and provide the first implementation of general purpose methods for finding good solutions to multi-stage, stochastic mixed integer (0,1) programming problems.

The solution method makes use of Rockafellar and Wets' progressive hedging algorithm that averages solutions rather than data, and then applies a tabu search algorithm to obtain solutions to the induced quadratic, (0,1) mixed-integer sub-problems. Computational experiments verify the effectiveness of the new method across a range of problem instances. The software that the authors have developed reads standard (SMPS) data files.

8.8.3 Portfolio Management

An important issue in portfolio management is how to measure and handle risk. The challenge is to solve large problems that typically cannot be attacked with non-linear programming methods. Rolland (1996) develops a tabu search method that handles real valued decision variables by discretizing the problem space in 1% and 0.1% increments, and by further incorporating a greedy search to adjust the decision variables to real numbers "finer" than the 0.1% accuracy level. The moves alter their focus to comply with minimum variance and the target return. The approach identified optimal solutions to all the random and real-world problems it was tested on. Computation time was also very modest, even for large problems with 100 or more assets. This approach recently has been extended by Rolland and Johnson (1996) to be able to handle skewness computations and targets.

8.9 Manufacturing

8.9.1 Task Assignment for Assembly Line Balancing

An assembly line consists of a sequence of m work stations which are connected by a conveyor belt. Each station repeatedly has to perform a subset of n tasks or operations on consecutive product units moving along the line at constant speed. Tasks are indivisible elements of work which have to be performed to assemble a product. Due to the uniform movement of the assembly line each product unit spends the same fixed time interval, called the cycle time c, in each work station. The cycle time c determines the production rate which is $1/c$. The execution of each task $j = 1,\dots,n$ requires a fixed task time t_j. Technological restrictions impose precedence constraints that partially specify the sequence of tasks to be performed. These constraints can be represented by a precedence graph containing nodes for all tasks and arcs (i,j) if task i has to be completed before task j can be started. In the simple assembly line balancing problem (SALBP) each task has to be assigned to exactly one station of the assembly line such that no precedence constraint is violated. The maximum station time determines the cycle time c, and all stations with less station time have a respective idle time.

Based on these observations a general problem (SALBP-G) may be formulated with the objective of finding a cycle time c, a number of stations m and a task assignment which minimizes the sum of idle times over all stations, while respecting bounds on the number of stations and the cycle time. In the literature usually the following variants of SALBP are considered:

SALBP-1: Given the cycle time c find an assignment of tasks to stations, subject to the precedence constraints, to minimize the number m of stations.

SALBP-2: Given m stations, find an assignment of tasks to stations, subject to the precedence constraints, to minimize the cycle time c.

A comprehensive survey on methods for SALBP-1 and SALBP-2, given in Scholl and Voss (1996). Voss (1997) investigates a general purpose tabu search method that combines local search with a lower bound search. The resulting method is applicable to problems where a suitable neighborhood is not obvious, and has applications in wider settings. The method also gives a new approach for using tabu search to overcome infeasibility. The tabu search method provides a considerable improvement over the best known results for SALBP-G from the literature, as measured by the average deviation from given lower bounds. Additional problems areas where this tabu search method can be applied include the bin packing problem and makespan minimization for identical parallel machines.

8.9.2 Facility Layout in Manufacturing

The design of the facility layout of a manufacturing system is critically important for its effective utilization: 20 to 50 percent of the total operating expenses in manufacturing are attributed to material handling and layout related costs. Use of effective methods for facilities planning can reduce these costs often by as much as 30 percent, and sometimes more. In general, the facility layout problem has been formulated as a quadratic assignment problem (QAP). The QAP is to find the optimal assignment of n candidate facilities (departments, machines, workstations) to n candidate sites, for the goal of minimizing the total layout costs (which includes the material handling cost, expressed as the product of workflow and travel distance, and a fixed cost associated with locating a facility at a specific location).

Chiang and Kouvelis (1994b) provide a new implementation of the tabu search metaheuristic to solve the QAP, with particular emphasis on facility layout problems, utilizing recency-based and long term memory structures, dynamic tabu size strategies, and intensification and diversification strategies. The tabu search algorithm quickly converges from an arbitrarily generated random solution. Computational experiments, including statistical analysis and library analysis, strongly support the superiority of the C & K tabu search method compared to other procedures for this facility layout application.

8.9.3 Two Dimensional Irregular Cutting

The two-dimensional cutting problem is an optimization problem in which two-dimensional elements of arbitrary specified shapes are to be cut out of a rectangular material. The objective is to determine a cutting pattern that will minimize the amount of material used. The importance of the problem is growing due to its relation to packing, loading and partitioning, which have applications in multiple branches of industry.

Blazewicz and Walkowiak (1995) apply three variants of tabu search to this problem, extending earlier work of Blazewicz, Hawryluk and Walkowiak (1993). The first approach is a simple short term memory version of tabu search that incorporates a tabu list, an aspiration function and a single criterion for optimization. The next variant introduces a new type of evaluating function which combines several criteria to be optimized, together with an associated tabu condition. In the last version a probabilistic approach is used which translates the evaluation criteria into probabilities of selection. An exact algorithm for the subproblem of finding the placement of the polygon is incorporated to enhance the quality of solutions. The final version is also embedded in a parallel version of the algorithm by taking into account various tabu method parallelization schemes and geometric features of the algorithm and problem space.

Extensive computational comparisons disclose that all variants of the tabu search method tested, but most particularly the advanced ones, obtain significantly improved patterns for cutting layouts. In addition, the parallel implementation demonstrates the ability of the method to be exploited highly effectively in a multiprocessor environment.

8.10 Financial Analysis

8.10.1 Computer Aided Design of Financial Products

Recent years have witnessed a rapid pace in *financial innovation*: the tradition of financing a firm's activities with a limited number of securities — equity, debt, preferred stock and convertibles — gave way to the introduction of innovative instruments, such as options, bonds with embedded options, securitized assets etc. This phenomenon has been the focus of a recent issue of *Financial Management*, where Allen (1993) wrote:

> "This [rapid pace of financial innovation] has highlighted the need to understand why the securities that are used have the particular form they do. As a result, financial economics has paid increasing attention to the issue of security design."

A model for designing a specific type of securities, namely of *callable bonds*, was proposed recently by Consiglio and Zenios (1997a). Callable bonds are a major borrowing instrument for corporations, utilities and government agencies with interest rate sensitive assets. Agencies like Fannie Mae and Freddie Mac fund more than 95% of their mortgage assets with the issue of callable bonds and indexed sinking funds debentures. As interest rates drop and mortgages prepay, the bonds are called. The problem faced by the agencies is to specify the kind of a callable bond that is sold to finance a particular asset. This is the problem of *integrated financial product management*, analyzed in Holmer and Zenios (1995). A nonlinear, global, optimization model was suggested by Consiglio and Zenios for this problem. Good solutions to the problem of integrated product management improve the efficiency of the financial intermediary, with potentially measurable rewards by the market.

The model was successfully solved using tabu search by coordinating the use of candidate lists with intensification and diversification search strategies. Empirical results show that the global optimization model produces better returns than the classical model of *portfolio immunization*. Computational experiments also establish that tabu search is both more efficient and more accurate than simulated annealing for solving these models.

The bond design model was used in a subsequent paper, Consiglio and Zenios (1997b), for developing parallel variants of the tabu search heuristics. A set of parallelization strategies was proposed, and evaluated empirically on a distributed network of workstations, reaching parallel efficiencies of up to 60% when using 12 workstations.

8.10.2 Dynamic Investment Problems

Financial decision makers often rely on dynamic investment rules. Traders search out patterns of stock and bond movements and try to anticipate these events in their investing programs. Portfolio managers attempt to surpass broad market indices, such as the S&P 500, when they reduce their exposure to stocks after sensing a market drop. In many of these cases, the trigger event can be modeled via a mathematical formula or rule. An area that is now attracting widespread attention involves the application of stochastic optimization models to improve the investor's performance.

Analyzing investment strategies becomes difficult, however, as dynamic and multi-stage elements are included in the model. Most dynamic investment strategies give rise to non-convex programs. A prominent example is the so-called fixed mix decision rule. Under this rule, the investor re-balances the portfolio by selling overperforming assets and buying underperforming assets. Each period, the investor must carry out the re-balancing decisions in order to achieve a determined asset mix. The proportions become the decision rules for the multi-stage investment model. The goal is to start the simulation clock at each period with a fixed proportions — say 60% stock, 30% bonds, and 10% cash.

The fixed mix model and its variants possess desirable theoretical properties even in the presence of transaction costs. It performs best in highly volatile but relatively trendless markets. Due to its stochastic optimization components, however, the model is a challenging non-convex optimization problem that cannot be adequately handled by traditional nonlinear optimization methods. An effective tabu search approach for this problem is developed by Glover, Mulvey, and Hoyland (1996). For small to moderate sized nonlinear problems, for which a reasonable bound on optimality can be established — ranging up to 20-30 variables and approximately 400 constraints — the TS approach uniformly obtains solutions within 1% of the optimality bound. Compared to a leading global optimization procedure, the method requires only a fraction of the computation time (from 2% to 5%) to yield solutions within the indicated range. As the size of the problems increases, the efficiency of the TS approach increasingly dominates, growing approximately linearly while the global optimization method becomes highly exponential at a size of roughly 40 variables.

An alternative investment decision rule is "constant proportional portfolio insurance." This approach depends upon a momentum strategy, whereby the investor purchases assets that are performing well. The dynamic investment model is non-convex but has a structure that can be taken advantage of by the same general form of tabu search strategy that has proved successful for the fixed mix problem.

Numerous other financial decision rules are similarly amenable to analysis by tabu search. Berger and Mulvey (1997) provide an application to long term financial planning for individual investors. This model addresses real world considerations, such as taxes, mortgage decisions, and investment strategies for paying off liabilities and goals (college tuition, retirement, house purchase). Tabu search makes it possible to handle both integer variables and stochastic elements in a natural manner.

8.11 Specialized Techniques and General Zero-One Solvers

8.11.1 Reactive Tabu Search for Combinatorial Optimization

Reactive Tabu Search (RTS) as developed by Battiti and Tecchiolli (1992, 1994a), has been applied to a considerable range of optimization problems. Combinatorial problems studied with this approach include:

1) Quadratic assignment problems
2) N-K Models (derived from biological inspiration)
3) 0-1 Knapsack and multi-knapsack problems
4) Max-clique problems
5) Biquadratic assignment problems.

In many cases the results obtained with alternative competitive heuristics have been duplicated with low computational complexity, and without intensive parameter and algorithm tuning. In some cases (e.g., in the Max-Clique and Biquadratic assignment problems) significantly better results have been obtained. A comparison of RTS with alternative heuristics (Repeated Local Minima Search, Simulated Annealing, Genetic Algorithms and Mean Field Neural Networks) is presented in Battiti and Tecchiolli (1995c) and a comparison with Simulated Annealing on QAP tasks is contained in Battiti and Tecchiolli (1994b), disclosing the effectiveness of the RTS approach relative to these alternative procedures.

8.11.2 Chunking in Tabu Search

Chunking — grouping basic units of information to create higher level units — is a critical component of human intelligence. The tendency for people to group information was described in the celebrated 1956 paper "The Magic Number Seven: Plus or Minus Two" by G. A. Mitchell. Human problem solvers, when faced with a hard problem, often proceed by linking and integrating features they perceive as related and germane to the solution process. However, people often prefer to organize problem data and problem solving methods in a hierarchical fashion. When possible, they decompose the problem into sub-problems and solve those. When the problem cannot be decomposed, a common strategy is to form groupings of solution attributes so that the search space can be reduced and higher level relationships can be discovered or exploited.

Woodruff (1996) identifies special types of chunking to enhance tabu search memory structures, producing improved problem solving ability and giving useful supporting information for decision makers. Although the proposals are most natural in the context of the tabu search paradigm, they can also be employed in genetic algorithms and simulated annealing by omitting links to memory based constructions. This work outlines theory and proposals for learning about chunks and using them. Computational experience to date is briefly summarized to support the contention that chunking can be an important part of effective optimization algorithms.

8.11.3 Zero-One Mixed Integer Programming

Tabu Search has been applied to solving a variety of zero-one mixed integer programming (MIP) problems with special structures, ranging from scheduling and routing to group technology and probabilistic logic. At the same time, there is a need for a more general heuristic approach that is capable of handling zero-one mixed integer programming problems as a class. The numerous important practical applications subsumed within the zero-one MIP domain suggest the value of a method capable of being applied to the domain as a whole, and thus which can avoid the necessity of relying on a different tailored approach for each different setting.

As a basis for handling this problem class, Tabu Search has also been applied to solving general zero-one MIP problems by taking advantage of the well known result that optimal zero-one solutions can be found at an extreme point of the linear programming feasible region. Several designs for doing this have been developed by superimposing the TS framework on the "Pivot and Complement" heuristic by Balas and Martin (1980).

Lokketangen and Glover (1996, 1997a) introduce a tabu search approach for solving general zero-one problems that directly exploits the extreme point property of zero-one solutions, without using another (previously designed) heuristic as a subroutine. This approach identifies specialized choice rules and aspiration criteria for the problems, expressed as functions of integer infeasibility measures and objective function values. The first-level TS mechanisms are integrated with advanced level strategies and learning. Probabilistic measures are incorporated in this framework, and the Target Analysis learning tool is applied to identify improved control structures and decision rules.

The resulting TS method is demonstrated to give better solutions than other adjacent extreme point methods on a portfolio of test problems. In addition, these solutions rival, and in some cases surpass, the best solutions obtained by special purpose methods that have been created to exploit the special structure of these problems.

The component strategies of this zero-one TS approach can easily be combined with other methods, as shown in Lokketangen and Woodruff (1996), where these strategies have been used to provide a subproblem (scenario) solver in a general tool based on the progressive hedging algorithm for solving mixed integer zero-one multistage stochastic programming problems.

8.12 Constraint Satisfaction and Satisfiability

8.12.1 Constraint Satisfaction Problem as a General Problem Solver

Many combinatorial problems, including a variety of combinatorial optimization problems, can be naturally formulated as a constraint satisfaction problem (CSP). Nonobe and Ibaraki (1997) develop a tabu search based algorithm for the CSP as a foundation for a general problem solver. Special features of the approach include an automatic control mechanism for the tabu tenure, a dynamically weighted penalty function to handle objective functions, and enlargement of the neighborhood by screened swap operations.

The versatility of the tabu search approach for the CSP is demonstrated by applying the method to problems selected from a wide range of applications, i.e., graph coloring, generalized assignment, set covering, timetabling and nurse scheduling. The results prove competitive with those of state-of-the-art algorithms specially developed for the respective problem domains. Findings from the study also indicate several directions for future research in creating general purpose solvers based on applying tabu search to the CSP.

8.12.2 Advances for Satisfiability Problems — Surrogate Constraint Analysis and Tabu Learning

A new approach to solving satisfiability (SAT) problems results by joining surrogate constraint analysis with a tabu learning procedure. The power of surrogate constraints to capture information from an original constraint system, which has been the foundation for a variety of exact methods for discrete and nonlinear optimization problems, is exploited heuristically in the setting of SAT problems by introducing adaptive memory and learning structures derived from tabu search (Lokketangen and

Glover, 1997b). The surrogate constraint information capturing function is achieved by special normalization strategies (as opposed to subgradient optimization strategies) for combining and processing the original constraints to create new ones, with the goal of analyzing and tracing consequences of the original system that may not otherwise be readily accessible.

The study joins both constructive and iterative approaches with the surrogate constraint and tabu learning design. The resulting SAT methods are tested on a portfolio of benchmark problems, and are shown to be more effective than methods based on probabilistic search designs, including variants that encompass probabilistic rules that have been highly favored in previous SAT approaches.

The surrogate constraint and learning based approach quickly found solutions to all problems in the standard benchmark test-set. On average the iterative surrogate/tabu learning approach took $n\log(n)$ or fewer flips to find these solutions, where n is the number of variables. By contrast, rival claims indicate that state-of-the-art iterative randomized methods require on the order of n^2 flips, while randomized constructive (multistart) approaches use up to n^2 complete constructive phases on some of the problems.

These outcomes challenge the widely held view that probabilistic search strategies provide the best heuristic methods for SAT problems. The surrogate constraint and tabu learning approaches used in the study represent only a first level of sophistication, and motivate a closer look at surrogate strategies and more advanced ways of integrating them with adaptive memory and learning procedures.

9 CONNECTIONS, HYBRID APPROACHES AND LEARNING

Relationships between tabu search and other procedures like simulated annealing and genetic algorithms provide a basis for understanding similarities and contrasts in their philosophies, and for creating potentially useful hybrid combinations of these approaches. We offer some speculation on preferable directions in this regard, and also suggest how elements of tabu search can add a useful dimension to neural network approaches.

From the standpoint of evolutionary strategies, we trace connections between population based models for combining solutions, as in genetic algorithms, and ideas that emerged from surrogate constraint approaches for exploiting optimization problems by combining constraints. We show how this provides the foundation for methods that give additional alternatives to genetic-based frameworks, specifically as embodied in the scatter search approach, which is the "primal complement" to the dual strategy of surrogate constraint approaches. Recent successes by integrating scatter search (and its path relinking extensions) with tabu search disclose potential advantages for evolutionary strategies that incorporate adaptive memory.

Finally, we describe the learning approach called target analysis, which provides a way to determine decision parameters for deterministic and probabilistic strategies — and thus affords an opportunity to create enhanced solution methods.

9.1 Simulated Annealing

The contrasts between simulated annealing and tabu search are fairly conspicuous, though undoubtedly the most prominent is the focus on exploiting memory in tabu search that is absent from simulated annealing. The introduction of this focus entails associated differences in search mechanisms, and in the elements on which they operate. Accompanying the differences directly attributable to the focus on memory, and also magnifying them, several additional elements are fundamental for understanding the relationship between the methods. We consider three such elements in order of increasing importance.

First, tabu search emphasizes scouting successive neighborhoods to identify moves of high quality, as by candidate list approaches of the form described in Chapter 3. This contrasts with the simulated annealing approach of randomly sampling among these moves to apply an acceptance criterion that disregards the quality of other moves available. (Such an acceptance criterion provides the sole basis for sorting the moves selected in the SA method.) The relevance of this difference in orientation is accentuated for tabu search, since its neighborhoods include linkages based on history, and therefore yield access to information for selecting moves that is not available in neighborhoods of the type used in simulated annealing.

Next, tabu search evaluates the relative attractiveness of moves not only in relation to objective function change, but in relation to additional factors that represent quality, which are balanced over time with factors that represent influence. Both types of measures are affected by the differentiation among move attributes, as embodied in tabu activation rules and aspiration criteria, and in turn by relationships manifested in recency, frequency, and sequential interdependence (hence, again, involving recourse to memory). Other aspects of the state of search also affect these measures, as reflected in the altered evaluations of strategic oscillation, which depend on the direction of the current trajectory and the region visited.

Finally TS emphasizes guiding the search by reference to multiple thresholds, reflected in the tenures for tabu-active attributes and in the conditional stipulations of aspiration criteria. This may be contrasted to the simulated annealing reliance on guiding the search by reference to the single threshold implicit in the temperature parameter. The treatment of thresholds by the two methods compounds this difference between them. Tabu search varies its thresholds nonmonotonically, reflecting the conception that multidirectional parameter changes are essential to adapt to different conditions, and to provide a basis for locating alternatives that might otherwise be missed. This contrasts with the simulated annealing philosophy of adhering to a temperature parameter that only changes monotonically.

Hybrids are now emerging that are taking preliminary steps to bridge some of these differences, particularly in the realm of transcending the simulated annealing reliance on a monotonic temperature parameter. A hybrid method that allows temperature to be strategically manipulated, rather than progressively diminished, has been shown to yield improved performance over standard SA approaches. A hybrid method that expands the SA basis for move evaluations also has been found to perform better than standard simulated annealing. Consideration of these findings invites the question of whether removing the memory scaffolding of tabu search and retaining its other features may yield a viable method in its own right, as by application of the tabu thresholding method described in Chapter 7. Experience cited in some of the studies reported in Chapter 8 suggests that, while memoryless tabu thresholding can outperform a variety of alternative heuristics, it generally does not match the performance of TS methods that appropriately exploit memory.

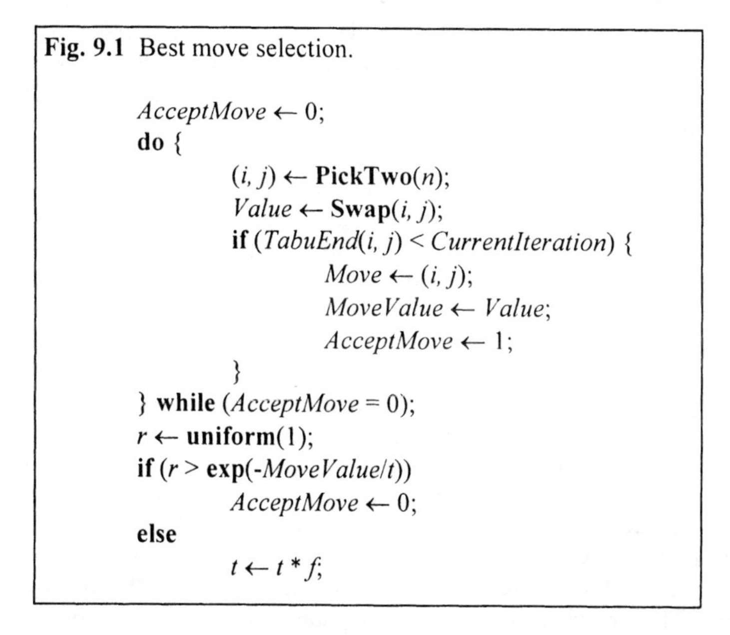

Fig. 9.1 Best move selection.

```
AcceptMove ← 0;
do {
        (i, j) ← PickTwo(n);
        Value ← Swap(i, j);
        if (TabuEnd(i, j) < CurrentIteration) {
                Move ← (i, j);
                MoveValue ← Value;
                AcceptMove ← 1;
        }
} while (AcceptMove = 0);
r ← uniform(1);
if (r > exp(-MoveValue/t))
        AcceptMove ← 0;
else
        t ← t * f;
```

For illustration purposes, suppose that it is desired to create a search method that employs a SA sampling procedure embedded in a TS framework. One way of implementing such hybrid procedure within the context of a sequencing problem (of size n) using a swap move neighborhood is illustrated by the *best move* selection shown in Figure 9.1. First note that the indicator variable *AcceptMove* is part of the memory structure that contains the information related to the "best" move. The "do-loop" is performed until an admissible move is found under a standard tabu activation rule. (The **if** statement within the loop may be replaced in order to enforce an alternative form of a tabu activation rule, or implement aspiration level criteria.) The **PickTwo** function is a simple procedure that returns two integers randomly selected between 1 and n, such that $j > i$. Once an admissible move is found, it is accepted with probability $e^{-\Delta/t}$, where Δ is the move value and t is the current temperature. The function **uniform**(x) returns a random real value uniformly distributed between 0 and x. If the move is not accepted the indicator variable *AcceptMove* is switched back to zero, otherwise its value remains as one, and the temperature is decrease by a factor *f*. In this context, typical values for the starting temperature *t* range from 1 to 2, while the values for *f* range between 0.9 and 0.99. The initial *t* value and the *f* factor value define what is known in the simulated annealing literature as the *temperature schedule.*

The hybrid approach presented above differs from the "pure" tabu search procedure in that an iteration may or may not result in a move being executed. When a move is not accepted, the main TS iteration is still performed (i.e., remaining tabu tenures for tabu moves are modified), however the trial solution is not altered. Due to the sampling nature of the hybrid approach, an iteration of this procedure generally requires less computational effort than in the "pure" TS case. Note that the hybrid procedure incorporates a "mildly strategic" diversification component by means of randomness, but this approach lacks the aggressiveness typical to TS methods (including probabilistic tabu search).

This an other hybrids of tabu search with simulated annealing are entirely possible, and it is an open question as to whether the introduction of the SA temperature parameter may add anything to the effectiveness of a TS design that is not achieved by TS choice strategies of the types indicated in preceding chapters.

9.2 Genetic Algorithms

Genetic algorithms offer a somewhat different set of comparisons and contrasts with tabu search. GAs are based on selecting subsets (traditionally pairs) of solutions from a population, called parents, and combining them to produce new solutions called children. Rules of combination to yield children are based on the genetic notion of crossover, which in the classical form consists of interchanging solution values of particular variables, together with occasional operations such as random value changes. Children that pass a survivability test, probabilistically biased to favor those of superior quality, are then available to be chosen as parents of the next generation. The choice of parents to be matched in each generation is based on random or biased random sampling from the population (in some parallel versions executed over separate subpopulations whose best members are periodically exchanged or shared). Genetic terminology customarily refers to solutions as chromosomes, variables as genes, and values of variables as alleles.

By means of coding conventions, the genes of genetic algorithms may be compared to attributes in tabu search. Introducing memory in GAs to track the history of genes and their alleles over subpopulations would provide an immediate and natural way to create a hybrid with TS.

Some important differences between genes and attributes are worth noting, however. The implicit differentiation of attributes into *from* and *to* components, each having different memory functions, does not have a counterpart in genetic algorithms. A *from* attribute is one that is part of the current solution but is not included in the next solution once a move is made. A *to* attribute is one that is not part of the current solution but becomes part of the next solution once a move is made. The lack of this type of differentiation in GAs results because these approaches are organized to operate without reference to moves (although, strictly speaking, combination by crossover can be viewed as a special type of move). Another distinction derives from differences in the use of coding conventions. Although an attribute change, from a state to its complement, can be encoded in a zero-one variable, such a variable does not necessarily provide a convenient or useful representation for the transformations provided by moves. Tabu activation rules and aspiration criteria handle the binary aspects of complementarity without requiring explicit reference to a zero-one x vector or two-valued functions. Adopting a similar orientation (relative to the special class of moves embodied in crossover) might yield benefits for genetic algorithms in dealing with issues of genetic representation, which currently pose difficult questions.

A contrast to be noted between genetic algorithms and tabu search arises in the treatment of context, i.e., in the consideration given to structure inherent in different problem classes. For tabu search, context is fundamental, embodied in the interplay of attribute definitions and the determination of move neighborhoods, and in the choice of conditions to define tabu restrictions. Context is also implicit in the identification of amended evaluations created in association with longer term memory, and in the regionally dependent neighborhoods and evaluations of strategic oscillation.

At the opposite end of the spectrum, GA literature has traditionally stressed the freedom of its rules from the influence of context. Crossover, in particular, is supposedly a *context neutral* operation, which assumes no reliance on conditions that solutions must obey in a particular problem setting, just as genes make no reference to the environment as they follow their instructions for recombination (except, perhaps, in the case of mutation). Practical application, however, generally renders this an inconvenient assumption, making solutions of interest difficult to find. Consequently, a good deal of effort in GA implementation is devoted to developing "special crossover" operations that compensate for the difficulties created by context, effectively reintroducing it on a case by case basis. The term *evolutionary algorithms* has come to encompass virtually any method that undertakes to improve a population of vectors (or at least the best members of the population) over time. Relative to this extremely broad classification, the scatter search and path relinking approaches linked with tabu search can be viewed as sharing the "evolutionary" label in common with GAs. The evolutionary components of TS, then, belong to a subcategory that is distinguished by rigorous dedication to exploiting context, as manifested in problem structure.

The chief method by which modern genetic algorithms handle structure is by relegating its treatment to some other method. For example, genetic algorithms combine solutions by their parent-children processes at one level, and then a descent method may be introduced to operate on the resulting solutions to produce new solutions. These new solutions in turn are submitted to be recombined by the GA processes. In these versions, genetic algorithms already take the form of hybrid methods. Hence there is a natural basis for marrying GA and TS procedures in such approaches. But genetic algorithms and tabu search also can be joined in a more fundamental way.

Specifically, tabu search strategies for intensification and diversification are based on the following question: how can information be extracted from a set of good solutions to help uncover additional (and better) solutions? GAs are generated in reverse, by posing a mechanism to provide an approach for answering this question, consisting of putting solutions together and interchanging components (in some loosely defined sense, if traditional crossover is not strictly enforced). Tabu search, by contrast, seeks an answer by utilizing processes that specifically incorporate neighborhood structures into their design.

Augmented by historical information, neighborhood structures are used as a basis for applying penalties and incentives to induce attributes of good solutions to become incorporated into current solutions. Consequently, although it may be meaningless to interchange or otherwise incorporate a set of attributes from one solution into another in a wholesale fashion, as attempted in traditional GA recombination operations, a stepwise approach to this goal through the use of neighborhood structures is entirely practicable. This observation provides a motive for creating *structured combinations* of solutions that embody desired characteristics such as feasibility — as is automatically achieved by the TS approach of path relinking discussed in Chapter 4. Instead of being compelled to create new types of crossover to remove deficiencies of standard operators upon being confronted by changing contexts, this approach addresses context directly and makes it an essential part of the design for generating combinations.

The current trend of genetic algorithms seems to be increasingly compatible with this perspective, and could provide a basis for a useful hybrid combination of genetic algorithm and tabu search ideas. However, a fundamental question emerges, as posed in the development of the next sections, about whether there is any advantage to introducing genetic crossover-based ideas over introducing the apparently more flexible and exploitable path relinking ideas.

If we assume that some merit may be attached to the genetic orientation, a type of hybrid approach that deserves consideration arises by the use of strategic oscillation. For the majority of cases, genetic algorithms do not allow crossover operations that would produce infeasible solutions. In restricted instances where infeasible solutions are permitted, they are usually transformed into feasible solutions before they are placed back into the population. By enriching the population with infeasible solutions, a genetic algorithm coupled with strategic oscillation gains the power to operate with increased diversity, which may improve its chances of finding optimal solutions.

A straightforward way to incorporate strategic oscillation into a genetic algorithm is as follows. First, a crossover operation that admits infeasible solutions is employed. Second, a feasibility measure is introduced. Typically, this measure quantifies the amount by which a solution violates its constraints. For example, a heuristic for a permutation problem that produces a solution with four subtours (viewing a permutation as a Hamiltonian path) may be assigned an infeasibility measure of three, perhaps modified by weighting each subtour as an inverse function of its size. Third, a suitable set of penalty factors must be obtained, to scale the infeasibility measure relative to the objective function. The fitness of a solution may then be evaluated as follows:

$$Fitness = ObjectiveValue + Penalty(t) * Infeasibility$$

The penalty and infeasibility measure terms may be conceived as vectors, hence producing a weighted sum of values, whose components terms are either linear or nonlinear. If a problem is a minimization problem, then "better" fitness levels are given by smaller *Fitness* values. *Penalty*(t) is controlled by strategic oscillation as a sequence of penalty factors that oscillate between large and small values (where t denotes the control parameter for the oscillation). The best strategy for manipulating t is probably problem dependent, although as in tabu search, relevant strategies may emerge that are relevant for special classes of problems. As changes to t cause *Penalty*(t) to decrease, infeasible solutions are permitted to enter the population, which then may in turn be gradually driven out by changes that cause *Penalty*(t) to increase. In this way the population will alternate between populations with different mixes of feasible and infeasible members.

9.2.1 Models of Nature — Beyond "Genetic Metaphors"

An aspect of tabu search that is often misunderstood concerns the relation between a subset of its strategies and certain approaches embodied in genetic algorithms. TS researchers have tended sometimes to overlook the part of the adaptive memory focus that is associated with strategies for combining sets of elite solutions. Complementing this, GA researchers have been largely unaware that such a collection of strategies outside their domain exists. This has quite possibly been due to the influence of the genetic metaphor, which on the one hand has helped to launch a number of useful problem solving ideas, and on the other hand has also sometimes obscured fertile connections to ideas that come from different foundations.

To understand the relevant ties, it is useful to go back in time to examine the origins of the GA framework and of an associated set of notions that became embodied in TS strategies. We will first sketch the original genetic algorithm design (see Figure 9.2), as characterized in Holland (1975). Our description is purposely somewhat loose, to be able to include approaches more general than the specific proposals that accompanied the introduction of GAs. Many variations and changes have come about over the years, as we subsequently observe.

Fig. 9.2 Genetic algorithm template.

1) Begin with a population of binary vectors.

2) Operate repeatedly on the current generation of vectors, for a selected number of steps, choosing two "parent vectors" at random. Then mate the parents by exchanging certain of their components to produce offspring. (The exchange, called "crossover," was originally designed to reflect the process by which chromosomes exchange components in genetic mating and, in common with the step of selecting parents themselves, was organized to rely heavily on randomization. In addition, a "mutation" operation is occasionally allowed to flip bits at random.)

3) Apply a measure of fitness to decide which offspring survive to become parents for the next generation. When the selected number of matings has been performed for the current generation, return to the start of Step 2 to initiate the mating of the resulting new set of parents.

4) Carry out the mating-and-survival operation of Steps 2 and 3 until the population becomes stable or until a chosen number of iterations has elapsed.

A somewhat different model for combining elements of a population comes from a class of relaxation strategies in mathematical optimization known as surrogate constraint methods (Glover, 1965). The goal of these approaches is to generate new constraints that capture information not contained in the original problem constraints taken independently, but which their union implies. We will see that some unexpected connections emerge between this development and that of genetic algorithms.

The information-capturing focus of the surrogate constraint framework has the aim of developing improved methods for solving difficult optimization problems by means of (a) providing better criteria for choice rules to guide a search for improved solutions, (b) inferring new bounds (constraints with special structures) to limit the space of solutions examined. (The basic framework and strategies for exploiting it are given in Glover (1965, 1968, 1975b), Greenberg and Pierskalla (1970, 1973), Karwan and Rardin (1976, 1979), and Freville and Plateau (1986, 1993).) Based on these objectives, the generation of new constraints proceeds as indicated in Figure 9.3.

Fig. 9.3 Surrogate constraint template.

1) Begin with an initial set of problem constraints (chosen to characterize all or a special part of the feasible region for the problem considered).

2) Create a measure of the relative influence of the constraints as basis for combining subsets to generate new constraints. The new (surrogate) constraints, are created from nonnegative linear combinations of other constraints, together with cutting planes inferred from such combinations. (The goal is to determine surrogate constraints that are most effective for guiding the solution process.)

3) Change the way the constraints are combined, based on the problem constraints that are not satisfied by trial solutions generated relative to the surrogate constraints, accounting for the degree to which different source constraints are violated. Then process the resulting new surrogate constraints to introduce additional inferred constraints obtained from bounds and cutting planes. (Weaker surrogate constraints and source constraints that are determined to be redundant are discarded.)

4) Change the way the constraints are combined, based on the problem constraints that are not satisfied by trial solutions generated relative to the surrogate constraints, accounting for the degree to which different source constraints are violated. Then process the resulting new surrogate constraints to introduce additional inferred constraints obtained from bounds and cutting planes. (Weaker surrogate constraints and source constraints that are determined to be redundant are discarded.)

A natural first impression is that the surrogate constraint design is quite unrelated to the GA design, stemming from the fact that the concept of combining constraints seems inherently different from the concept of combining vectors. However, in many types of problem formulations, including those where surrogate constraints were first introduced, constraints are summarized by vectors. More particularly, over time, as the surrogate constraint approach became embedded in both exact and heuristic methods, variations led to the creation of a "primal counterpart" called *scatter search*. The scatter search approach combines solution vectors by rules patterned after those that govern the generation of new constraints, and specifically inherits the strategy of exploiting linear combinations and inference (Glover, 1977). The motivation for this development goes beyond the goal of producing a primal analog for the surrogate constraint approach. In situations where surrogate constraint relaxations yield a duality gap, the natural response is to combine elite solutions that "rim" this gap as a basis for exploiting information that may be contained in their union.

9.3 Scatter Search

We have alluded to the form of scatter search in Chapter 7, but now examine it in greater detail. Following the principles that underlie the surrogate constraint design, the scatter search process is organized to (1) capture information not contained separately in the original vectors, (2) take advantage of auxiliary heuristic solution methods (to evaluate the combinations produced and to actively generate new vectors), (3) make dedicated use of strategy instead of randomization to carry out component steps.

Fig. 9.4 Scatter search procedure.

1) Apply heuristic processes to generate a starting set of solution vectors (trial points). Designate a subset of the best vectors to be *reference points.* (Subsequent iterations of this step, transferring from Step 3 below, incorporate advanced starting solutions and best solutions from previous history as candidates for the reference points.)

2) Form linear combinations of subsets of the current reference points to create new points. The linear combinations are:

   chosen to produce points both inside and outside the convex region spanned by the reference points.

   modified by generalized rounding processes to yield integer values for integer-constrained vector components.

3) Extract a collection of the best points generated in Step 2 to be used as starting points for a new application of the heuristic processes of Step 1. Repeat these steps until reaching a specified iteration limit.

Figure 9.4 sketches the scatter search approach in its original form and identifies some novel connections and contrasts to GA methods. Based on this sketch, we will examine extensions that additionally take advantage of the memory-based designs of tabu search.

Two particular features of the scatter search proposal, which will be elaborated later, deserve mention. The use of clustering strategies is suggested for selecting subsets of points in Step 2, which allows different blends of intensification and diversification by generating new points "within clusters" and "across clusters." Also, solutions selected in Step 3 as starting points for re-applying heuristic processes are not required to be feasible, since heuristics proposed to accompany scatter search include those capable of starting from an infeasible solution.

In sum, scatter search is founded on the following premises.

(P1) Useful information about the form or location of optimal solutions is typically contained in a (sufficiently diverse) collection of elite solutions. Useful information may also be contained in bad solutions. However, such solutions are usually much more numerous and varied than good ones, and consequently there is less advantage in trying to make use of them. On the other hand, valuable information can be contained in *trajectories* from bad solutions to good solutions (or from good solutions to other good solutions), and such trajectory-based information is one of the elements that tabu search seeks to exploit.

(P2) When solutions are combined as a strategy for exploiting such information, it is important to provide for combinations that can extrapolate beyond the regions spanned by the solutions considered, and further to incorporate heuristic processes to map combined solutions into new points.

(P3) Taking account of multiple solutions simultaneously, as a foundation for creating combinations, enhances the opportunity to exploit information contained in the union of elite solutions.

The fact that the heuristic processes of scatter search (as referred to in (P2)) are not restricted to a single uniform design, but represent a varied collection of procedures, affords strategic possibilities whose implications are examined in the next chapter. This theme also shares a link with the original surrogate constraint proposal, where heuristics for surrogate relaxations are introduced to improve the application of exact solution methods. In combination, the heuristics are used to generate strengthened surrogate constraints and, iteratively applied, to generate trial solutions for integer programming problems.

The catalog in Figure 9.5 traces the links between the conceptions underlying scatter search and conceptions that have been introduced over time as amendments to the GA framework.

These innovations in the GA domain, which have subsequently been incorporated in a wide range of studies, are variously considered to be advances or heresies according to whether they are viewed from liberal or traditional perspectives. Significantly, their origins are somewhat diffuse, rather than integrated within a single framework. The "press" for the GA approach suggests by contrast that it is not subject to such variation, but represents a manifestation of immutable natural law. An amusing quote from the January 16, 1996 issue of the *Wall Street Journal* is illustrative: "Three billion years of evolution can't be wrong," [according to a genetic algorithm pioneer]…. "It's the most powerful algorithm there is."

Fig. 9.5 Scatter search features (1977) incorporated into non-traditional GA approaches.

- Introduction of "flexible crossover operations." (Scatter search combinations include all possibilities generated by the early GA crossover operations, and also include all possibilities embedded in the more advanced "uniform" and "Bernoulli" crossovers (Ackley (1987), Spears and DeJong (1991)). Path relinking descendants of scatter search provide further possibilities, noted subsequently.)

- Use of heuristic methods to improve solutions generated from processes for combining vectors (Mühlenbein et al. (1988), Ulder et al. (1991)), (Whitley, Gordon and Mathias (1994)).

- Exploitation of vector representations that are not restricted to binary representations (Davis (1989), Eschelman and Schaffer (1992)).

- Introduction of special cases of linear combinations for operating on continuous vectors (Davis (1989), Wright (1990), Bäck et al. (1991), Michalewicz and Janikow (1991)).

- Use of combinations of more than two parents simultaneously to produce offspring (Eiben et al. (1994), Mühlenbein and Voigt (1996)).

- Introduction of strategies that subdivide the population into different groupings (Mühlenbein and Schlierkamp-Voosen (1994)).

It is clear that a number of the elements of the scatter search approach remain outside of the changes brought about by these proposals. A simple example is the approach of introducing adaptive rounding processes for mapping fractional components into integers. There also has conspicuously been no GA counterpart to the use of clustering to create strategic groupings of points, nor (as a result) to the notion of combining points according to distinctions between membership in different clusters. (The closest approximation to this has been the use of "island populations" that evolve separately, but without concern for analyzing or subdividing populations based on inference and clustering. The relevance of such matters, and of related conditional analyses, is discussed in Chapter 10.)

The most important distinction, however, is the link between scatter search and the theme of exploiting history. The prescriptions for combining solutions within scatter search are part of a larger design for taking advantage of information about characteristics of previously generated solutions to guide current search. In retrospect, it is perhaps not surprising that such a design should share an intimate association with the surrogate constraint framework, with its emphasis on extracting and coordinating information across different solution phases. This orientation, which takes account of elements such as the recency, frequency and quality of particular value assignments, clearly shares a common foundation with notions incorporated within tabu search. (The same reference on surrogate constraint strategies that is the starting point for scatter search is also often cited as a source of early TS conceptions.) By this means, the link between tabu search and so-called "evolutionary" approaches also becomes apparent. The term *evolutionary* has undergone an interesting evolution of its own. By a novel turn, the term "mutation" in the GA terminology has become reinterpreted to refer to any form of change, including the purposeful change produced by a heuristic process. As a result, all methods that apply heuristics to multiple solutions, whether or not they incorporate strategies for combining solutions, are now considered kindred to genetic algorithms, and the enlarged collection is labeled "evolutionary methods." (This terminology accordingly has acquired the distinction of embracing nearly every kind of method conceivable.)

9.3.1 Modern Forms and Applications of Scatter Search

Recent implementations of scatter search (cited below) have taken advantage of the implicit learning capabilities provided by the tabu search framework, leading to refined methods for determining reference points and for generating new points. Current scatter search versions have also introduced more sophisticated mechanisms to map fractional values into integer values. This work is reinforced by new theorems about searches over spaces of zero-one integer variables, as described in the star-path development of Chapter 6. Special models have also been developed to allow both heuristic and exact methods to transform infeasible trial points into feasible points. Finally, scatter search is the source of the broader class of *path relinking* methods, as described in Chapter 4, which offer a wide range of mechanisms for creating productive combinations of reference solutions. A brief summary of some of these developments appears in Figure 9.6.

Implementation of various components of these extensions have provided advances for solving general nonlinear mixed discrete optimization problems with both linear and nonlinear constraints, as noted in the references cited.

In the Appendix of Section 9.8 we elaborate a particular approach for nonlinear scatter search to provide an understanding of how tabu search, and associated intensification and diversification strategies, can be used to guide and enhance the process.

Fig. 9.6 Scatter Search Extensions.

- Tabu search memory is used to select current reference points from a historical pool (Glover, 1989, 1994a).
- Tabu search intensification and diversification strategies guide the generation of new points (Mulvey, 1995; Zenios, 1996; Fleurent et al. 1996).
- Solutions generated as "vector combinations" are further improved by explicit tabu search guidance (Trafalis and Al-Harkan, 1995; Glover, Kelly and Laguna, 1996; Fleurent et al., 1996; Cung, et al. 1997).
- Directional rounding processes focus the search for feasible zero-one solutions allowing them to be mapped into convex subregions of hyperplanes produced by valid cutting plane inequalities (Glover, 1995a).
- Neural network learning is applied to filter out promising and unpromising points for further examination, and pattern analysis is used to predict the location of promising new solutions (Glover, Kelly and Laguna, 1996).
- Mixed integer programming models generate sets of diversified points, and yield refined procedures for mapping infeasible points into feasible points (Glover, Kelly and Laguna, 1996).
- Structured combinations of points take the role of linear combinations, to expand the range of alternatives generated (Glover, 1994a).

9.3.2 Scatter Search and Path Relinking Interconnections

The relation between scatter search and path relinking sheds additional light on the character of these approaches. As already remarked, path relinking is a direct extension of scatter search. The way this extension comes about is as follows.

From a spatial orientation, the process of generating linear combinations of a set of reference points may be characterized as generating paths between and beyond these points (where points on such paths also serve as sources for generating additional points). This leads to a broader conception of the meaning of *combinations* of points. That is, by natural extension, we may conceive such combinations to arise by generating paths between and beyond selected points in neighborhood space, rather than in Euclidean space.

The form of these paths in neighborhood space is easily specified by reference to attribute-based memory, as used in tabu search. The path relinking strategy thus emerges as a direct consequence. Just as scatter search encompasses the possibility to generate new solutions by weighting and combining more than two reference solutions at a time, path relinking includes the possibility to generate new solutions by multi-parent path constructions that incorporate attributes from a set of guiding solutions, where these attributes are weighted to determine which moves are given higher priority (as we have seen in Chapter 4). The name *path relinking* comes from the fact that the generation of such paths in neighborhood space characteristically "relinks" previous points in ways not achieved in the previous search history.

It is useful to note that re-expressing scatter search relative to neighborhood space also leads to more general forms of scatter search in Euclidean space. The form of path relinking manifested in vocabulary building (which results by using constructive and destructive neighborhoods to create and reassemble components of solutions), also suggests the relevance of combining solutions in Euclidean space by allowing different linear combinations to be created for different solution components. The design considerations that underlie vocabulary building generally, as described in earlier chapters, carry over to this particular instance.

The broader conception of solution combinations provided by path relinking has useful implications for evolutionary procedures. The exploitation of neighborhood space and attribute-based memory gives specific, versatile mechanisms for achieving such combinations, and provides a further interesting connection between tabu search proposals and genetic algorithm proposals. In particular, many recently developed "crossover operators," which have no apparent relation between each other in the GA setting, can be shown to arise as special instances of path relinking, by restricting attention to two reference points (taken as parents in GAs), and by replacing the strategic neighborhood guidance of path relinking with a reliance on randomization. In short, the options afforded by path relinking for combining solutions are more unified, more systematic and more encompassing than those provided by the "crossover" concept, which changes from instance to instance and offers no guidance for how to take advantage of any given context.

9.4 Greedy Randomized Adaptive Search Procedures (GRASP)

The GRASP methodology was developed in the late 1980s, and the acronym was coined by Tom Feo (Feo and Resende, 1995). It was first used to solve computationally difficult set covering problems (Feo and Resende, 1989). Each GRASP iteration consists of constructing a trial solution and then applying an exchange procedure to find a local optimum (i.e., the final solution for that iteration). The construction phase is iterative, greedy, and adaptive. It is iterative because the initial solution is built considering one element at a time. It is greedy because the addition of each element is guided by a greedy function. It is adaptive because the element chosen at any iteration in a construction is a function of those previously chosen. (That is, the method is adaptive in the sense of updating relevant information form iteration to iteration, as in most constructive procedures.) The improvement phase typically consists of a local search procedure.

For illustration purposes, consider the design of a GRASP for the 2-partition problem (see, e.g., Laguna et al., 1994). This problem consists of clustering the nodes of a weighted graph into two equal sized sets such that the weight of the edges between the two sets is minimized. In this context, the iterative, greedy, and adaptive elements of the GRASP construction phase may be interpreted as follows. The initial solution is built considering one node at a time. The addition of each node is guided by a greedy function that minimizes the augmented weight of the partition. The node chosen at any iteration in the construction is a function of the adjacencies of previously chosen nodes. There is also a probabilistic component in GRASP, that is applied to the selection of elements during the construction phase. After choosing the first node for one set, all non-adjacent nodes are of equal quality with respect to the given greedy function. If one of those nodes is chosen by some deterministic rule, then every GRASP iteration will repeat this selection. In such stages within a construction where there are multiple greedy choices, choosing any one of them will not compromise the greedy approach, yet each will often lead to a very different solution.

To generalize this strategy, consider forming a *candidate list* (at each stage of the construction) consisting of high quality elements according to an adaptive greedy function. Then, the next element to be included in the initial solution is randomly selected from this list. A similar strategy has been categorized as a cardinality-based semi-greedy heuristic.

The solution generated by a greedy randomized adaptive construction can generally be improved by the application of an improvement phase following selected construction phases, as by using a descent method based on an exchange mechanism, since usually the result of the construction phase is not a local minimum with respect to simple exchange neighborhoods. There is an obvious computational tradeoff between the construction and improving phases. An intelligent construction requires fewer improving exchanges to reach a local optimum, and therefore, it results in a reduction of the total CPU time required per GRASP iteration. The exchange mechanism can also be used as a basis for a hybrid method, as by incorporating elements of other methodologies such as simulated annealing or tabu search. In particular, given that the GRASP constructions inject a degree of diversification to the search process, the improvement phase may consist of a short term memory tabu search that is fine tuned for intensification purposes. Other connections may be established with methods such as scatter search or the path relinking strategy of tabu search, by using the GRASP constructions (or their associated local optima) as reference points.

Performing multiple GRASP iterations may be interpreted as a means of strategically sampling the solution space. Based on empirical observations, it has been found that the sampling distribution generally has a mean value that is inferior to the one obtained by a deterministic construction, but the best over all trials dominates the deterministic solution with a high probability. The intuitive justification of this phenomenon is based on the ordering statistics of sampling. GRASP implementations are generally robust in the sense that it is difficult to find or devise pathological instances for which the method will perform arbitrarily bad. The robustness of this method has been well documented in applications to production, flight scheduling, equipment and tool selection, location, and maximum independent sets.

An interesting connection exists between GRASP and probabilistic tabu search (PTS). If PTS is implemented in a memoryless form, and restricted to operate only in the constructive phase of a multistart procedure (stripping away memory, and even probabilistic choice, from the improving phase), then a procedure resembling GRASP results. The chief difference is that the probabilities used in PTS are rarely chosen to be uniform over members of the candidate list, but generally seek to capture variations in the evaluations, whenever these variations reflect anticipated differences in the effective quality of the moves considered.

This connection raises the question of whether a multistart variant of probabilistic tabu search may offer a useful alternative to memoryless multistart approaches like GRASP. A study of this issue for the quadratic assignment problem, where GRASP has been reported to perform well, was conducted by Fleurent and Glover (1996). To provide a basis for comparison, the improving phases of the PTS multistart method excluded the use of TS memory and guidance strategies, and were restricted to employ a standard descent procedure. Probabilistic tabu search mechanisms were used in the constructive phases, incorporating frequency-based intensification and applying the principle of proximate optimality (as described in Chapter 7) to improve the effectiveness of successive constructions. The resulting multistart method proved significantly superior to other multistart approaches previously reported for the quadratic assignment problem. However, it also turned out to be not as effective as the leading tabu search methods that use memory in the improving phases as well as (or instead of) in the constructive phases. Nevertheless, it seems reasonable to conjecture that classes of problems exist where increased reliance on re-starting will prove advantageous, and where the best results may be obtained from appropriately designed multistart strategies such as based on greedy randomized search and multistart variants of PTS.

9.5 Neural Networks

Neural networks have a somewhat different set of goals than tabu search, although some overlaps exist. We indicate how tabu search can be used to extend certain neural net conceptions, yielding a hybrid that may have both hardware and software implications. The basic transferable insight from tabu search is that memory components with dimensions such as recency and frequency can increase the efficacy of a system designed to evolve toward a desired state. We suggest the merit of fusing neural network memory with tabu search memory as follows. (A rudimentary acquaintance with neural network ideas is assumed.)

Recency based considerations can be introduced from tabu search into neural networks by a *time delay feedback loop* from a given neuron back to itself (or from a given synapse back to itself, by the device of interposing additional neurons). This permits firing rules and synapse weights to be changed only after a certain time threshold, determined by the length of the feedback loop. Aspiration thresholds of the form conceived in tabu search can be embodied in inputs transmitted on a secondary level, giving the ability to override the time delay for altering firing thresholds and synaptic weights. Frequency based effects employed in tabu search similarly may be incorporated by introducing a form of cumulative averaged feedback.

Time delay feedback mechanisms for creating recency and frequency effects also can have other functions. In a problem solving context, for example, it may be convenient to disregard one set of options to concentrate on another, while retaining the ability to recover the suppressed options after an interval. This familiar type of human activity is not a customary part of neural network design, but can be introduced by the time dependent functions previously indicated. In addition, a threshold can be created to allow a suppressed option to "go unnoticed" if current activity levels fall in a certain range, effectively altering the interval before the option reemerges for consideration. Neural network designs to incorporate those features may directly make use of the TS ideas that have made these elements effective in the problem solving domain.

Tabu search strategies that introduce longer term intensification and diversification concerns are also relevant to neural network processes. As a foundation for blending these approaches, it is useful to adopt an orientation where a collection of neurons linked by synapses with various activation weights is treated as a set of attribute variables which can be assigned alternative values. Then the condition that synapse j (from a specified origin neuron to a specified destination neuron) is assigned an activation weight in interval p can be coded by the assignment $y_j = p$, where y_j is a component of an attribute vector y as identified in the discussion of attribute creation processes in connection with vocabulary building, in Chapter 7. A similar coding identifies the condition under which a neuron fires (or does not fire) to activate its associated synapses. As a neural network process evolves, a sequence of these attribute vectors is produced over time. The association between successive vectors may be imagined to operate by reference to a neighborhood structure implicit in the neural architecture and associated connection weights. There also may be an implicit association with some (unknown) optimization problem, or a more explicit association with a known problem and set of constraints. In the latter case, attribute assignments (neuron firings and synapse activation) can be evaluated for efficacy by transformation into a vector x, to be checked for feasibility by $x \in \mathbf{X}$. (We maintain a distinction between y and x since there may not be a one-one association between them.)

Time records identifying the quality of outcomes produced by recent firings, and identifying the frequency particular attribute assignments produce the highest quality firing outcomes, yield a basis for delaying changes in certain weight assignments and for encouraging changes in others. The concept of influence, in the form introduced in tabu search, should be considered in parallel with quality of outcomes.

Early designs to incorporate tabu search into neural networks are provided in the work of de Werra and Hertz (1989) and Beyer and Ogier (1991). These applications, which respectively treat visual pattern identification and nonconvex optimization, are reported to significantly reduce training times and increase the reliability of outcomes generated. More recent uses of tabu search to enhance the function of neural networks are provided by the studies reported in Chapter 8.

9.6 Target Analysis

Target analysis (Glover and Greenberg, 1989) links artificial intelligence and operation research perspectives to give heuristic or exact solution procedures the ability to learn what rules are best to solve a particular class of problems. Many existing solution methods have evolved by adopting, a priori, a somewhat limited characterization of appropriate rules for evaluating decisions. An illustration is provided by restricting the definition of a "best" move to be one that produces the most attractive objective function change. However, this strategy does not guarantee that the selected move will lead the search in the direction of the optimal solution. In fact, in some settings it has been shown that the merit of such a decision rule diminishes as the number of iterations increases during a solution attempt.

As seen earlier, the tabu search philosophy is to select a best admissible move (from a strategically controlled candidate list) at each iteration, interpreting best in a broad sense that goes beyond the use of objective function measures, and relies upon historical parameters to aid in composing an appropriate evaluation. Target analysis provides a means to exploit this broader view. For example, target analysis can be used to create a dynamic evaluation function that incorporates a systematic process for diversifying the search over the longer term.

A few examples of the types of questions that target analysis can be used to answer are:

(1) Which decision rule from a collection of proposed alternatives should be selected to guide the search? (In an expanded setting, how should the rules from the collection be combined? By interpreting "decision rule" broadly, this encompasses the issue of selecting a neighborhood, or a combination of neighborhoods, as the source of a move at a given stage.) Similarly, which parameter

values should be chosen to provide effective instances of the decision rules?

(2) What attributes are most relevant for determining tabu status, and what associated tabu restrictions, tabu tenures and aspiration criteria should be used?

(3) What weights should be assigned to create penalties or inducements (e.g., as a function of frequency-based memory), and what thresholds should govern their application?

(4) Which measures of quality and influence are most appropriate, and which combinations of these lead to the best results in different search phases?

(5) What features of the search trajectory disclose when to focus more strongly on intensification and when to focus more strongly on diversification? (In general, what is the best relative emphasis between intensification and diversification, and under what conditions should this emphasis change?)

Motivation for using target analysis to answer such questions is provided by contrasting target analysis with the way answers are normally determined. Typically, an experimenter begins with a set of alternative rules and decision criteria which are intended to capture the principal elements of a given method, often accompanied by ranges of associated parameters for implementing the rules. Then various combinations of options are tried, to see how each one works for a preliminary set of test problems. However, even a modest number of rules and parameters may create a large number of possibilities in combination, and there is usually little hope of testing these with any degree of thoroughness. As a result, such testing for preferred alternatives generally amounts to a process of blind groping. Where methods boast the lack of optional parameters and rules, typically it is because the experimenter has already done the advance work to settle upon a particular combination that has been hard-wired for the user, at best with some degree of adaptiveness built in, but the process that led to this hard-wiring still raises the prospect that another set of options may be preferable.

More importantly, in an adaptive memory approach, where information from the history of the search is included among the inputs that determine current choices, a trial and error testing of parameters may overlook key elements of timing and yield no insights about relationships to be exploited. Such a process affords no way to uncover or characterize the circumstances encountered during the search that may cause a given rule to perform well or badly, and consequently gives no way to anticipate the nature of rules that may perform better than those originally envisioned. Target analysis replaces this by a systematic approach to create *hindsight before the fact*, and then undertakes to "reverse engineer" the types of rules that will lead to good solutions.

9.6.1. Target Analysis Features

The main features of target analysis may briefly be sketched by viewing the approach as a five phase procedure (see Figure 9.7). *Phase* 1 of target analysis is devoted to applying existing methods to determine optimal or exceptionally high quality solutions to representative problems of a given class. In order to allow subsequent analysis to be carried out more conveniently, the problems are often selected to be relatively small, provided this can be done in a way to assure these problems will exhibit features expected to be encountered in hard problems from the class examined.

Fig. 9.7 Overview of the target analysis methodology.

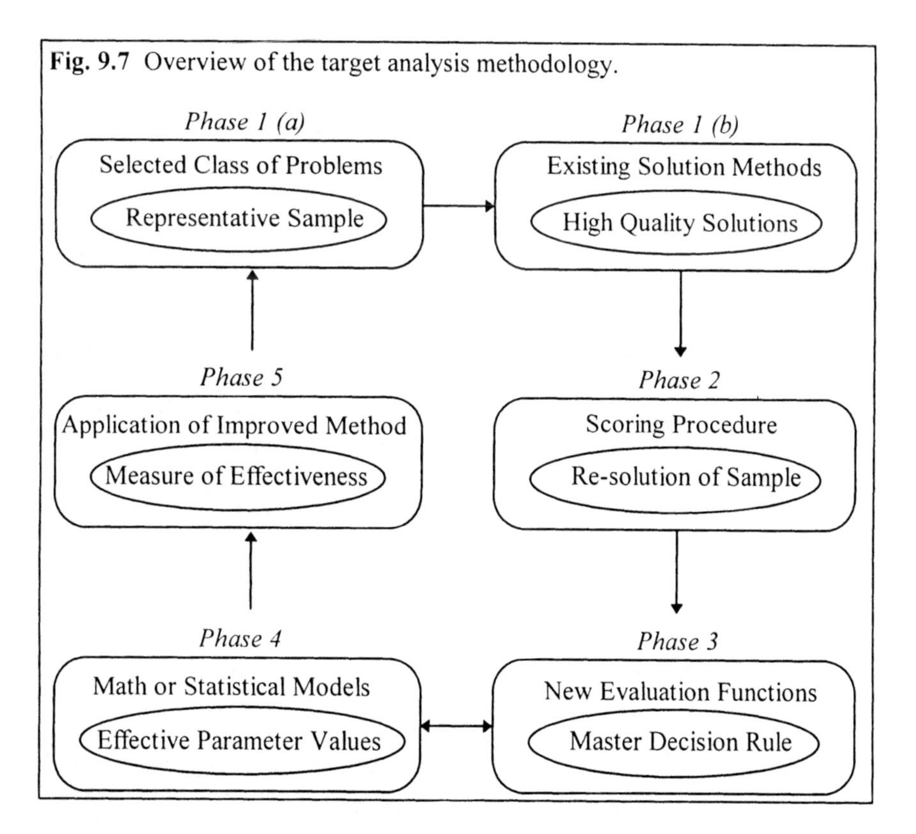

Although this phase is straightforward, the effort allotted to obtaining solutions of the specified quality will generally be somewhat greater than would be allotted during the normal operation of the existing solution procedures, in order to assure that the solutions have the quality sought. (Such an effort may be circumvented in cases where optimal solutions to a particular testbed of problems are known in advance.)

Phase 2 uses the solutions produced by Phase 1 as *targets,* which become the focus of a new set of solution passes. During these passes, each problem is solved again, this time scoring all available moves (or a high-ranking subset) on the basis of their ability to progress effectively toward the target solution. The scoring can be a simple classification, such as "good" or "bad," or it may capture more refined gradations. In the case where multiple best or near best solutions may reasonably qualify as targets, the scores may be based on the target that is "closest to" the current solution.

In some implementations, choices during Phase 2 are biased to select moves that have high scores, thereby leading to a target solution more quickly than the customary choice rules. In other implementations, the method is simply allowed to make its regular moves. In either case, the goal is to generate information during this solution effort which may be useful in inferring the solution scores. That is, the scores provide a basis for creating modified evaluations — and more generally, for creating new rules to generate such evaluations in order to more closely match them with the measures that represent "true goodness" (for reaching the targets).

In the case of tabu search intensification strategies such as elite solution recovery approaches, scores can be assigned to parameterized rules for determining the types of solutions to be saved. For example, such rules may take account of characteristics of clustering and dispersion among elite solutions. In environments where data bases can be maintained of solutions to related problems previously encountered, the scores may be assigned to rules for recovering and exploiting particular instances of these past solutions, and for determining which new solutions will be added to the data bases as additional problems are solved. (The latter step, which is part of the target analysis and not part of the solution effort can be performed "off line.") An integration of target analysis with a generalized form of sensitivity analysis for these types of applications has been developed and implemented in financial planning and industrial engineering by Glover, Mulvey and Bai (1996) and Glover, et al. (1997). Such designs are also relevant, for example, in applications of linear and nonlinear optimization based on simplex method subroutines, to identify sets of variables to provide crash-basis starting solution. (A very useful framework for recovering and exploiting past solutions is provided by Kraay and Harker (1996, 1997).)

In path relinking strategies, scores can be applied to rules for matching initiating solutions with guiding solutions. As with other types of decision rules produced by target analysis, these will preferably include reference to parameters that distinguish different problem instances. The parameter-based rules similarly can be used to select initiating and guiding solutions from pre-existing solutions pools. Tunneling applications of path relinking, which allow traversal of infeasible regions, and strategic oscillation designs that purposely drive the search into and out of such regions, are natural accompaniments for handling recovered solutions that may be infeasible.

Phase 3 constructs parameterized functions of the information generated in Phase 2, with the goal of finding values of the parameters to create a *master decision rule*. This rule is designed to choose moves that score highly, in order to achieve the goal that underlies Phase 2. It should be noted that the parameters available for constructing a master decision rule depend on the search method employed. Thus, for example, tabu search may include parameters that embody various elements of recency-based and frequency-based memory, together with measures of influence linked to different classes of attributes or to different regions from which elite solutions have been derived.

Phase 4 transforms the general design of the master decision rule into a specific design by applying a model to determine effective values for its parameters. This model can be a simple set of relationships based on intuition, or can be a more rigorous formulation based on mathematics or statistics (such as a goal programming or discriminant analysis model, or even a "connectionist" model based on neural networks).

The components of phases 2, 3 and 4 are not entirely distinct, and may be iterative. On the basis of the outcomes of these phases, the master decision rule becomes the rule that drives the solution method. In the case of tabu search, this rule may use feedback of outcomes obtained during the solution process to modify its parameters for the problem being solved.

Phase 5 concludes the process by applying the master decision rule to the original representative problems and to other problems from the chosen solution class to confirm its merit. The process can be repeated and nested to achieve further refinement.

Target analysis has an additional important function. On the basis of the information generated during its application, and particularly during its final confirmation phase, the method produces empirical frequency measures for the probabilities that choices with high evaluations will lead to an optimal (or near-optimal) solution within a certain number of steps. These decisions are not only at tactical levels but also at strategic levels, such as when to initiate alternative solution phases, and which sources of information to use for guiding these phases (e.g., whether from processes for tracking solution trajectories or for recovering and analyzing solutions). By this means, target analysis can provide inferences concerning expected solution behavior, as a supplement to classical "worst case" complexity analysis. These inferences can aid the practitioner by indicating how long to run a solution method to achieve a solution desired quality, according to specified empirical probability.

One of the useful features of target analysis is its capacity for taking advantage of human interaction. The determination of key parameters, and the rules for connecting them, can draw directly on the insight of the observer as well as on supplementary analytical techniques. The ability to derive inferences from pre-established knowledge of optimal or near optimal solutions, instead of manipulating parameters blindly (without information about the relation of decisions to targeted outcomes), can save significant investment in time and energy. The key, of course, is to coordinate the phases of solution and guided re-solution to obtain knowledge that has the greatest utility. Many potential applications of target analysis exist, and recent applications suggest the approach holds considerable promise for developing improved decision rules for difficult optimization problems.

9.6.2 Illustrative Application and Implications

An application of target analysis to a production scheduling problem (Laguna and Glover, 1993) provides a basis for illustrating some of the relevant considerations of the approach. In this study, the moves consisted of a combination of swap and insert moves, and scores were generated to identify the degree to which a move brought a solution closer to the target solution (which consisted of the best known solution before improving the method by means of target analysis). In the case of a swap move, for example, a move might improve, worsen (or, by the measure used, leave unchanged) the "positional value" of each component of the swap, and by the simplification of assigning scores of 1, 0 or -1 to each component, a move could accordingly receive a score ranging from 2 to -2. The application of target analysis then proceeded by tracking the scores of the 10 highest evaluation moves at each iteration, to determine the circumstances under which the highest evaluations tended to correspond to the highest scores. Both tabu and non-tabu moves were included in the analysis, to see whether tabu status was also appropriately defined.

At an early stage of the analysis a surprising relationship emerged. Although the scores of the highest evaluation non-tabu moves ranged across both positive and negative values, the positive values were largely associated with moves that improved the schedule while the negative values were largely associated with moves that worsened the schedule. In short, the highest evaluations were significantly more "accurate" (corresponded more closely to high scores) during phases where the objective function value of the schedule improved than during phases when it deteriorated.

A simple diversification strategy was devised to exploit this discovery. Instead of relying on the original evaluations during "disimproving phases," the strategy supplemented the evaluations over these intervals by assigning penalties to moves whose component jobs had been moved frequently in the past. The approach was initiated at a local optimum after the progress of the search began to slow (as measured by how often a new best solution was found), and was de-activated as soon as a move was executed that also was an improving move (to become re-activated the next time that all available moves were disimproving moves). The outcome was highly effective, producing new solutions that were often superior to the best previously found, especially for larger problems, and also finding the highest quality solutions more quickly.

The success of this application, in view of its clearly limited scope, provides an incentive for more thorough applications. For example, a more complete analysis would reasonably proceed by first seeking to isolate the high scoring moves during the disimproving phases and to determine how frequency-based memory and other factors could be used to identify these moves more effectively. (Designs to determine appropriate combinations of factors could include reference to considerations such as persistent voting and persistent attractiveness, as discussed in Chapter 5.) Comparisons between evaluations proposed in this manner and their associated move scores would then offer a foundation for identifying more intelligent choices. Classifications to segregate the moves based on criteria other than "improving" and "disimproving" could also be investigated. Additional relevant factors that may profitably be taken into account are examined in the illustration of the next subsection.

A Hypothetical Illustration. The following hypothetical example embodies a pattern related to the one uncovered in the scheduling application cited above. However, the pattern in this case is slightly more ambiguous, and less clearly points to options that it may be exploited.

For simplicity in this illustration, suppose that moves are scored to be either "good" or "bad." (If each move changes the value of a single 0-1 variable, for instance, a move may be judged good or bad depending on whether the assigned value is the same as in the target solution. More generally, a threshold can be used to differentiate the two classifications.)

Table 9.1 indicates the percent of time each of the five highest evaluation moves, restricting attention in this case to those that are non-tabu, receives a good score during the search history. (At a first stage of conducting the target analysis, this history could be for a single hard problem, or for a small collection of such problems.) The *Move Rank* in the table ranges from 1 to 5, corresponding to the highest evaluation move, the 2nd highest evaluation move, and so on to the 5th highest evaluation move.

The indicated percent values do not total 100 because good scores may also be assigned to moves with lower evaluations, whose ranks are not included among those shown. Also, it may be expected that some non-tabu moves will also receive good scores. (A fuller analysis would similarly show ranks and scores for these moves.)

Table 9.1 Moves throughout the search history.

Move Rank	1	2	3	4	5
Percent of moves with "good" scores	22	14	10	20	16

At first glance, the table appears to suggest that the fourth and fifth ranked moves are almost as good as the first ranked move, although the percent of moves that receive good scores is not particularly impressive for any of the ranks. Without further information, a strategy might be contemplated that allocates choices probabilistically among the first, fourth and fifth ranked moves (though such an approach would not be assured to do better than choosing the first ranked move at each step). Tables 9.2 and 9.3 below provide more useful information about choices that are potentially favorable, by dividing the iterations into improving and disimproving phases as in the scheduling study previously discussed.

Table 9.2 Moves during improving phases.

Move Rank	1	2	3	4	5
Percent of moves with "good" scores	34	21	9	14	7

Table 9.3 Moves during disimproving phases.

Move Rank	1	2	3	4	5
Percent of moves with "good" scores	8	7	11	26	25

These tables are based on a hypothetical situation where improving and disimproving moves are roughly equal in number, so that the percent values shown in Table 9.1 are the average of the corresponding values in Tables 9.2 and 9.3. (For definiteness, moves that do not change the problem objective function may be assumed to be included in the improving phase, though a better analysis might treat them separately.)

The foregoing outcomes to an extent resemble those found in the scheduling study, though with a lower success rate for the highest evaluation improving moves. Clearly Tables 9.2 and 9.3 give information that is more exploitable than the information in Table 9.1. According to these latter tables, it would be preferable to focus more strongly on choosing one of the two highest evaluation moves during an improving phase, and one of the fourth or fifth highest evaluation moves during a disimproving phase. This conclusion is still weak in several respects, however, and we examine considerations that may lead to doing better.

Refining the Analysis. The approach of assigning scores to moves, as illustrated in Tables 9.1, 9.2 and 9.3, disregards the fact that some solution attributes (such as assignments of values to particular 0-1 variables) may be fairly easy to choose "correctly," while others may be somewhat harder. Separate tables of the type illustrated should therefore be created for easy and hard attributes (as determined by how readily their evaluations lead to choices that would generate the target solution), since the preferred rules for evaluating moves may well differ depending on the types of attributes the moves contain. Likewise, an effective strategy may require that easy and hard attributes become the focus of different search phases. The question therefore arises as to how to identify such attributes.

As a first approximation, we may consider an easy attribute to be one that often generates an evaluation that keeps it out of the solution if it belongs out, or that brings it into the solution if it belongs in. A hard attribute behaves oppositely. Thus, a comparison between frequency-based memory and move scores gives a straightforward way to differentiate these types of attributes. Both residence and transition frequencies are relevant, though residence measures are probably more usually appropriate. For example, an attribute that belongs to the current solution a high percentage of the time, and that also belongs to the target solution, would evidently qualify as easy. On the other hand, the number of times the attribute is accepted or rejected from the current solution may sometimes be less meaningful than how long it stays in or out. The fact that residence and transition frequencies are characteristically used in tabu search makes them conveniently available to assist in differentiations that can improve the effectiveness of target analysis.

9.6.3 Conditional Dependencies Among Attributes

Tables 9.1, 9.2 and 9.3 suggest that the search process that produced them is relatively unlikely to find the target solution. Even during improving phases, the highest evaluation move is almost twice as likely to be bad as good. However, this analysis is limited, and discloses a limitation of the tables themselves. In spite of first appearances, it is entirely possible that these tables could be produced by a search process that successfully obtains the target solution (by a rule that chooses a highest evaluation move at each step). The reason is that the relation between scores and evaluations may change over time. While there may be fairly long intervals where choices are made poorly, there may be other shorter intervals where the choices are made more effectively – until eventually one of these shorter intervals succeeds in bringing all of the proper attributes into the solution.

Such behavior is likely to occur in situations where correctly choosing some attributes may pave the way for correctly choosing others. The interdependence of easy and hard attributes previously discussed is carried a step farther by these conditional relationships, because an attribute that at one point deserves to be classified hard may later deserve to be classified easy, once the appropriate foundations are laid.

Instead of simply generating tables that summarize results over long periods of the search history, therefore, it can be important to look for blocks of iterations where the success rate of choosing good moves may differ appreciably from the success rate overall. These blocks provide clues about intermediate solution compositions that may transform hard attributes into easy ones, and thus about preferred sequences for introducing attributes that may exploit conditional dependencies. The natural step then is to see which additional types of evaluation information may independently lead to identifying such sequences.

A simple instance of this type of effect occurs where the likelihood that a given attribute will correctly be selected (to enter or leave the solution) depends roughly on the number of attributes that already correctly belong to the solution. In such situations, the appropriate way to determine a "best choice" is therefore also likely to depend on this number of attributes correctly in solution. Even though such information will not generally be known during the search, it may be possible to estimate it and adjust the move evaluations accordingly. Such relationships, as well as the more general ones previously indicated, are therefore worth ferreting out by target analysis.

9.6.4 Differentiating Among Targets

In describing the steps of target analysis, it has already been noted that scores should not always be rigidly determined by only one specific target, but may account for alternative targets, and in general may be determined by the target that is closest to the current solution (by a metric that depends on the context). Acknowledging that there may be more than one good solution that is worth finding, such a differentiation among targets can prove useful. Yet even in the case where a particular solution is uniquely the one to be sought (as where its quality may be significantly better than that of all others known), alternative targets may be still be valuable to consider in the role of intermediate solutions, and may provide a springboard to finding additional solutions that are better. Making reference to intermediate targets is another way of accounting for the fact that conditional dependencies may exist among the attributes, as previously discussed. However, such dependencies in some cases may be more easily exploited by explicitly seeking constructions at particular stages that may progressively lead to a final destination.

Some elite solutions may provide better targets than others because they are easier to obtain — completely apart from developing strategies to reach ultimate targets by means of intermediate ones. However, some care is needed in making the decision to focus on such easier targets as a basis for developing choice rules. As in the study of Lokketangen and Glover (1997b), it may be that focusing instead on the harder targets will yield rules that likewise cause the easier targets to be found more readily, and these rules may apply to a wider spectrum of problems than those derived by focusing on easier targets.

9.6.5 Generating Rules by Optimization Models

Target analysis can use optimization models to generate decision rules by finding weights for various decision criteria to create a composite (master) rule. To illustrate, let G and B respectively denote index sets for good moves and bad moves, as determined from move scores, as in the classification embodied in Tables 9.1, 9.2, and 9.3. Incorporate the values of the different decision criteria in a vector A_i for $i \in G$ and $i \in B$; i.e., the jth component a_{ij} of A_i is the value assigned to move i by the decision criterion j. These components need not be the result of rules, but can simply correspond to data considered relevant to constructing rules. In the tabu search setting, such data can include elements of recency-based and frequency-based memory. Then we may consider a master rule which is created by applying a weight vector w to each vector A_i to produce a composite decision value $A_i w = \sum_j a_{ij} w_j$.

An ambitious objective is to find a vector w that yields

$$A_i w > 0 \qquad \text{for } i \in G$$
$$A_i w \leq 0 \qquad \text{for } i \in B$$

If such a weight vector w could be found, then all good moves would have higher evaluations by the composite criterion than all bad moves, which of course is normally too much to ask. A step toward formulating a more reasonable goal is as follows. Let $G(iter)$ and $B(iter)$ identify the sets G and B for a given iteration *iter*. Then an alternative objective is to find a w so that, at each such iteration, at least one $i \in G(iter)$ would yield

$$A_i w > A_k w \qquad \text{for all } k \in B(iter)$$

or equivalently

$$Max\{A_i w\text{: } i \in G(iter)\} > Max\{A_k w\text{: } k \in B(iter)\}$$

This outcome would insure that a highest evaluation move by the composite criterion will always be a good move. Naturally, this latter goal is still too optimistic. Nevertheless, it is possible to devise goal programming models (related to LP discriminant analysis models) that can be used to approximate this goal. A model of this type has proved to be effective for devising branching rules to solve a problem of refueling nuclear reactors (Glover, Klingman and Phillips, 1990).

A variety of opportunities exist for going farther in such strategies. For example, issues of creating nonlinear and discontinuous functions to achieve better master rules can be addressed by using trial functions to transform components of A_i vectors into new components, guided by LP sensitivity and postoptimality analysis. Target analysis ideas previously indicated can also be useful in this quest.

The range of possibilities for taking advantage of target analysis is considerable, and for the most part only the most rudimentary applications of this learning approach have been initiated. The successes of these applications make further exploration of this approach attractive.

9.7 Discussion Questions and Exercises

1. Develop a tabu search mechanism to control the temperature in simulated annealing, such that the resulting annealing schedule is nonmonotonic.

2. Suggest a form of a GA/TS hybrid in which recency and frequency information are used to create the right balance between search intensification and diversification.

3. Establish a connection between the path relinking approach within tabu search and scatter search.

4. Devise a probabilistic tabu search Multistart strategy as an alternative to GRASP and other "randomized restarting" approaches, which has the following features:

   (a) Candidate lists using designs such as described in Chapter 3 are used, instead of relying solely on an evaluation threshold. Also, probabilities are assigned to choosing moves from the candidate list that employ biases to reflect move evaluations, instead of using a uniform distribution.

   (b) Attributes of elite solutions found during improvement phases are given inducements to be selected during subsequent construction phases (using frequency-based memory to give increased weight to attributes that occur more frequently in these solutions).

   (c) Critical event memory is used — as by conceiving a re-start as an extreme form of strategic oscillation, where an entire solution is "de-constructed" before initiating a reconstruction. (Also, consider options for identifying critical events as discussed in Chapter 2.)

   (d) The Proximate Optimality Principle is applied by allowing the solution to be improved at intermediate stages of construction.

   Given the inherent ability to diversify based on the use of critical event memory, and the additional source of probabilistic bias derived by reference to elite solutions, would you anticipate that candidate list sizes might be made smaller than in randomized restarting approaches that also use candidate lists?

5. Indicate how you might apply target analysis to uncover, and take advantage of, relationships of the form suggested by the Principle of Persistent Voting and the Principle of Persistent Attractiveness, as described in Chapter 5.

6. What types of statistical analysis might be applied to relate scores to evaluations in target analysis? (What roles could be played by clustering and discrimination analysis?) How might such analysis be used to create alternative decision rules? Could thresholds that divide moves into classes other than "improving" and "disimproving" be uncovered in this way? (What types of classes might you be prompted to look for?)

7. Consider a type of neighborhood whose solutions include many that are infeasible (e.g., as in the case of an adjacent extreme point neighborhood for 0-1 integer programming problems, where not all extreme points satisfy integer feasibility conditions). In such a neighborhood, even simple evaluations may reasonably have two dimensions, one based on changes in the objective function value and one based on changes in infeasibility. Suppose each dimension is divided into two improving intervals and two non-improving intervals, designated "greatly improving" and "slightly improving," and "greatly non-improving" and "slightly non-improving." (For simplicity, treat a move that begins and ends with a feasible solution to be classed as "slightly improving" relative to the infeasibility dimension.)

   The relevance of the improving/non-improving (disimproving) distinction found by target analysis for scheduling problems suggests the relevance of a more general distinction in the present case. If any moves currently exist that fall within the improving intervals both for the objective function and for infeasiblity, then the relationship between move scores and evaluations may be different than if no such moves currently exist. From this standpoint, create a 4×4 matrix, whose rows correspond to the 2 improving and 2 non-improving intervals for the objective function, and whose columns correspond to the 2 improving and 2 non-improving intervals for infeasibility, progressing in the order from greatly improving, to slightly improving, to slightly non-improving, to greatly non-improving. Then the previously indicated distinction corresponds to subdividing the matrix into the two cases where the first quadrant (consisting of the first two rows and first two columns) is respectively non-empty or empty, relative to moves available on the current iteration. If each cell of the matrix records, say, the scores of the 5 highest evaluation moves that belong to this classification, then a basis would be provided to identify possible correlations between scores and evaluations in this context.

   What other ways of subdividing the matrix may be useful for uncovering additional distinctions? (For example, one way might be to subdivide the case where the where the first quadrant is empty into subcases where the second and third quadrants are both non-empty, versus where one or both may be empty. What is the meaning of such subdivisions?)

9.8 Appendix: Illustrative Version of Nonlinear Scatter Search

Nonlinear optimization applications are conveniently suited to the use of specialized forms of scatter search. We sketch a version of such an approach that selects each new set of reference points from trial points and reference points previously created. Tabu restrictions and aspiration criteria are applied to determine the admissible composition of reference points at each stage.

By narrowing the focus of the method to generating new points only from previous ones, this implementation of a scatter search and tabu search combination constitutes an intensification strategy. To be most effective, therefore, the approach should feed back into a process that creates new reference points at a global level (as from a separate ongoing tabu search approach at this level). Such an integration with a higher level process provides a natural basis for parallelization.

9.8.1 The Method in Overview

Let M be a mapping that transforms x into a trial solution $M(x)$ for the specified problem to be solved. A criterion of merit, such as an objective function, is then applied to the solution $M(x)$ to give the evaluation of x. For example, the mapping M can incorporate a post-processing operation such as generalized rounding to produce integer trial solutions from fractional vectors. M should be designed to exploit gradient information (exact or approximate) where possible. As in customary scatter search approaches, it is generally important to use mappings that transform solutions into improved solutions such as local optima, which then provide the x vectors that continue the process.

The method relies on parameters that define the sizes of particular sets and intervals for exploring search lines. Following the intensification theme, the values of these constants can be kept relatively small, which conveniently limits calibration possibilities. In addition, we indicate specific values of the constants as tentative "defaults," to remove the burden of selecting these values where the setting does not immediately prompt different choices.

Distance measures used by the method can be based on a variety of norms, of which the L1 (absolute value) norm is generally convenient.

9.8.2 Steps of the Nonlinear Scatter Search

1. Initialization and Classification. The method starts with a set S of initial trial points, from which all other sets derive. The generation of such points can be based on a strategy for creating *diverse subsets*, subsequently described.

Reference points, initially selected from the more attractive elements of S, are used to generate new elements that replace previous members of S. Elite elements (those of especially high quality) become part of a special historical collection that also contributes to the choice of current reference points. The following sets are referenced and updated at each iteration.

$H =$ a set of elite *historical generators*, consisting of h best (highest evaluation) points generated historically throughout the search (e.g., $h = 20$). On the first iteration, H consists of h best elements of S.

$T =$ a set of *tabu generators*, composing a subset of H currently excluded from consideration. Initially T is empty.

H^* = a set of *selected historical generators*, consisting of h^* best elements of H - T (e.g., $h^* = 4$).

S^* = a set of *selected current generators*, consisting of s^* best elements of S (e.g., $s^* = 5$).

On the first iteration, S^* and H^* overlap (H^* is a subset of S^*). Subsequently they overlap only as current elements of S have higher evaluations than some of those in H^* (when H changes to incorporate elements of S).

R = a set of *current reference points*, defined to equal the union of H^* and S^*. The number r of elements in R, which equals $h^* + s^*$ when H^* and S^* are disjoint, is given a lower bound r^* which is slightly less than this sum, to apply when H^* and S^* overlap. The bound is achieved by adjusting s^* upward so that the number of elements in S^* - H^* will be large enough to compensate for elements shared with H^*. (For example, s^* will be increased to r^* on the first iteration, when H^* lies wholly within S^*.)

2. Outline of an Iteration. In overview, the method creates centers of gravity for S^* and for each s^* - 1 element subset of S^*. These newly created trial points are paired with points of R to create search lines for generating additional trial points. Elements of S^* are also paired with each other to create new search lines and trial points in the same manner. Finally, diverse subsets, $D(S^*)$ and $D(S\text{-}S^*)$ of S^* and $S - S^*$ are identified by rules indicated in Step 4 below, and used to create additional search lines by pairing their elements with those of H^* and R, respectively.

3. Evaluation and Updates. Throughout an iteration, each trail point with an evaluation superior to the mean evaluation of the points in S^* initiates an intensified search on the line that produced it. At the same time, elements that are candidates to compose S^* for the next iteration are identified, together with candidates for inclusion in H. Finally, T is updated by reference to the tabu restrictions and aspiration criteria, and the method repeats.

To provide a complete description of the method, we now indicate the rules for generating the trial points and for creating the tabu restrictions defined over elements of H.

9.8.3 Generating Trial Points

Trial points are produced by the following steps.

Step 1. Denote the elements of S^* by $x(k)$, $k = 1,\ldots, s^*$. To begin, generate the center of gravity of S^*

$$y(0) = \frac{\sum x(k)}{s^*}$$

and the centers of gravity of each s^* - 1 element subset of S^*

$$y(k) = y(0) + \frac{y(0) - x(k)}{\left(s^* - 1\right)}, \qquad k = 1, \ldots, s^*.$$

These points provide the initial set of trial points.

Step 2. Let $x(s^* + 1), \ldots, x(r)$ denote the reference points in R that are not in S^*, and let (x,y) successively refer to each of the r pairs $(x(k), y(k))$ for $k = 1, \ldots, s^*$ and $(x(k), y(0))$ for $k = s^* + 1, \ldots, r$ (thereby matching each reference point in R with one of the trial points created in Step 1).

Consider the line through x and y given the representation $z(w) = x + w(y-x)$, where w is a weight. Subject to refinements indicated later, we begin by restricting attention to the points $z(1/2)$, $z(1/3)$, $z(2/3)$, $z(-1/3)$. (By definition, the point $z(1/2)$ is the midpoint of the line segment joining x and y, $z(1/3)$ and $z(2/3)$ are the associated exterior points on the x side and y side of the segment.)

The computational effort of transforming such points into solutions by the mapping M influences the number that will be examined as trial points. E.g., when it is expensive to examine all points, preference is given to examining points closer to x or y, depending on which of x and y has a higher evaluation. (Alternative approaches for generating such points are provided subsequently.)

Step 3. Apply the rule of Step 2 to create trial points from the pairs (x,y), $x \neq y$, both of whose members are in S^*.

Step 4. Let $D(X)$ denote a (small) diverse subset of the set X, for $X = S^*$ and $X = S - S^*$, generated as follows. The first element is selected to maximize its distance from the center of gravity of X. The second element is selected to maximize its distance from the first element, and the third element is selected to maximize its minimum distance from the first and second elements. The rule of Step 2 is then applied to the pairs (x, y) in the two cross product sets $H^* \times D(S^*)$ and $R \times D(S - S^*)$, to generate search lines and trial points.

Step 5. *Intensification Component* — During the execution of Steps 2 to 4 identify the highest evaluation trial point z on each search line (including trial points generated in Step 1 as candidates). If the evaluation of z is higher than the average of the evaluations over S^*, then intensify the search around z by generating two new trial points at the midpoints of the segments that join z to its nearest neighbors already examined. (If z is one of the exterior points $z(-1/3)$ or $z(4/3)$, add the point $z(-2/3)$ or $z(5/3)$, respectively, to serve as its "missing" nearest neighbor.) Then if a newly created point has an evaluation that exceeds the evaluation of z by an appropriate threshold increment the process is repeated, designating the best new point as z, and generating analogous missing neighbors as necessary. In the special case where z is the point $y(0)$ from Step 1, this step should be executed only for the line joining z to the best element of H^*. (For problems at risk of unbounded optimality, a safeguarding limit may be imposed on the number of improving iterations.)

Refinements of the foregoing procedure will be indicated after describing the rules for maintaining and updating T.

9.8.4 Tabu Restrictions and Aspiration Criteria

The implementation of tabu restrictions and aspiration criteria to guide successive iterations of the scatter search process likewise relies on parameters which will be expressed as small constants. First, tabu restrictions are introduced by transferring elements of the historical set H to the set T, and thus temporarily excluding them from membership in H^*. Such a restriction is initiated as a result of belonging to H^* (i.e., as a result of using an element to generate trial points).

Two types of aspiration criteria are applied to postpone or mitigate this restriction, based on the use of a small iteration parameter i^* (e.g., $i^* = 3$). First, when an element x is incorporated into H^*, a "minimum level of exposure" is provided by allowing x to remain in H^* for i^* iterations. At this point x automatically becomes tabu and enters T unless it satisfies the second aspiration criterion. This criterion permits x to remain in H^* for up to i^* iterations after generating a trial point z which entered the set S^* (i.e., where x belonged to a pair (x, y) that produced a search line containing z). Once a limit of i^*h^* consecutive iterations is accumulated in H^*, then x enters T regardless of the operation of the second aspiration criterion.

Progressive Tabu Restrictions. The first time an element becomes tabu, its tabu status remains in force for the same duration (i^* iterations) that the element is initially guarded from becoming tabu. However, this tabu restriction is progressively increased by incrementing the tabu tenure of an element x each additional time that x enters T, i.e., increasing by 1 by the number of iterations it must remain tabu. The tabu tenure for x returns to a value of i^* whenever x contributes to generating a trial point that qualifies for entry into H. (This is a variant of the second aspiration criterion.) In addition, the maximum duration that x remains in T is limited to h/i^* iterations.

Diversification. The final component of the tabu search process assures that all elements of H eventually have an opportunity to belong to H^*. Let *count*(x) denote the number of consecutive iterations an element x of H does not belong to either H^* or T, and let *frequency*(x) be the total cumulative number of iterations (not necessarily consecutive) x has been a member of H^*. Then the stipulation that H^* consists of h^* best elements of H-T is modified by interpreting "best" to include reference to the values of *count*(x) and *frequency*(x). Whenever the maximum *count*(x) value exceeds a threshold of i^*h^*, then each x yielding this maximum *count*(x) value receives first priority for inclusion in H^*, breaking ties to favor the smallest accompanying value of *frequency*(x) (and automatically favoring higher ranking elements of H subject to this).

9.8.5 Refinements

There are a variety of possibilities for refining the method. Evidently, for example, the centers of gravity produced in Step 1 can be weighted, as by reference to evaluations of the elements in S^*. An important refinement is to create a "parallel scatter" approach by simultaneously generating several different solution streams, each with its own sets S and H. Periodically (e.g., as the new admissions to H become infrequent) the elements from the union of the different sets H may be reallocated to form new sets. This may be done by applying a rule to generate diverse subsets, as elaborated in item (6) below, to provide a more selective foundation for generating new points than relying on simple randomization. A parallel processing approach coordinated in this way also can result in smaller preferred sizes for the sets H than otherwise would be appropriate. Specific alternatives for refinement are as follows.

(1) The previously indicated values of w used to generate trial points (1/2, 1/3, 2/3, etc.) are initial estimates of appropriate values, and assume the behavior of the evaluation function in the region examined is unknown. If knowledge of this behavior exists or can be inferred from the search then it should be used to select other values for w.

(2) During an intensified line search, a preferred strategy is to iteratively construct a quadratic or cubic to "fit" the evaluations for successive trial points, and then to calculate the location of a hypothetical maximum. (Variants of Golden Section and Fibonacci search may similarly be applied.) In conjunction with such an approach, a line on which to intensify the search may alternately be chosen according to the progression of evaluations of its trial points, rather than according to the quality of its "best" trial point. If there is a degree of uniformity in the parameters used to create a fit by polynomial interpolation, then the resulting function can be used without recalculation to give first estimates of appropriate w values in (1). However, this issue is complicated slightly in the case where the mapping $M(x)$ is designed to create a locally optimal point which then replaces x. Depending on the effort to apply such an $M(x)$, the method may

proceed for selected numbers of iterations without generating local optima, alternated with generating local optima from the solutions that result.

(3) The method may be improved by recognizing *approximate duplications*. When a point considered for inclusion in S^* or H^* has an evaluation close to that of another point in the set, then the distance between these two points may be examined. If this distance is sufficiently small the point with the lower evaluation may be discarded. The degree of separation that characterizes "sufficiently small" depends on the problem context. The choice of this degree can be a heuristic device to prevent points from clustering too tightly.

(4) When a new element x is added to H, it is generally preferable to pair it with other elements of H, to assure that all representative search lines generated by members of H have been explored. (If H is not large, as in a parallel approach, then the total computation will not be greatly affected, particularly since the composition of H changes less frequently than that of S.)

(5) A simple form of diversification may be achieved by restricting the number of trial points from any single search line permitted to be included in S^* or H. The goal is related to that of avoiding "approximate duplications," as proposed in (3), but requires less effort to check. For example, a rule may be imposed to prevent more than two points from entering S^* or H from any given line. These points can also be required to be separated by a minimum distance, as in (3), or else only one of them may be selected.

(6) Finally, more ambitiously, the definition of a diverse subset may be usefully broadened. Given a parent set X and a seed point u (where u need not belong to X), a diverse subset $D(X,u)$ of X may be characterized as a subset generated by the following rules: (a) the first element is selected from x to maximize its distance from u; (b) each subsequent element is selected from X to maximize its minimum distance from all elements previously chosen (excluding u); (c) ties are broken by maximizing the minimum distance to the last k-1 elements chosen, and then the last k-2 elements, etc., where k is the total number of elements currently selected to compose $D(X,u)$. Note that when a selected point x is first mapped into a local optimum, this new point immediately replaces x in defining the current composition of $D(x,u)$.

The definition of (6) can become computationally expensive to implement if $D(X,u)$ is allowed to grow beyond a modest size. In this case, a relaxation of the definition can be used to significantly reduce the computation: select a small value of v, and choose each successive point to maximize the minimum distance from the v elements most recently added to $D(X,u)$, breaking ties by maximizing the minimum distance from the first v (or $v/2$ elements) elements added to $D(X,u)$. (Ties that remain may be broken by randomization.) This rule is similar to the use of a simple form of short term memory in tabu search. Another approximation is to assign weights to elements of $D(X,u)$, so that the weight of each element is somewhat greater than that of the element added immediately before it. Then using these weights, the new element to add is selected to maximize the weighted sum of distances from the other elements.

The notion of a diverse subset has a broader relevance. As noted earlier, the initial set S to start the process may be generated by reference to this concept. Thus, for example, once a seed point u is selected, remaining points may be chosen by the diversification approach, retaining for S a subset consisting of the more attractive elements found. Since this subset will establish different diversification criteria than the full set, the process may be repeated relative to the points retained to create a more appropriate set of diverse elements. In this second phase (and in the first, if enough information is available), new points that do not qualify as acceptably attractive may be immediately screened out.

9.8.6 Parallel Processing Implementation.

Two variants of this approach are relevant to initiate a *parallel scatter search* procedure. One is to generate several initial diverse subsets, each to provide a different starting S. The other is instead to make each element of the initial diverse subset a "first member" of a separate S. Then these single element sets can be expanded by a series of steps that augment each one again by a single element, where the criterion for selecting this element is first to maximize the minimum distance from centers of gravity of other sets, and subject to this to maximize the minimum distance from elements of the set under consideration. (Maximizing weighted sums of distances alternately can replace the max-min criterion.) Such an approach assumes that points representative of the universe to be considered are already specified, in order to give candidates to augment the sets under consideration. When this is not the case, candidates may be generated by conditional random sampling as the process evolves.

Once the parallel scatter approach is thus initiated, the process for generating diverse subsets also can be periodically applied in order to reallocate elements from the current H sets to form new H sets, as previously noted. Within the operation of the scatter search method itself, related options are to select additional diverse subsets of the elements generated, including subsets of H and S^*, and to define H^* to be a diverse subset of $H - T$.

10 NEGLECTED TABU SEARCH STRATEGIES

This chapter briefly reviews several key strategies in tabu search that are often neglected (especially in beginning studies), but which are important for producing the best results.

Our purpose is to call attention to the relevance of particular elements that are mutually reinforcing, but which are not always discussed "side by side" in the literature, and which deserve special emphasis. In addition, observations about useful directions for future research are included.

A comment regarding implementation: first steps do not have to include the most sophisticated variants of the ideas discussed in the following sections, but the difference between "some inclusion" and "no inclusion" can be significant. Implementations that incorporate simple instances of these ideas will often disclose the manner in which refined implementations can lead to improved performance.

The material that follows is largely, though not exclusively, a review of particular ideas described in preceding chapters, but which are brought together to provide an additional perspective on how they interrelate. Later sections of this chapter introduce observations that have not previously been made.

10.1 Candidate List Strategies

Efficiency and quality can be greatly affected by using intelligent procedures for isolating effective candidate moves, rather than trying to evaluate every possible move in a current neighborhood of alternatives. This is particularly true when such a neighborhood is large or expensive to examine. The gains to be achieved by using candidate lists have been widely documented, yet many TS studies overlook their relevance.

Careful organization in applying candidate lists, as by saving evaluations from previous iterations and updating them efficiently, can also be valuable for reducing overall effort. Time saved in these ways allows a chance to devote more time to higher level features of the search.

While the basic theme of candidate lists is straightforward, there are some subtleties in the ways candidate list strategies may be used. Considerable benefit can result by being aware of fundamental candidate list approaches, such as the *Subdivision Strategy*, the *Aspiration Plus Strategy*, the *Elite Candidate List Strategy*, the *Bounded Change Strategy* and the *Sequential Fan Strategy* (as discussed in Chapter 3).

An effective integration of a candidate list strategy with the rest of a tabu search method will typically benefit by using TS memory designs to facilitate functions to be performed by the candidate lists. This applies especially to the use of frequency based memory. A major mistake of some TS implementations, whether or not they make use of candidate lists, is to consider only the use of recency based memory. Frequency based memory — which itself takes different forms in intensification phases and diversification phases — can not only have a dramatic impact on the performance of the search in general but also can often yield gains in the design of candidate list procedures.

10.2 Probabilistic Tabu Search

Several studies have suggested the value of a probabilistic version of TS, where evaluations (including reference to tabu status) are translated into probabilities of selection, strongly skewed to favor higher evaluations. Findings from such studies support the notion that probabilities may partly substitute for certain functions of memory (hence reduce the amount of memory needed) but also suggest that probabilities may have a role in counteracting "noise" in the evaluations.

In well designed TS implementations, the gains of probabilistic TS over deterministic TS are chiefly in accelerating the rate at which good solutions are discovered in earlier stages of search. Overall, some settings appear more exploitable by probabilistic TS and others appear more exploitable by deterministic TS. The most effective forms of each type of approach depend strongly on identifying and implementing TS strategies of the type described below.

10.3 Intensification Approaches

Intensification strategies, which are based on recording and exploiting elite solutions or, characteristically, specific features of these solutions, have proved very useful in a variety of applications. Some of the relevant forms of such strategies and considerations for implementing them are as follows.

10.3.1 Restarting with Elite Solutions

The simplest intensification approach is the strategy of recovering elite solutions in some order, each time the search progress slows, and then using these solutions as a basis for re-initiating the search. The list of solutions that are candidates to be recovered is generally limited in size, often in the range of 20 to 40 (although in parallel processing applications the number is characteristically somewhat larger). The size chosen for the list in serial TS applications also corresponds roughly to the number of solution recoveries anticipated to be done during the search, and so may be less or more depending on the setting. When an elite solution is recovered from the list, it is removed, and new elite solutions are allowed to replace less attractive previous solutions — usually dropping the worst of the current list members. However, if a new elite solution is highly similar to a solution presently recorded, instead of replacing the current worst solution, the new solution will compete directly with its similar counterpart to determine which solution is saved.

This approach has been applied very effectively in job shop and flow shop scheduling, in vehicle routing, and in telecommunication design problems. One of the best approaches for scheduling applications keeps the old TS memory associated with the solution, but makes sure the first new move away from this solution goes to a different neighbor than the one visited after encountering this solution the first time. Another effective variant does not bother to save the old TS memory, but uses a probabilistic TS choice design.

The most common strategy is to go through the list from best to worst, but in some cases it has worked even better to go through the list in the other direction. In this approach, it appears effective to allow two passes of the list. On the first pass, when a new elite solution is found that falls below the quality of the solution currently recovered, but which is still better than the worst already examined on the list, the method still adds the new solution to the list and displaces the worst solution. Then a second pass, after reaching the top of the list, recovers any added solutions not previously recovered.

10.3.2 Frequency of Elite Solutions

Another primary intensification strategy is to examine elite solutions to determine the frequency in which particular solution attributes occur (where the frequency is typically weighted by the quality of the solutions in which the attributes are found).

This strategy was originally formulated in the context of identifying "consistent" and "strongly determined" variables — where, loosely speaking, consistent variables are those more frequently found in elite solutions, while strongly determined variables are those that would cause the greatest disruption by changing their values (as sometimes approximately measured by weighting the frequencies based on solution quality). The idea is to isolate the variables that qualify as more consistent and strongly determined (according to varying thresholds), and then to generate new solutions that give these variables their "preferred values." This can be done either by rebuilding new solutions in a multistart approach or by modifying the choice rules of an ongoing solution effort to favor the inclusion of these value assignments.

Keeping track of the frequency that elite solutions include particular attributes (such as edges of tours, assignments of elements to positions, narrow ranges of values taken on by variables, etc.) and then favoring the inclusion of the highest frequency elements, effectively allows the search to concentrate on finding the best supporting uses and values of other elements. A simple variant is to "lock in" a small subset of the most attractive attributes (value assignments) — allowing this subset to change over time or on different passes.

A Relevant Concern: In the approach that starts from a current (good) solution, and tries to bring in favored elements, it is important to introduce an element that yields a best outcome from among the current contenders (where, as always, best is defined to encompass considerations that are not solely restricted to objective function changes). If an attractive alternative move shows up during this process, which does not involve bringing in one of these elements, aspiration criteria may determine whether such a move should be taken instead. Under circumstances where the outcome of such a move appears sufficiently promising, the approach may be discontinued and allowed to enter an improving phase (reflecting a decision that enough intensification has been applied, and it is time to return to searching by customary means).

Intensification of this form makes it possible to determine what percent of "good attributes" from prior solutions should be included in the solution currently generated. It also gives information about which subsets of these attributes should go together, since it is preferable not to choose attributes during this process that cause the solution to deteriorate compared to other choices. This type of intensification strategy has proved highly effective in the settings of vehicle routing and zero-one mixed integer optimization.

10.3.3 Memory and Intensification

It is clearly somewhat more dangerous to hold elements "in" solution than to hold them "out" (considering that a solution normally is composed of a small fraction of available elements — as where a tree contains only a fraction of the edges of a graph). However, there is an important exception, previously intimated. As part of a longer term intensification strategy, elements may be selected very judiciously to be "locked in" on the basis of having occurred with high frequency in the best solutions found. In that case, choosing different mutually compatible (and mutually reinforcing) sets to lock in can be quite helpful. This creates a *combinatorial implosion* effect (opposite to a combinatorial explosion effect) that shrinks the solution space to a point where best solutions over the reduced space are likely to be found more readily.

The key to this type of intensification strategy naturally is to select an appropriate set of elements to lock in, but the chances appear empirically to be quite high that some subset of those with high frequencies in earlier best solutions will be correct. Varying the subsets selected gives a significant likelihood of picking a good one. (More than one subset can be correct, because different subsets can still be part of the same complete set.) Aspiration criteria make it possible to drop elements that are supposedly locked in, to give this approach more flexibility.

10.3.4 Relevance of Clustering for Intensification

A search process over a complex space is likely to produce clusters of elite solutions, where one group of solutions gives high frequencies for one set of attributes and another group gives high frequencies for a different set. It is important to recognize this situation when it arises. Otherwise there is a danger that an intensification strategy may try to compel a solution to include attributes that work against each other. This is particularly true in a strategy that seeks to generate a solution by incorporating a collection of attributes "all at once," rather than using a step by step evaluation process that is reapplied at each move through a neighborhood space. (Stepping through a neighborhood has the disadvantage of being slower, but may compensate by being more selective. Experimentation to determine the circumstances under which each of these alternative intensification approaches may be preferable would be quite valuable.)

A strategy that incorporates a block of attributes together may yield benefits by varying both the size and composition of the subsets of high frequency "attractive" attributes, even if these attributes are derived from solutions that lie in a common cluster, since the truly best solutions may not include them all. Threshold based forms of logical restructuring, as discussed in Chapter 3, may additionally lead to identifying elements to integrate into solutions that may not necessarily belong to solutions previously encountered. The vocabulary building theme becomes important in this connection. The relevance of clustering analysis for logical restructuring and vocabulary building is reinforced by the use of a related conditional analysis, which is examined subsequently in Section 10.6.

10.4 Diversification Approaches

Diversification processes in tabu search are sometimes applied in ways that limit their effectiveness, due to overlooking the fact that diversification is not just "random" or "impulsive," but depends on a purposeful blend of memory and strategy. As noted in Chapter 3, recency and frequency based memory are both relevant for diversification. Historically, these ideas stem in part from proposals for exploiting surrogate constraint methods. In this setting, the impetus is not simply to achieve diversification, but to derive appropriate weights in order to assure that evaluations will lead to solutions that satisfy required conditions (see Chapter 9). Accordingly, it is important to account for elements such as how often, to what extent, and how recently, particular constraints have been violated, in order to determine weights that produce more effective valuations.

The implicit *learning effects* that underlie such uses of recency, frequency and influence are analogous to those that motivate the procedures used for diversification (and intensification) in tabu search. Early strategic oscillation approaches exploited this principle by driving the search to various depths outside (and inside) feasibility boundaries, and then employing evaluations and directional search to move toward preferred regions.

In the same way that these early strategies bring diversification and intensification together as part of a continuously modulated process, it is important to stress that these two elements should be interwoven in general. A common mistake in many TS implementations is to apply diversification without regard for intensification. "Pure" diversification strategies are appropriate for truly long term strategies, but over the intermediate term, diversification is generally more effective if it is applied by heeding information that is also incorporated in intensification strategies. In fact, intensification by itself can sometimes cause a form of diversification, because intensifying over part of the space allows a broader search of the rest of the space. A few relevant concerns are as follows.

10.4.1 Diversification and Intensification Links

A simple and natural diversification approach is to keep track of the frequency that attributes occur in non-elite solutions, as opposed to solutions encountered in general, and then to periodically discourage the incorporation of attributes that have modest to high frequencies (giving greater penalties to larger frequencies). The reference to non-elite solutions tends to avoid penalizing attributes that would be encouraged by an intensification strategy.

More generally, for a "first level" balance, an Intermediate Term Memory matrix may be used, where the high frequency items in elite solutions are not penalized by the long term values, but may even be encouraged. The tradeoffs involved in establishing the degree of encouragement, or the degree of reducing the penalties, represents an area where a small amount of preliminary testing can be valuable. This applies as well to picking thresholds to identify high frequency items. (Simple guesses about appropriate parameter values can often yield benefits, and tests of such initial guesses can build an understanding that leads to increasingly effective strategies.)

By extension, if an element has never or rarely been in a solution generated, then it should be given a higher evaluation for being incorporated in a diversification approach if it was "almost chosen" in the past but didn't make the grade. This observation has not been widely heeded, but is not difficult to implement, and is relevant to intensification strategies as well. The relevant concerns are illustrated in the discussion of "Persistent Attractiveness" and "Persistent Voting" in Chapter 7.

10.4.2 Implicit Conflict and the Importance of Interactions

Current evaluations also should not be disregarded while diversification influences are activated. Otherwise, a diversification process may bring elements together that conflict with each other, make it harder rather than easier to find improved solutions.

For example, a design that gives high penalties to a wide range of elements, without considering interactions, may drive the solution to avoid good combinations of elements. Consequently, diversification — especially in intermediate term phases — should be carried out for a limited number of steps, accompanied by watching for and sidestepping situations where indiscriminately applying penalties would create incompatibilities or severe deterioration of quality. To repeat the theme: even in diversification, attention to quality is important. And as in "medical remedies," sometimes small doses are better than large ones. Larger doses (i.e., more radical departures from previous solutions) which are normally applied less frequently, can still benefit by coordinating the elements of quality and change.

10.4.3 Reactive Tabu Search

An approach called Reactive Tabu Search (RTS) developed by Battiti and Tecchiolli (1992, 1994a) and briefly reviewed in Chapter 8, deserves additional consideration as a way to achieve a useful blend of intensification and diversification. RTS incorporates hashing in a highly effective manner to generate attributes that are very nearly able to differentiate among distinct solutions. That is, very few solutions contain the same hashed attribute, applying techniques such as those discussed in Chapter 7. Accompanying this, Battiti and Tecchiolli use an automated tabu tenure, which begins with the value of 1 (preventing a hashed attribute from being reinstated if this attribute gives the "signature" of the solution visited on the immediately preceding step). This tenure is then increased if examination shows the method is possibly cycling, as indicated by periodically generating solutions that produce the same hashed attribute.

The tabu tenure, which is the same for all attributes, is increased exponentially when repetitions are encountered, and decreased gradually when repetitions disappear. Under circumstances where the search nevertheless encounters an excessive number of repetitions within a given span (i.e., where a moving frequency measure exceeds a certain threshold), a diversification step is activated, which consists of making a number of random moves proportional to a moving average of the cycle length.

The reported successes of this approach invite further investigations of its underlying ideas and related variants. As a potential bases for generating such variants, attributes created by hashing may be viewed as *fine grain* attributes, which give them the ability to distinguish among different solutions. By contrast, "standard" solution attributes, which are the raw material for hashing, may be viewed as *coarse grain* attributes, since each may be contained in (and hence provide a signature for) many different solutions. Experience has shown that tabu restrictions based on coarse grain attributes are often advantageous for giving increased vigor to the search. (There can exist a variety of ways of defining and exploiting attributes, particularly at coarser levels, which complicates the issue somewhat.) This raises the question of when particular degrees of granularity are more effective than others.

It seems reasonable to suspect that fine grain attributes may yield greater benefits if they are activated in the vicinity of elite solutions, thereby allowing the search to scour "high quality terrain" more minutely. This effect may also be achieved by reducing tabu tenures for coarse grain attributes — or basing tabu restrictions on attribute conjunctions — and using more specialized aspiration criteria. Closer scouring of critical regions can also be brought about by using strongly focused candidate list strategies, such as a sequential fan candidate list strategy. (Empirical comparisons of such alternatives to hashing clearly would be of interest.) On the other hand, as documented by Nonobe and Ibaraki (1997), the use of "extra coarse grain" attributes (those that prohibit larger numbers of moves when embodied in tabu restrictions) can prove advantageous for solving large problems over a broadly defined problem domain.

Another type of alternative to hashing also exists, which is to create new attributes by processes that are not so uniform as hashing. A potential drawback of hashing is its inability to distinguish the relative importance (and appropriate influence) of the attributes that it seeks to map into others that are fine grained. A potential way to overcome this drawback is to make use of vocabulary building (Chapter 7) and of conditional analysis (Section 10.6).

10.4.4 Ejection Chain Approaches

Ejection chain methods provide an implicit blending of diversification and intensification by generating compound moves out of simpler components. Such approaches have provided breakthroughs in handling certain types of tough problems, particularly those related to optimization over graphs (as discussed in Chapters 7 and 9). TS memory structures can be used at two levels with ejection chains, both at a simple internal level which operates primarily as a bookkeeping function to avoid duplicate patterns (as complex moves are woven from simpler ones), and at an external level that guides the successively generated compound moves to go beyond conditions of local optimality. So far ejection chain studies have chiefly focused on internal as opposed to external levels of control. New discoveries may be expected by broadening this focus. In addition, opportunities exist in many settings for applications of ejection chains where such forms of compound neighborhoods have so far remained uninvestigated.

10.5 Strategic Oscillation

A considerable amount has been written on strategic oscillation and its advantages. However, one of the uses of this approach that is frequently overlooked involves the idea of oscillating among alternative choice rules and neighborhoods. As stressed in Chapter 4, an important aspect of strategic oscillation is the fact that there naturally arise different types of moves and choice rules that are appropriate for negotiating different regions and different directions of search. Thus, for example, there are many constructive methods in graph and scheduling problems, but strategic oscillation further leads to the creation of complementary "destructive methods" which can operate together with their constructive counterparts. Different criteria emerge as relevant for selecting a move to take on a constructive step versus one to take on a destructive step. Similarly, different criteria apply according to whether moves are chosen within a feasible region or outside a feasible region (and whether the search is moving toward or away from a feasibility boundary).

The variation among moves and evaluations introduces an inherent vitality into the search that provides one of the sources underlying the success of strategic oscillation approaches. This reinforces the motivation to apply strategic oscillation to the choice of moves and evaluation criteria themselves, selecting moves from a pool of possibilities according to rules for transitioning from one choice to another. In general, instead of picking a single rule, a process of invoking multiple rules provides a range of alternatives that run all the way from "strong diversification" to "strong intensification."

This form of oscillation has much greater scope than may at first be apparent, because it invokes the possibility of simultaneously integrating decision rules and neighborhoods, rather than only visiting them in a strategically determined sequence. Basic considerations of such integration are described in Section 1.8 of Chapter 1, and are elaborated in the discussion of Persistent Voting in Chapter 7.

Such concepts are beginning to find counterparts in investigations being launched by the computer science community. The "agent" terminology is being invoked in such applications to characterize different choice mechanisms and neighborhoods as representing different agents. Relying on this representation, different agents then are assigned to work on (or "attack") the problem serially or in parallel. The CS community has begun to look upon this as a significant innovation — unaware of the literature where such ideas were introduced a decade or more ago — and the potential richness and variation of these ideas still seems not to be fully recognized. For example, there have not yet been any studies that consider the idea of "strategically sequencing" rules and neighborhoods, let alone those that envision the notion of parametric integration. The further incorporation of adaptive memory structures to enhance the application of such concepts also lies somewhat outside the purview of most current CS proposals. At the same time, however, TS research has also neglected to conduct empirical investigations of the broader possibilities. This is clearly an area that deserves fuller study.

10.6 Clustering and Conditional Analysis

To reinforce the theme of identifying opportunities for future research, we provide an illustration to clarify the relevance of clustering and conditional analysis, particularly as a basis for intensification and diversification strategies in tabu research.

An Example: Suppose 40 elite solutions have been saved during the search, and each solution is characterized as a vector $\boldsymbol{x}$ of zero-one variables x_j, for $j \in N = \{1,\ldots,n\}$. Assume the variables that receive positive values in at least one of the elite solutions are indexed x_1 to x_{30}. (Commonly in such circumstances, n may be expected to be somewhat larger than the number of positive valued variables, e.g., in this case, reasonable values may be $n = 100$ or 1000.)

For simplicity, we restrict attention to a simple weighted measure of consistency which is given by the frequency that the variables x_1 to x_{30} receive the value 1 in these elite solutions. (We temporarily disregard weightings based on solution quality and other aspects of "strongly determined" assignments.) Specifically, assume the frequency measures are as shown in Table 10.1.

Since each of x_1 to x_{15} receives a value of 1 in 24 of the 40 solutions, these variables tie for giving "most frequent" assignments. An intensification strategy that favors the inclusion of some number of such assignments would give equal bias to introducing each of x_1 to x_{15} at the value 1. (Such a bias would typically be administrated either by creating modified evaluations or by incorporating probabilities based on such evaluations.)

Table 10.1 Frequency measures.

Variables $x_j = 1$	Number of Solutions
x_1 to x_{15}	24
x_{16} to x_{20}	21
x_{21} to x_{25}	17
x_{26} to x_{30}	12

To illustrate the relevance of clustering, suppose the collection of 40 elite solutions can be partitioned into two subsets of 20 solutions each, whose characteristics are summarized in Table 10.2.

Table 10.2 Frequency measures for two subsets.

Subset 1 (20 solutions)		Subset 2 (20 solutions)	
Variables $x_j = 1$	No. of Solutions	Variables $x_j = 1$	No. of Solutions
x_{11} to x_{15}	20	x_{16} to x_{20}	20
x_{21} to x_{25}	16	x_6 to x_{10}	16
x_1 to x_5	12	x_1 to x_5	12
x_6 to x_{10}	8	x_{26} to x_{30}	8
x_{26} to x_{30}	4	x_{11} to x_{15}	4
x_{16} to x_{20}	1	x_{21} to x_{25}	1

A very different picture now emerges. The variables x_1 to x_{15} no longer appear to deserve equal status as "most favored" variables. Treating them with equal status may be a useful source of diversification, as opposed to intensification, but the clustered data provide more useful information for diversification concerns as well. In short, clustering gives a relevant contextual basis for determining the variables (and combinations of variables) that should be given special treatment.

10.6.1 Conditional Relationships

To go a step beyond the level of differentiation provided by cluster analysis, it is useful to sharpen the focus by referring explicitly to interactions among variables. Such interactions can often be identified in a very straightforward way, and can form a basis for more effective clustering. In many types of problems, the number of value assignments (or the number of "critical attributes") needed to specify a solution is relatively small compared to the total number of problem variables. (For example, in routing, distribution and telecommunication applications, the number of links contained in feasible constructions is typically a small fraction of those contained in the underlying graph.) Using a 0-1 variable representation of possibilities, it is not unreasonable in such cases to create a *cross reference* matrix, which identifies variables (or coded attributes) that simultaneously receive a value of 1 in a specific collection of elite solutions.

To illustrate, suppose the index set $P = \{1,...,p\}$ identifies the variables x_j that receive a value of 1 in at least r solutions from the collection of elite solutions under consideration. (Apart from other strategic considerations, the parameter r can also be used to control the size of p, since larger values of r result in smaller values of p.)

Then create a $p \times p$ symmetric matrix **M** whose entries m_{ij} identify the number of solutions in which x_i and x_j are both 1. (Thus, row M_i of **M** represents the sum of the solution vectors in which $x_i = 1$, restricted to components x_j for $j \in P$.) The value m_{ii} identifies the total number of elite solutions in which $x_i = 1$, and the value m_{ij}/m_{ii} represents the "conditional probability" that $x_j = 1$ in this subset of solutions. Because p can be controlled to be of modest size, as by the choice of r and the number of solutions admitted to the elite set, the matrix **M** is not generally highly expensive to create or maintain.

By means of the conditional probability interpretation, the entries of **M** give a basis for a variety of analyses and choice rules for incorporating preferred attributes into new solutions. Once an assignment $x_j = 1$ is made in a solution currently under consideration (which may be either partly or completely constructed), an updated conditional matrix **M** can be created by restricting attention to elite solution vectors for which $x_j = 1$. (Restricted updates of this form can also be used for look-ahead purposes.) Weighted versions of **M**, whose entries additionally reflect the quality of solutions in which specific assignments occur, likewise can be used.

Critical event memory as described in Chapter 2, provides a convenient mechanism to maintain appropriate variation when conditional influences are taken into account. The "critical solutions" associated with such memory in the present case are simply those constituting a selected subset of elite solutions. Frequency measures for value assignments can be obtained by summing these solution vectors for problems with 0-1 representations and the critical event control mechanisms can then assure assignments are chosen to generate solutions that differ from those of previous elite solutions.

Conditional analysis, independent of such memory structures, can also be a useful foundation for generating solution fragments to be exploited by vocabulary building processes, as discussed in Chapter 7.

10.7 Referent-Domain Optimization

Referent-domain optimization is based on introducing one or more optimization models to strategically restructure the problem or neighborhood, accompanied by auxiliary heuristic or algorithmic process to map the solutions back to the original problem space. The optimization models are characteristically devised to embody selected heuristic goals (e.g., of intensification, diversification or both), within the context of particular classes of problems.

There are several ways to control the problem environment as a basis for applying referent-domain optimization. A natural control method is to limit the structure and range of parameters that define a neighborhood (or the rules used to navigate through a neighborhood), and to create an optimization model that operates under these restricted conditions.

The examples that follow assume the approach starts from a current trial solution, which may or may not be feasible. The steps described yield a new solution, and then the step is repeated, using tabu search as a master guiding strategy to avoid cycling, and to incorporate intensification and diversification.

Example 1. A heuristic selects k variables to change values, holding other variables constant. An exact method determines the (conditionally) optimal new values of the k selected variables.

Example 2. A heuristic identifies a set of restrictive bounds that bracket the values of the variables in the current trial solution (where the bounds may compel some variables to take on a single value). An exact method determines an optimal solution to the problem as modified to include these bounds. (This approach is related to the one in Observation 7 of Chapter 6 for the cut search method.)

Example 3. A heuristic selects a restructured and exploitable region around the current solution to search for an alternative solution. An exact method finds the best solution in this region.

Example 4. For add/drop neighborhoods, a heuristic chooses k elements to add (or to drop). For example, the heuristic may operate by both adding and dropping k specific elements, as in k-opt moves for the TSP or k-swap moves for graph bipartitioning that add and drop k nodes. Then, attention is restricted to consider only the subset of elements added or the subset of elements dropped (and further restricted in the case of a bipartitioning problem to just one of the two sets). Then an exact method identifies the remaining k elements to drop (or to add), that will complete the move optimally.

Example 5. A heuristic chooses a modified problem formulation, that also admits the current trial solution as a trial solution. (For example, the heuristic may relax some part of the formulation and/or restrict another part.) An exact method then finds an optimal solution to the modified formulation. An illustration occurs where a two phase exact algorithm first finds an optimal solution to a relaxed portion of the problem, and then finds an optimal solution to a restricted portion. Finally, a small part of the feasible region of the original problem close to or encompassing this latter solution is identified, and an exact solution method finds an optimal solution in this region.

Example 6. The use of specially constructed neighborhoods (and aggregations or partitions of integer variables) permits the application of mixed integer programming (MIP) models to identify the best options from all moves of depth at most k (or from associated collections of at most k variables). When k is sufficiently small, such MIP models can be quite tractable, and produce moves considerably more powerful than those provided by lower level heuristics.

Example 7. In problems with graph-related structures, the imposition of directionality or non-looping conditions gives a basis for devising generalized shortest path (or dynamic programming) models to generate moves that are optimal over a significant subclass of possibilities. This type of approach gives rise to a combinatorial leverage phenomenon, where a low order effort (e.g., linear or quadratic) can yield solutions that dominate exponential numbers of alternatives. (See, e.g., Glover (1992), Rego (1996a), Punnen and Glover (1997).)

Example 8. A broadly applicable control strategy, similar to that of a relaxation procedure but more flexible, is to create a proxy model that resembles the original problem of interest, and which is easier to solve. Such an approach must be accompanied with a method to transform the solution to the proxy model into a trial solution for the original problem. A version of such an approach, which also induces special structure into the proxy model, can be patterned after layered surrogate/Lagrangean decomposition strategies for mixed integer optimization.

Referent-domain optimization can also be applied in conjunction with target analysis (Chapter 9) to create more effective solution strategies. In this case, a first stage learning model, based on controlled solution attempts, identifies a set of desired properties of good solutions, together with target solutions (or target regions) that embody these properties. Then a second stage model is devised to generate neighborhoods and choice rules to take advantage of the outcomes of the learning model. Useful strategic possibilities are created by basing these two models on a proxy model for referent-domain optimization, to structure the outcomes so that they may be treated by one of the control methods indicated in the foregoing examples.

10.8 Discussion Questions and Exercises

1. Which strategies do you consider most appropriate for balancing intensification and diversification in a beginning study? In a more advanced study? How would you bridge the implementation aspects of these studies? (Formulate your answer to account for two specific types of applications. Would your answer differ for these two applications, or would your strategies apply to both of them?)

2. Indicate criteria you might use to decide when to employ fine grained attributes and when to employ coarse grain attributes as a foundation for recency-based and frequency-based memory.

3. What kinds of applications do you think might benefit most usefully from applying clustering and conditional analysis as described in Section 10.6?

4. Which aspects of logical restructuring may be relevant to referent domain optimization? Can threshold based restructuring be used to create restricted problems to solve optimally in this approach?

5. Taking account of considerations stressed in this chapter, design a “first stage” tabu search approach for a problem of your choice. What additional types of problems do you think would be especially susceptible to being treated by your approach?

6. For each subsection of this chapter, create a specific exercise that draws on one or more of the concepts of the subsection in order to "answer" (or "solve") the exercise.

7. What research areas suggested by observations of previous chapters do you think would offer the greatest opportunities for devising improved TS procedures? Which of these areas do you think might be explored most easily? (How would you restrict the exploration initially to make it additionally manageable?)

REFERENCES

Aarts, E. H. L. and J. H. Korst (1989a) *Simulated Annealing and Boltzmann Machines: A Stochastic Approach to Combinatorial Optimization and Neural Computing*, Wiley, New York.

Aarts, E. H. L. and J. H. M. Korst (1989b), "Boltzmann Machines for Traveling Salesman Problems," *European Journal of Operational Research,* Vol. 39, No. 1, pp. 79-95.

Ackley, D. (1987) "*A Connectionist Model for Genetic Hillclimbing*," Kluwer, Dordrecht. Academic Publishers.

Aiex, R.M., S. L. Martins, C. C. Ribeiro and N. R. Rodriguez (1996) "Asynchronous parallel strategies for tabu search applied to the partitioning of VLSI circuits," Research report, Department of Computer Science, Catholic University of Rio de Janeiro.

Allen, F. (1993) "Security Design Special Issue: Introduction," *Financial Management*, pp. 32-33.

Al-Mahmeed, A. S. (1996) "Tabu Search Combination and Integration," *Meta-Heuristics: Theory and Applications,* I. H. Osman and J. P. Kelly (eds.), Kluwer Academic Publishers, pp. 319-330.

Andreatta, A. A. and C. C. Ribeiro (1994) "A Graph Partitioning Heuristic for the Parallel Pseudo-Exhaustive Logical Test of VLSI Combinational Circuits," *Annals of Operations Research,* Vol. 50, pp. 1-36.

Anzellotti, G., R. Battiti, I. Lazzizzera, G. Soncini, A. Zorat, A. Sartori, G. Tecchiolli and P. Lee (1995) "A Highly Parallel Chip for Triggering Applications with Inductive Learning Based on the Reactive Tabu Search," *International Journal of Modern Physics C,* Vol. 6, No. 4, pp. 555-560.

Bäck, T., F. Hoffmeister and H. Schwefel (1991) "A Survey of Evolution Strategies," *Proceedings of the Fourth International Conference on Genetic Algorithms*, R. Belew and L. Booker (eds.), pp. 2-9.

Balas, E. (1971) "The Intersection Cut – A New Cutting Plane for Integer Programming," *Operations Research,* Vol. 19, pp. 19-39.

Balas, E. and C. Martin (1980) "Pivot and Complement — A Heuristic for 0-1 Programming," *Management Science,* Vol. 26, No. 1, pp. 86-96.

Barnes, J. W. and M. Laguna (1993) "Solving the Multiple-Machine Weighted Flow Time Problem Using Tabu Search," *IIE Transactions*, Vol. 25, No. 2, pp. 121-128.

Barnes, J. W. and W. B. Carlton (1995) "Solving the Vehicle Routing Problem with Time Windows Using Reactive Tabu Search," presented at the Fall INFORMS National Meeting in New Orleans, Louisiana.

Battiti, R., P. Lee, A. Sartori, and G. Tecchiolli (1994a) "TOTEM: A Digital Processor for Neural Networks and Reactive Tabu Search," *Fourth International Conference on Microelectronics for Neural Networks and Fuzzy Systems*, MICRONEURO 94, pp. 17-25.

Battiti, R., P. Lee, A. Sartori, and G. Tecchiolli (1994b) "Combinatorial Optimization for Neural Nets: RTS Algorithm and Silicon," Technical Report no. 9406-04, IRST, Trento, IT.

Battiti, R., A. Sartori, G. Tecchiolli and A. Zorat (1995) "Neural Compression: An Integrated Approach to eeg Signals, *International Workshop on Applications of Neural Networks in Telecommunications* (IWANNT 95), J. Alsoector, R. Goodman and T. X. Brown, (eds.), Stockholm, Sweden, pp. 210-217.

Battiti, R. and G. Tecchiolli (1992) "Parallel Biased Search for Combinatorial Optimization: Genetic Algorithms and Tabu Search," *Microprocessor and Microsystems*, Vol. 16, pp. 351-367.

Battiti, R. and G. Tecchiolli (1994a) "The Reactive Tabu Search," *ORSA Journal on Computing*, Vol. 6, No. 2, pp. 126-140.

Battiti, R. and G. Tecchiolli (1994b) "Simulated Annealing and Tabu Search in the Long Run: A Comparison on QAP Tasks," *Computers and Mathematics with Applications*, Vol. 28, No. 6, pp. 1-8.

Battiti, R. and G. Tecchiolli (1995a) "The Continuous Reactive Tabu Search: Blending Combinatorial Optimization and Stochastic Search for Global Optimization," to appear in *Annals of Operations Research.*

Battiti, R. and G. Tecchiolli (1995b) "Training Neural Nets with the Reactive Tabu Search," *IEEE Transactions on Neural Networks*, Vol. 6, No. 5, pp. 1185-1200.

Battiti, R. and G. Tecchiolli (1995c) "Local Search with Memory: Benchmarking RTS," *Operations Research Spektrum*, Vol. 17, No. 2/3, pp. 67-86.

Beasley, J. E. (1996) *Advances in Linear and Integer Programming,* Oxford Lecture Series in Mathematics and Its Applications, Vol. 4, Oxford University Press.

Berge, C. (1992) *Theory of Graphs and its Applications,* Methuen, London.

Berger, A. J. and J. M. Mulvey (1997) "The Home Account Advisor, Asset and Liability for Individual Investors," to appear in *World Wide Asset and Liability Modeling*, W. T. Ziemba and J. M. Mulvey (eds.), Cambridge University Press.

Beyer, D. and R. Ogier (1991) "Tabu Learning: A Neural Network Search Method for Solving Nonconvex Optimization Problems," *Proceedings of the International Conference in Neural Networks,* IEEE and INNS, Singapore.

Blacewicz, J., P. Hawryluk and R. Walkowiak (1993) "Using a Tabu Search Approach for Solving the Two-dimensional Irregular Cutting Problem," *Annals of Operations Research*, Vol. 41, pp. 313-325.

Blacewicz, J. and R. Walkowiak (1995) "A Local Search Approach for Two-dimensional Irregular Cutting," *OR Spektrum*, Vol. 17, pp. 93-98.

Brumelle, S., D. Granot, M. Halme, and I. Vertinsky (1996) "A Tabu Search Algorithm for Finding a Good Forest Harvest Schedule Satisfying Green-Up Constraints," Forest Economics and Policy Analysis Research Unit, The University of British Columbia Vancouver, BC, Canada.

Carlton, W. B. and J. W. Barnes (1995) "A Note on Hashing Functions and Tabu Search Algorithms," to appear in *European Journal of Operational Research.*

Chakrapani, J. and J. Skorin-Kapov (1992) "Connectionist Approaches to the Quadratic Assignment Problem," *Computers and Operations Research*, Vol. 19, No. 3/4, pp. 287-295.

Chakrapani, J. and J. Skorin-Kapov (1993a) "Massively Parallel Tabu Search for the Quadratic Assignment Problem," *Annals of Operations Research*, Vol. 41, pp. 327-341.

Chakrapani, J. and J. Skorin-Kapov (1993b) "Connection Machine Implementation of a Tabu Search Algorithm for the Traveling Salesman Problem," *Journal of Computing and Information Technology*, Vol. 1, No. 1, pp. 29-36.

Chakrapani, J. and J. Skorin-Kapov (1995) "Mapping Tasks to Processors to Minimize Communication Time in a Multiprocessor System," *The Impact of Emerging Technologies of Computer Science and Operations Research*, Kluwer Academic Publishers, pp. 45-64.

Charon, I. and O. Hudry (1996) "Mixing Different Components of Metaheuristics," *Meta-Heuristics: Theory and Applications,* I. H. Osman and J. P. Kelly (eds.), Kluwer Academic Publishers, pp. 589-604.

Chiang, W-C. and P. Kouvelis (1994a) "Simulated Annealing and Tabu Search Approaches for Unidirectional Flowpath Design for Automated Guided Vehicle Systems," *Annals of Operation Research*, Vol. 50.

Chiang, W-C. and P. Kouvelis (1994b) "An Improved Tabu Search Heuristic for Solving Facility Layout Design Problems," to appear in *International Journal of Production Research.*

Chiang, W-C. and R. Russell (1995) "A Reactive Tabu Search Metaheuristic for the Vehicle Routing Problem with Time Windows," Working Paper, Department of Quantitative Methods and Management Information Systems.

Committee on the Next Decade of Operations Research (Condor 1988) "Operations Research: The Next Decade," *Operations Research,* Vol. 36, pp. 619-637.

Consiglio, A. and S. A. Zenios (1997a) "Optimal Design of Callable Bonds Using Tabu Search," to appear in *Journal of Economic Dynamics and Control*, Vol. 21.

Consiglio, A. and S. A. Zenios (1997b) "High-performance Computing for the Computer Aided Design of Financial Products," to appear in *High Performance Computing: Technology and Applications,* J. S. Kowalik and L. Grandinetti (eds.), Kluwer Academic Publishers, Dordrecht.

Costamagna, F., A. Fanni and G. Giacinto (1995) "Tabu Search for the Optimization of B-ISDN Telecommunication Networks," Tech. Report No. 60, Dept. of Electrical Eng., University of Cagliari.

Crainic, T. G., M. Gendreau and J. M. Farvolden (1996) "Simplex-based Tabu Search for the Multicommodity Capacitated Fixed Charge Network Design Problem," publication CRT-96-8, Centre de recherche sur les transports, University of Montreal.

Crainic, T. G., M. Toulouse and M. Gendreau (1997) "Toward a Taxonomy of Parallel Tabu Search Heuristics," *INFORMS Journal on Computing,* Vol. 9, No. 1, pp. 61-72.

Crowston, W. B., F. Glover, G. L. Thompson and J. D. Trawick (1963) "Probabilistic and Parametric Learning Combinations of Local Job Shop Scheduling Rules," ONR Research Memorandum No. 117, GSIA, Carnegie-Mellon University, Pittsburgh, PA.

Cung, V-T, T. Mautor, P. Michelo and A. Tavares (1997) "A Scatter Search Based Approach for the Quadratic Assignment Problem," Laboratoire PR*i*SM-CNRS URA 1525, University of Versailles.

Daduna, J. R. and S. Voss (1995) "Practical Experiences in Schedule Synchronization," *Lecture Notes in Economics and Mathematical Systems,* No. 430, pp. 39-55.

Dammeyer, F. and S. Voss (1993) "Dynamic Tabu List Management Using the Reverse Elimination Method," *Annals of Operations Research*, Vol. 41, pp. 31-46.

Davis, L. (1989) "Adapting Operator Probabilities in Genetic Algorithms," *Proceedings of the Third International Conference on Genetic Algorithms*, Morgan Kaufmann, San Mateo, CA, pp. 61-69.

Dell'Amico, M. and F. Maffioli (1996) "A New Tabu Search Approach to the 0-1 Equicut Problem," *Meta-Heuristics: Theory and Applications*, I. H. Osman and J. P. Kelly (eds.), Kluwer Academic Publishers, pp. 361-378.

Dell'Amico, M. and M. Trubian (1993) "Applying Tabu Search to the Job-Shop Scheduling Problem," *Annals of Operations Research*, Vol. 41, 231-252.

Dell'Amico, M. and M. Trubian (1997) "Solution of Large Weighted Equicut Problems," to appear in *European Journal of Operational Research.*

Denardo, E. V. and B. L. Fox (1979) "Shortest Route Methods: 2 Group Knapsacks, Expanded Networks and Branch and Bound," *Operations Research,* Vol. 27, pp. 548-566.

Dodin, B., A. A. Elimam and E. Rolland (1995) "Tabu Search in Audit Scheduling," to appear in *European Journal of Operational Research.*

Domschke, W., P. Forst and S. Voss (1992) "Tabu Search Techniques for the Quadratic Semi-Assignment Problem," *New Directions for Operations Research in Manufacturing*, G. Fandel, T. Gulledge and A. Jones (eds.), Springer, Berlin, pp. 389-405.

Dorndorf, U. and E. Pesch (1994) "Fast Clustering Algorithms," *ORSA Journal on Computing*, Vol. 6, pp. 141-153.

Eiben, A. E., P-E Raue and Z. Ruttkay (1994) "Genetic Algorithms with Multi-Parent Recombination," *Proceedings of the Third International Conference on Parallel Problem Solving from Nature* (PPSN), Y. Davidor, H-P Schwefel and R. Manner (eds.), New York: Springer-Verlag, pp. 78-87.

Eschelman, L. J. and J. D. Schaffer (1992) "Real-Coded Genetic Algorithms and Interval-Schemata," Technical Report, Phillips Laboratories.

Fanni, A., G. Giacinto and M. Marchesi (1996) "Tabu Search for Continuous Optimization of Electromagnetic Structures," *Int. Workshop on Optimization and Inverse Problems in Electromagnetism*, Brno, Czech Republic.

Feo, T. and M. G. C. Resende (1989) "A Probabilistic Heuristic for a Computationally Difficult Set Covering Problem," *Operations Research Letters,* Vol. 8, pp. 67-71.

Feo, T. and M. G. C. Resende (1995) "Greedy Randomized Adaptive Search Procedures," *Journal of Global Optimization,* Vol. 2, pp. 1-27.

Fiechter, C-N (1994) "A Parallel Tabu Search Algorithm for Large Traveling Salesman Problems," *Discrete Applied Mathematics,* Vol. 51, pp. 243-267.

Fisher, H. and G. L. Thompson (1963) "Probabilistic Learning Combinations of Local Job-Shop Scheduling Rules," *Industrial Scheduling*, J. F. Muth and G. L. Thompson (eds.), Prentice-Hall, pp. 225-251.

Fleurent, C. and F. Glover (1996) "Improved Constructive Multistart Strategies for the Quadratic Assignment Problem," Graduate School of Business, University of Colorado at Boulder.

Fleurent, C., F. Glover, P. Michelon and Z. Valli (1996) "A Scatter Search Approach for Unconstrained Continuous Optimization," *Proceedings of the 1996 IEEE International Conference on Evolutionary Computation*, pp. 643-648.

Ford, L. R. and D. R. Fulkerson (1962) *Flows in Networks,* Princeton University Press.

Fréville, A. and G. Plateau (1986) "Heuristics and Reduction Methods for Multiple Constraint 0-1 Linear Programming Problems," *European Journal of Operational Research*, Vol. 24, pp. 206-215.

Fréville, A. and G. Plateau (1993) "An Exact Search for the Solution of the Surrogate Dual of the 0-1 Bidimensional Knapsack Problem," *European Journal of Operational Research,* Vol. 68, pp. 413-421.

Gendreau, M., A. Hertz and G. Laporte (1994) "A Tabu Search Heuristic for the Vehicle Routing Problem," *Management Science,* Vol. 40, No. 10, pp. 1276-1290.

Gendreau, M., G. Laporte and J-Y, Potvin (1995) "Metaheuristics for the Vehicle Routing Problem," *Local Search Algorithms*, J. K. Lenstra and E. H. L. Aarts (eds.), John Wiley & Sons, Chichester.

Gendreau, M., P. Soriano, and L. Salvail (1993) "Solving the Maximum Clique Problem Using a Tabu Search Approach," *Annals of Operations Research,* Vol. 41, pp. 385-403.

Glover, F. (1963) "Parametric Combinations of Local Job Shop Rules," Chapter IV, ONR Research Memorandum No. 117, GSIA, Carnegie-Mellon University, Pittsburgh, PA.

Glover, F. (1964) "A Bound Escalation Method for the Solution of Integer Linear Programs," *Cahiers de Recherche Opérationelle*, Vol. 6, No. 3, pp. 131-168.

Glover, F. (1965) "A Multiphase-dual Algorithm for the Zero-one Integer Programming Problem," *Operations Research,* Vol. 13, pp. 879-919.

Glover, F. (1966) "An Algorithm for Solving the Linear Integer Programming Problem over a Finite Additive Group, with Extensions to Solving General and Certain Non-linear Integer Programs," CRC 66-29, University of California at Berkeley.

Glover, F. (1967) "A Pseudo Primal-Dual Integer Programming Algorithm," *Journal of Research of the National Bureau of Standards — B, Mathematics and Mathematical Physics*, Vol. 71B, No. 4, pp. 187-195.

Glover, F. (1968) "Surrogate Constraints," *Operations Research*, Vol. 16, pp. 741-749.

Glover, F. (1969) "Integer Programming Over a Finite Additive Group," *SIAM Journal on Control,* Vol. 7, pp. 213-231.

Glover, F. (1972) "Cut Search Methods in Integer Programming," *Mathematical Programming,* Vol. 3, No. 1, pp. 86-100.

Glover, F. (1975a) "Polyhedral Annexation in Mixed Integer and Combinatorial Programming," *Mathematical Programming,* Vol. 8, pp. 161-188.

Glover, F. (1975b) "Surrogate Constraint Duality in Mathematical Programming," *Operations Research,* Vol. 23, pp. 434-451.

Glover, F. (1977) "Heuristics for Integer Programming Using Surrogate Constraints," *Decision Sciences,* Vol. 8, pp. 156-166.

Glover, F. (1986) "Future Paths for Integer Programming and Links to Artificial Intelligence," *Computers and Operations Research,* Vol. 13, pp. 533-549.

Glover, F. (1989) "Tabu Search — Part I," *ORSA Journal on Computing,* Vol. 1, pp. 190-206.

Glover, F. (1992) "Ejection Chains, Reference Structures and Alternating Path Methods for Traveling Salesman Problems," University of Colorado. Shortened version published in *Discrete Applied Mathematics,* Vol. 65, pp. 223-253, 1996.

Glover, F. (1994a) "Tabu Search for Nonlinear and Parametric Optimization (with Links to Genetic Algorithms)," *Discrete Applied Mathematics,* Vol. 49, pp. 231-255.

Glover, F. (1994b) "Genetic Algorithms and Scatter Search: Unsuspected Potentials," *Statistics and Computing,* Vol. 4, pp. 131-140.

Glover, F. (1995a) "Scatter Search and Star-Paths: Beyond the Genetic Metaphor," *OR Spektrum,* Vol. 17, pp. 125-137.

Glover, F. (1995b) "Tabu Search Fundamentals and Uses," Graduate School of Business, University of Colorado. Condensed version published in *Mathematical Programming: State of the Art,* J. Birge and K. Murty (eds.), pp. 64-92, 1994.

Glover, F. (1995c) "Tabu Thresholding: Improved Search by Nonmonotonic Trajectories," *ORSA Journal on Computing,* Vol. 7, No. 4, pp. 426-442.

Glover, F. (1996) "Tabu Search and Adaptive Memory Programming — Advances, Applications and Challenges", *Interfaces in Computer Science and Operations Research,* R. Barr, R. Helgason and J. Kennington (eds.), Kluwer Academic Publishers, pp. 1-75.

Glover, F. and H. Greenberg (1989) "New Approaches for Heuristic Search: A Bilateral Linkage with Artificial Intelligence," *European Journal of Operational Research,* Vol. 39, No. 2, pp. 119-130.

Glover, F., D. Karney, D. Klingman and A. Napier (1974) "A Computational Study on Start Procedures, Basis Change Criteria and Solution Algorithms for Transportation Problems," *Management Science,* Vol. 20, No. 5, pp. 793-813.

Glover, F., J. P. Kelly and M. Laguna (1996) "New Advances and Applications of Combining Simulation and Optimization," *Proceedings of the 1996 Winter Simulation Conference,* J. M. Charnes, D. J. Morrice, D. T. Brunner, and J. J. Swain (eds.), pp. 144-152.

Glover, F., D. Klingman and N. Phillips (1990) "A Network-Related Nuclear Power Plant Model with an Intelligent Branch and Bound Solution Approach," *Annals of Operations Research,* Vol. 21, pp. 317-332.

Glover, F. and G. Kochenberger (1996) "Critical Event Tabu Search for Multidimensional Knapsack Problems," *Meta-Heuristics: Theory and Applications,* I. H. Osman and J. P. Kelly (eds.), Kluwer Academic Publishers, pp. 407-427.

Glover, F., G. Kochenberger and B. Alidaee (1997) "Adaptive Memory Tabu Search for Binary Quadratic Programs," to appear in *Management Science.*

Glover, F. and M. Laguna (1993) "Tabu Search," *Modern Heuristic Techniques for Combinatorial Problems,* C. Reeves (ed.), Blackwell Scientific Publishing, Oxford, pp. 70-150.

Glover, F. and M. Laguna (1997a) "Properties of Optimal Solutions to the Min *k*-Tree Problem," Graduate School of Business, University of Colorado at Boulder.

Glover, F. and M. Laguna (1997b) "General Purpose Heuristics for Integer Programming — Part I," *Journal of Heuristics,* Vol. 2, No. 4, pp. 343-358.

Glover, F. and M. Laguna (1997c) "General Purpose Heuristics for Integer Programming — Part II," *Journal of Heuristics,* Vol. 3, No. 2, pp. 161-179.

Glover, F. and C. McMillan (1986) "The General Employee Scheduling Problem: An Integration of Management Science and Artificial Intelligence," *Computers and Operations Research,* Vol. 15, No. 5, pp. 563-593.

Glover, F., C. McMillan and B. Novick (1985) "Interactive Decision Software and Computer Graphics for Architectural and Space Planning," *Annals of Operations Research,* Vol. 5, pp. 557-573.

Glover, F., J. M. Mulvey and D. Bai (1996) "Improved Approaches to Optimization Via Integrative Population Analysis," Graduate School of Business, University of Colorado at Boulder.

Glover, F., J. M. Mulvey, D. Bai and M. Tapia (1997) "Integrative Population Analysis for Better Solutions and What-If Analysis in Industrial Engineering Applications," to appear in *Industrial Applications of Combinatorial Optimization,* G. Yu (ed.).

Glover, F., J. M. Mulvey and K. Hoyland (1996) "Solving Dynamics Stochastic Control Problems in Finance using Tabu Search with Variable Scaling," *Meta-Heuristics: Theory and Applications*, I. H. Osman and J. P. Kelly (eds.), Kluwer Academic Publishers, pp. 429-448.

Gomory, R. E. (1958) "Outline of an Algorithm for Integer Solutions to Linear Programs," *Bulletin of the American Mathematical Society,* Vol. 64, pp. 275-278.

Gomory, R. E. (1960) "An Algorithm for the Mixed Integer Problem," Research Memorandum RM-2597, Rand Corporation, Santa Monica.

Gomory, R. E. (1965) "On the Relation Between Integer and Non-Integer Solutions to Linear Programs," *Proceedings of the National Academy of Science,* Vol. 53, pp. 260-265.

Gomory, R. E. (1996) Interview, *Optima,* No. 50, June.

Gomory, R. E. and E. L. Johnson (1972) "Some Continuous Functions Related to Corner Polyhedra," *Mathematical Programming*, Vol. 3, pp. 23-85.

Greenberg, H. J. and W. P. Pierskalla (1970) "Surrogate Mathematical Programs," *Operations Research*, Vol. 18, pp. 924-939.

Greenberg, H. J. and W. P. Pierskalla (1973) "Quasi-conjugate Functions and Surrogate Duality," *Cahiers du Centre d'Etudes de Recherche Operationelle*, Vol. 15, pp. 437-448.

Hansen, P., B. Jaumard, and Da Silva (1992) "Average Linkage Divisive Hierarchical Clustering," to appear in *Journal of Classification.*

Hansen, P., B. Jaumard, and M. Poggi di Aragao (1992) "Mixed Integer Column Generation Algorithms and the Probabilistic Maximum Satisfiability Problem," Proceedings of the 2nd Integer Programming and Combinatorial Optimization Conference, Carnegie Mellon.

Hertz, A., B. Jaumard, and M. Poggi di Aragao (1991) "Topology of Local Optima for the K-Coloring Problem," to appear in *Discrete Applied Mathematics.*

Hertz, A. and D. De Werra (1991) "The Tabu Search Metaheuristic: How We Used It," *Annals of Mathematics and Artificial Intelligence*, Vol. 1, pp. 111-121.

Holland, J. H. (1975) *Adaptation in Natural and Artificial Systems*, University of Michigan Press, Ann Arbor, MI.

Holmer, M. R. and S. A. Zenios (1995) "The Productivity of Financial Intermediation and the Technology of Financial Product Management," *Operations Research*, Vol. 43, No. 6, pp. 970-982.

Hong, I., A. B. Kahng and B-R Moon (1997) "Improved Large-Step Markov Chain Variants for Symmetric TSP," to appear in *Journal of Heuristics.*

Hubscher, R. and F. Glover (1994) "Applying Tabu Search with Influential Diversification to Multiprocessor Scheduling," *Computers and Operations Research,* Vol. 21, No. 8, pp. 877-884.

Johnson, D. S. (1990) "Local Optimization and the Traveling Salesman Problem," *Proc. 17th Intl. Colloquium on Automata, Languages and Programming,* pp. 446-460.

Karwan, M. H. and R. L. Rardin (1976) "Surrogate Dual Multiplier Search Procedures in Integer Programming," School of Industrial Systems Engineering, Report Series No. J-77-13, Georgia Institute of Technology.

Karwan, M. H. and R. L. Rardin (1979) "Some Relationships Between Lagrangean and Surrogate Duality in Integer Programming," *Mathematical Programming*, Vol. 17, pp. 230-334.

Kelly, J. P. (1995) "Determination of Market Niches Using Tabu Search-Based Cluster Analysis," Graduate School of Business, University of Colorado at Boulder.

Kelly, J. P., M. Laguna and F. Glover (1994) "A Study on Diversification Strategies for the Quadratic Assignment Problem," *Computers and Operations Research*, Vol. 22, No. 8, pp. 885-893.

Kelly, J. P. and J. Xu (1995) "Tabu Search and Vocabulary Building for Routing Problems," Graduate School of Business Administration, University of Colorado at Boulder.

Kernighan, B. W. and S. Lin (1970) "An Efficient Heuristic Procedure for Partitioning Graphs, *Bell Syst. Tech. J.,* Vol. 49.

Kincaid, R. K. (1995) "Actuator Placement for Active Sound and Vibration Control of Cylinders," NASA ASEE Report.

Kincaid, R. K. and R. T. Berger (1993) "The Damper Placement Problem on Space Truss Structures," *Location Science*, Vol. 1, No. 3, pp. 219-234.

Kincaid, R. K., A. D. Martin and J. A. Hinkley (1995) "Heuristic Search for the Polymer Straightening Problem," *Computational Polymer Science*, Vol. 5, pp. 1-5.

Kirkpatrick, S., C. D. Gelatt Jr. and M. P. Vecchi (1983) "Optimization by Simulated Annealing," *Science,* Vol. 220, pp. 671-680.

Kraay, D. and P. Harker (1996) "Case-Based Reasoning for Repetitive Combinatorial Optimization Problems, Part I: Framework," *Journal of Heuristics,* Vol. 2, No. 1, pp. 55-86.

Kraay, D. and P. Harker (1997) "Case-Based Reasoning for Repetitive Combinatorial Optimization Problems, Part II: Numerical Results," to appear in *Journal of Heuristics.*

Laguna, M., T. Feo and H. Elrod (1994) "A Greedy Randomized Adaptive Search Procedure for the 2-Partition Problem," *Operations Research*, Vol. 42, No. 4, pp. 677-687.

Laguna, M., J. P. Kelly, J. L. Gonzalez-Velarde, and F. Glover (1995) "Tabu Search for the Multilevel Generalized Assignment Problem," *European Journal of Operational Research*, Vol. 82, pp. 176-189.

Laguna M., R. Marti and V. Valls (1995) "Arc Crossing Minimization in Hierarchical Digraphs with Tabu Search," to appear in *Computers and Operations Research.*

Laporte, G. and I.H. Osman (1995) "Metaheuristics in Combinatorial Optimization," *Annals of Operations Research*, Vol. 60, J. C. Baltzer Science Publishers, Basel, Switzerland.

Lin, S. and B. W. Kernighan (1973) "An Efficient Heuristic Algorithm for the Traveling Salesman Problem," *Operations Research,* Vol. 21, pp. 498-516.

Lokketangen, A. and F. Glover (1996) "Probabilistic Move Selection in Tabu Search for 0/1 Mixed Integer Programming Problems," *Meta-Heuristics: Theory and Applications*, I. H. Osman and J. P. Kelly (eds.), Kluwer Academic Publishers, pp. 467-488.

Lokketangen, A. and F. Glover (1997a) "Solving Zero/One Mixed Integer Programming Problems using Tabu Search," to appear in the *European Journal of Operational Research.*

Lokketangen, A. and F. Glover (1997b) "Surrogate Constraint Analysis — New Heuristics and Learning Schemes for Satisfiability Problems," Proceedings of the DIMACS workshop on Satisfiability Problems: Theory and Applications, D-Z. Du, J. Gu and P. Pardalos (eds.).

Lokketangen, A. and G. Hasle (1997) "Solving the Forest Treatment Scheduling Problem using Abstraction and Iterative Improvement Techniques," Working Paper, Molde College.

Lokketangen, A. K. Jornsten and S. Storoy (1994) "Tabu Search within a Pivot and Complement Framework," *International Transactions in Operations Research,* Vol. 1, No. 3, pp. 305-316.

Lopez, L., M. W. Carter and M. Gendreau (1996) "The Hot Strip Mill Production Scheduling Problem: A Tabu Search Approach," Centre de recherche sur les transports, Université de Montréal.

Lourenco, H. R. and M. Zwijnenburg (1996) "Combining the Large-Step Optimization with Tabu Search: Application to the Job Shop Scheduling Problem," *Meta-Heuristics: Theory and Applications*, I. H. Osman and J. P. Kelly (eds.), Kluwer Academic Publishers, pp. 219-236.

Magee, T. M. and F. Glover (1995) "Integer Programming," *Mathematical Programming for Industrial Engineers*, M. Avriel and B. Golany (eds.), Marcel Dekker, Inc., New York, pp. 123-270.

Malek, M., M. Guruswamy, M. Pandya and H. Owens (1989) "Serial and Parallel Simulated Annealing and Tabu Search Algorithms for the Traveling Salesman Problem," *Annals of Operations Research,* Vol. 21, pp. 59-84.

Marti, R. (1996) "An Aggressive Search Procedure for the Bipartite Drawing Problem," *Meta-Heuristics: Theory and Applications*, I. H. Osman and J. P. Kelly (eds.), Kluwer Academic Publishers, pp. 97-113.

Martin, O., S. W. Otto and E. W. Felten (1991) "Large-Step Markov Chains for the Traveling Salesman Problem," *Complex Systems,* Vol. 5, No. 3, pp. 299-326.

Martin, O., S. W. Otto and E. W. Felten (1992) "Large-Step Markov Chains for TSP Incorporating Local Search Heuristics," *Operations Research Letters,* Vol. 11, No. 4, pp. 219-224.

Martins, S. L., C. C. Ribeiro and N. R. Rodriguez (1996) "Parallel Programming Tools for Distributed Memory Environments," Research report MCC-01/96, Department of Computer Science, Catholic University of Rio de Janeiro.

Mazzola, J. B., A. W. Neebe, and C.M. Rump (1995) "Multiproduct Production Planning in the Presence of Work-Force Learning," Working Paper, Fuqua School of Business, Duke University.

Mazzola, J. B. and R. H. Schantz (1995a) "Single-facility Resource Allocation Under Capacity-based Economies and Diseconomies of Scope," *Management Science*, Vol. 41, No. 4, pp. 669-689.

Mazzola, J. B. and R. H. Schantz (1995b) "Multiple-Facility Loading Under Capacity-Based Economies of Scope," Working Paper, Fuqua School of Business, Duke University.

Michalewicz, Z. and C. Janikow (1991) "Genetic Algorithms for Numerical Optimization," *Statistics and Computing*, Vol. 1, pp. 75-91.

Mühlenbein, H., M. Gorges-Schleuter, and O. Krämer (1988) "Evolution Algorithms in Combinatorial Optimization," *Parallel Computing*, Vol. 7, pp. 65-88.

Mühlenbein, H. and D. Schlierkamp-Voosen (1994) "The Science of Breeding and its Application to the Breeder Genetic Algorithm," *Evolutionary Computation,* Vol. 1, pp. 335-360.

Mühlenbein, H. and H-M Voigt (1996) "Gene Pool Recombination in Genetic Algorithms," *Meta-Heuristics: Theory and Applications*, I. H. Osman and J. P. Kelly (eds.), Kluwer Academic Publishers, pp. 53-62.

Mulvey, J. (1995) "Generating Scenarios for the Towers Perrin Investment Systems," to appear in *Interfaces*.

Nemati, H. R. and M. Sun (1994) "A Neural Network Tabu Machine for the Traveling Salesman Problem," presented at the TIMS/ORSA Joint National Meeting, Detroit, MI.

Nemhauser, G. L. and L. A. Wolsey (1988) *Integer and Combinatorial Optimization*, John Wiley & Sons, New York.

Nonobe, K. and T. Ibaraki (1997) "A Tabu Search Approach to the CSP (Constraint Satisfaction Problem) as a General Problem Solver," to appear in *European Journal of Operational Research.*

Nowicki, E. and C. Smutnicki (1993) "A Fast Taboo Search Algorithm for the Job Shop," to appear in *Management Science*.

Nowicki, E. and C. Smutnicki (1994) "A Fast Taboo Search Algorithm for the Flow Shop," to appear in *European Journal of Operational Research.*

Nowicki, E. and C. Smutnicki (1995) "The Flow Shop with Parallel Machines: A Tabu Search Approach," Report ICT PRE 30/95, Technical University of Wroclaw.

Osman, I. H. and J. P. Kelly (1996) *Metaheuristics: Theory and Applications*, Kluwer Academic Publishers, Norwell, MA.

Papadimitriou, C.H. and K. Steiglitz (1982) *Combinatorial Optimization: Algorithms and Complexity*, Prentice Hall, New York.

Pardalos, P. M., X. Liu and G. Xue (1995) "Protein Conformation of Lattice Using Tabu Search," Unpublished.

Parker, G. and R. Rardin (1988) *Discrete Optimization*, Academic Press, New York.

Pesch, E. and F. Glover (1995) "TSP Ejection Chains," to appear in *Discrete Applied Mathematics.*

Piwakowski, K. (1996) "Applying Tabu Search to Determine New Ramsey Graphs," *Electronic Journal of Combinatorics,* Vol. #R6, No. 3, http:// ejc.math.gatech.edu:8080 /Journal/ejc-wce.html.

Porto, S. C. S. and C. C. Ribeiro (1995a) "A Tabu Search Approach to Task Scheduling on Heterogeneous Processors Under Precedence Constraints," *International Journal of High-Speed Computing,* Vol. 7, pp. 45-71.

Porto, S. C. S. and C. C. Ribeiro (1995b) "Parallel Tabu Search Message-Passing Synchronous Strategies for Task Scheduling Under Precedence Constraints," *Journal of Heuristics*, Vol. 1, No. 2, pp. 207-225.

Punnen, A. P. and F. Glover (1997) "Ejection Chains with Combinatorial Leverage for the Traveling Salesman Problem," Graduate School of Business, University of Colorado at Boulder.

Rego, C. (1996a) "Relaxed Tours and Path Ejections for the Traveling Salesman Problems," to appear in *European Journal of Operational Research.*

Rego, C. (1996b) "A Subpath Ejection Method for the Vehicle Routing Problem," to appear in *Managment Science.*

Rego, C. and C. Roucairol (1996) "A Parallel Tabu Search Algorithm Using Ejection Chains for the Vehicle Routing Problem," *Meta-Heuristics: Theory and Applications*, I. H. Osman and J. P. Kelly (eds.), Kluwer Academic Publishers, pp. 661-675.

Rochat, Y. and F. Semet (1994) "A Tabu Search Approach for Delivering Pet Food and Flour," *Journal of the Operational Research Society*, Vol. 45, No. 11, pp. 1233-1246.

Rochat, Y. and E. Taillard (1995) "Probabilistic Diversification and Intensification in Local Search for Vehicle Routing," *Journal of Heuristics,* Vol. 1, No. 1, pp. 147-167.

Rolland, E. (1996) "A Tabu Search Method for Constrained Real-Number Search: Applications to Portfolio Selection," Working paper, The A. Gary Anderson Graduate School of Management, University of California, Riverside.

Rolland, E. and H. Johnson (1996) "Skewness and the Mean-Variance Frontier: A Tabu Search Approach," Working paper, The A. Gary Anderson Graduate School of Management, University of California, Riverside.

Rolland, E., H. Pirkul and F. Glover (1995) "Tabu Search for Graph Partitioning," to appear in *Annals of Operations Research.*

Rolland, E., D. A. Schilling and J. R. Current (1996) "An Efficient Tabu Search Procedure for the p-Median Problem," *European Journal of Operational Research,* Vol. 96, No. 2, pp. 329-342.

Savelsbergh, M. W. P. (1994) "Preprocessing and Probing Techniques for Mixed Integer Programming Problems," *ORSA Journal on Computing*, Vol. 6, No. 4, pp. 445-454.

Scholl, A. and S. Voss (1996) "Simple Assembly Line Balancing — Heuristic Approaches" *Journal of Heuristics,* Vol. 2, No. 3, pp. 217-244.

Sexton, R. S, B. Alidaee, R. E. Dorsey and J. D. Johnson (1997) "Global Optimization for Artificial Neural Networks: A Tabu Search Application," to appear in *European Journal of Operational Research.*

Sherali, H. D. and C. M. Shetty (1980) *Optimization with Disjunctive Constraints,* Lecture Notes in Economics and Mathematical Systems, No. 181, Springer-Verlag, New York.

Simon, H. A. and A. Newell (1958) "Heuristic Problem Solving: The Next Advance in Operations Research", *Operations Research*, Vol. 6, No. 1.

Skorin-Kapov, D. and J. Skorin-Kapov (1994) "On Tabu Search for the Location of Interacting Hub Facilities," *European Journal of Operational Research*, Vol. 73, No. 3, pp. 501-508.

Skorin-Kapov, D., J. Skorin-Kapov and M. O'Kelly (1995) "Tight Linear Programming Relaxations of p-Hub Median problems," to appear in *European Journal of Operational Research.*

Skorin-Kapov, J. (1990) "Tabu Search Applied to the Quadratic Assignment Problem," *ORSA Journal on Computing*, Vol. 2, No. 1, pp. 33-45.

Skorin-Kapov, J. (1994) "Extensions of a Tabu Search Adaptation to the Quadratic Assignment Problem," *Computers and Operations Research*, Vol. 21, No. 8, pp. 855-865.

Skorin-Kapov, J. and J-F Labourdette (1995) "On Minimum Congestion in Logically Rearrangeable Multihop Lightwave Networks," *Journal of Heuristics*, Vol. 1, No. 1, pp. 129-146.

Skorin-Kapov, J. and A. Vakharia (1993) "Scheduling a Flow-Line Manufacturing Cell: A Tabu Search Approach," *International Journal of Production Research*, Vol. 31, No. 7, pp. 1721-1734.

Sondergeld, L. and S. Voss (1996a) "A Star-Shaped Diversification Approach in Tabu Search," I. H. Osman and J. P. Kelly (eds.), *Meta-Heuristics: Theory and Applications*, Kluwer Academic Publishers, pp. 489-502.

Sondergeld, L. and S. Voss (1996b) "A Multi-Level Star-Shaped Intensification and Diversification Approach in Tabu Search for the Steiner Tree Problem in Graphs," Working Paper, TU Braunschweig.

Soriano, P. and M. Gendreau (1996a) "Diversification Strategies in Tabu Search Algorithms for the Maximum Clique Problem," *Annals of Operations Research,* Vol. 63, pp. 189-207.

Soriano, P. and M. Gendreau (1996b) "Tabu Search Algorithms for the Maximum Clique Problem," *Cliques, Coloring and Satisfiability*, D. S. Johnson and M. A. Trick (eds.), DIMACS Series in Discrete Mathematics and Theoretical Computer Science, Vol. 26, American Mathematical Society, pp. 221-242.

Spears, W. M. and K. A. DeJong (1991) "On the Virtues of Uniform Crossover," 4th International Conference on Genetic Algorithms, La Jolla, CA.

Sun, M. (1996) "A Tabu Search Heuristic Procedure for Solving the Transportation Problem with Exclusionary Side Constraints," Working Paper, Vol. 18, No. 1, Institute for Studies in Business, College of Business, The University of Texas at San Antonio.

Sun, M., J. E. Aronson, P. G. McKeown, and D. Drinka (1995) "A Tabu Search Heuristic Procedure for the Fixed Charge Transportation Problem," Working Paper No. 95-414, Terry College of Business, The University of Georgia.

Sun, M., and H. R. Nemati (1994) "TABU MACHINE: A New Method of Solving Combinatorial Optimization Problems," presented at the TIMS/ORSA Joint National Meeting, Boston, MA.

Taillard, E. (1991) "Robust Taboo Search for the Quadratic Assignment Problem," *Parallel Computing,* Vol. 17, pp. 443-455.

Taillard, E. (1993) "Parallel Iterative Search Methods for Vehicle Routing Problems," *Networks*, Vol. 23, pp. 661-673.

Taillard, E. (1994) "Parallel Taboo Search Techniques for the Job Shop Scheduling Problem," *ORSA Journal on Computing,* Vol. 6, pp. 108-117.

Taillard, E., P. Badeau, M. Gendreau, F. Guertin and J-Y. Potvin (1995) "A New Neighborhood Structure for the Vehicle Routing Problem with Time Windows," Centre de Recherche sur les Transports, Publication CRT-95-66.

Talukdar, S., L. Baerentzen, A. Gove and P. de Souza (1996) "Asynchronous Teams: Cooperation Schemes for Autonomous Agents," Carnegie Mellon University.

Toulouse, M., Teodor G. C. and M. Gendreau (1996) "Communication Issues in Designing Cooperative Multi-Thread Parallel Searches," *Meta-Heuristics: Theory and Applications*, I. H. Osman and J. P. Kelly (eds.), Kluwer Academic Publishers, pp. 503-522

Trafalis, T., and I. Al-Harkan (1995) "A Continuous Scatter Search Approach for Global Optimization," Extended Abstract in: *Conference in Applied Mathematical Programming and Modeling (APMOD'95),* London, UK, 1995.

Tuy, H. (1964) "Concave Programming Under Linear Constraints," *Doklady Akademii Nauk SSSR* (in Russian), *Soviet Mathematics* (English translation), pp. 1437-1440.

Ulder, N. L. J., E. Pesch, P. J. M. van Laarhoven, H. J. Bandelt and E. H. L. Aarts (1991) "Genetic Local Search Algorithm for the Traveling Salesman Problem," *Parallel Problem Solving from Nature*, R. Maenner and H. P. Schwefel (eds.), Springer-Verlag, Berlin, pp. 109-116.

Vaessens, R. J. M, E. H. L. Aarts and J. K. Lenstra (1995) "Job Shop Scheduling by Local Search," Memorandum COSOR 94-05, Eindhoven University of Technology.

Valls V., R. Marti and P. Lino (1995) "A Tabu Thresholding Algorithm for Arc Crossing Minimization in Bipartite Graphs," *Annals of Operations Research,* Vol. 60, Metaheuristics in Combinatorial Optimization.

Veeramani, D. and A. H. Stinnes (1996) "Optimal Process Planning for Four-Axis Turing Centers," Proceedings of the 1996 ASME Design Engineering Technical Conferences and Computers in Engineering Conference, Irvine, CA.

Veeramani, D., A. H. Stinnes, and D. Sanghi (1996) "Application of Tabu Search for Process Plan Optimization of Four-Axis CNC Turning Centers," Technical Report, Department of Industrial Engineering, University of Wisconsin-Madison.

Voss, S. (1992) "Network Design Formulations in Schedule Synchronization," *Lecture Notes in Economics and Mathematical Systems*, No. 386, pp. 137-152.

Voss, S. (1993) "Tabu Search: Applications and Prospects," *Network Optimization Problems: Algorithms, Applications and Complexity,* D-Z Du and P. M. Pardalos (eds.), World Scientific, Singapore, pp. 333-353.

Voss, S. (1997) "Optimization by Strategically Solving Feasibility Problems Using Tabu Search," *Modern Heuristics for Decision Support Proceedings,* Unicom, Uxbridge, pp. 29-47.

De Werra, D. and A. Hertz (1989) " Tabu Search Techniques: A Tutorial and Applications to Neural Networks," *OR Spektrum,* Vol. 11, pp. 131-141.

Whitley, D. (1993) *Foundations of Genetic Algorithms 2,* Morgan Kaufmann.

Whitley, D., V. S. Gordon and K. Mathias (1994) "Lamarckian Evolution, the Baldwin Effect and Function Optimization," *Proceedings of the Parallel Problem Solving from Nature,* Vol. 3, New York: Springer-Verlag, pp. 6-15.

Woodruff, D. L. (1996) "Proposals for Chunking and Tabu Search," Working Paper, University of California at Irvine.

Woodruff, D. L. and E. Zemel (1993) "Hashing Vectors for Tabu Search," *Annals of Operations Research,* Vol. 41, pp. 123-138.

Wright, A. H. (1990) "Genetic Algorithms for Real Parameter Optimization," *Foundations of Genetic Algorithms*, G. Rawlins (ed.), Morgan Kaufmann, Los Altos, CA, pp. 205-218.

Xu, J., S. Y. Chiu and F. Glover (1996a) "Fine-Tuning a Tabu Search Algorithm with Statistical Tests," Graduate School of Business, University of Colorado, Boulder, CO.

Xu, J., S. Y. Chiu and F. Glover (1996b) "Tabu Search for Dynamic Routing Communications Network Design," Graduate School of Business, University of Colorado, Boulder, CO.

Young, R. D. (1971) "Hypercylindrically Deduced Cuts in Zero—One Integer Programming," *Operations Research*, Vol. 19, pp. 1393-1405.

Zenios, S. (1996) "Dynamic Financial Modeling and Optimizing the Design of Financial Products," Presented at the National INFORMS Meeting, Washington, D.C.

INDEX

A

B

C

D

E

F

G

H

I

J

L

M

N

O

P

Q

R

S

T

U

V

W

The cover design was inspired by the painting "God is at Home" from the "Geometry of the Heart" series by Horst Becker (www.cybec.com)

BIRKBECK COLLEGE
1911867691